OCEAN SUSTAINABILITY IN THE 21st CENTURY

The world's oceans are an essential source of food and other resources, as well as providing an important means of transportation, trade, and recreation. Covering more than two thirds of the Earth's surface, our oceans are intricately linked to our climate system and require careful management to ensure their continued sustainability.

Describing the emerging and unresolved issues related to the oceans and the marine environment, this book presents the developments made in marine science and policy since the implementation of the United Nations Convention on the Law of the Sea (UNCLOS), and implications for the sustainable management of ocean areas and resources. This comprehensive volume also provides a number of scientific, policy, and legal tools to address such issues, and to ensure better science-based management of the oceans. Topics covered include the impacts of human-induced climate change on the oceans, the marine genetic resources debate, the current legal framework for the oceans, and a comparative study of the legal issues associated with outer space.

Including practical examples and worldwide case studies, this book is a valuable resource for policy makers, students, and academics, in marine science and policy, ocean affairs, and the law of the sea.

SALVATORE ARICÒ is Senior Programme Specialist for biodiversity science and policy in the Natural Sciences Sector at UNESCO, and coordinator of UNESCO's Biodiversity Initiative. He is also Visiting Researcher and Professor at the University of Delaware, the Parthenope University of Naples, and UNU's Institute of Advanced Studies in Yokohama. He has co-authored approximately 70 scientific publications related to oceans and biodiversity.

OCEAN SUSTAINABILITY IN THE 21ST CENTURY

EDITED BY SALVATORE ARICÒ

United Nations
Educational, Scientific and
Cultural Organization

CAMBRIDGE
UNIVERSITY PRESS

University Printing House, Cambridge CB2 8BS, United Kingdom

One Liberty Plaza, 20th Floor, New York, NY 10006, USA

477 Williamstown Road, Port Melbourne, VIC 3207, Australia

4843/24, 2nd Floor, Ansari Road, Daryaganj, Delhi - 110002, India

79 Anson Road, #06-04/06, Singapore 079906

Cambridge University Press is part of the University of Cambridge.

It furthers the University's mission by disseminating knowledge in the pursuit of
education, learning and research at the highest international levels of excellence.

Published jointly by the United Nations Educational, Scientific and Cultural Organization (UNESCO),
7, place de Fontenoy, 75007 Paris, France, and Cambridge University Press,
University Printing House, Shaftesbury Road, Cambridge CB2 8BS, United Kingdom.

www.cambridge.org
Information on this title: www.cambridge.org/9781108447867

First published 2015
First paperback edition 2017

A catalogue record for this publication is available from the British Library

Library of Congress Cataloging in Publication data
Ocean sustainability in the 21st century / edited by Salvatore Aricò.
pages   cm
ISBN 978-1-107-10013-8 (Hardback)
1. Marine resources conservation.   2. Marine ecosystem health.   3. Sustainability.   4. Ocean
5. Ocean–atmosphere interaction.   6. Climatic changes.   7. Law of the sea.   I. Aricò, Salvatore, editor of
compilation.   II. Title: Ocean sustainability in the 21st century.
GC1018.O238 2015
333.91´6416–dc23   2014039486

ISBN Cambridge 978-1-107-10013-8
ISBN UNESCO 978-92-3-100055-3
ISBN 978-1-108-44786-7 Paperback

The designations employed and the presentation of material throughout this publication do not imply the expression of any
opinion whatsoever on the part of UNESCO concerning the legal status of any country, territory, city or area or of its
authorities, or the delimitation of its frontiers or boundaries.

The authors are responsible for the choice and the presentation of the facts contained in this book and for the opinions
expressed therein, which are not necessarily those of UNESCO and do not commit the Organization.

Contents

Contributors

Salvatore Aricò is a senior programme specialist at the Natural Sciences Sector of UNESCO. He has served at the Intergovernmental Oceanographic Commission and the Convention on Biological Diversity. He holds affiliations with the University of Delaware and the United Nations University and a PhD from the Stazione Zoologica Anton Dohrn in Naples.

Stefano Belfiore has been a programme specialist at the Intergovernmental Oceanographic Commission of the United Nations Educational, Scientific and Cultural Organization (UNESCO/IOC). He has published on ocean governance, especially at the regional level, and coastal zone management. In 2012 he joined the World Meteorological Organization (WMO).

Professor **Peter Bridgewater** is currently a visiting professor at UNU-IAS and Beijing Forestry University. He has also served as Chair of the UK Joint Nature Conservation Committee, Secretary General of the Ramsar Convention, Director of the Division of Ecological Sciences in UNESCO, and as Commissioner of the World Commission on the Oceans.

Eddy Carmack is an emeritus senior research scientist with Fisheries and Oceans Canada with over 40 years of working in high-latitude waters and a veteran of over 80 field missions. He is a fellow of the American Geophysical Institute and was the recipient of the RCGS Massey Medal in 2007 and the CMOS Tully Medal in 2010.

Claudio Chiarolla is a senior advisor/negotiator with the Italian Biodiversity EU Presidency Team. He is also a research fellow on International Biodiversity Governance at the Institute for Sustainable Development and International Relations and *maître de conférences* at the Paris School of International Affairs, Sciences Po. He holds a PhD in law from the Queen Mary University of London.

Dr **Biliana Cicin-Sain** is the Director of the Gerard J. Mangone Center for Marine Policy and Professor of Marine Policy at the University of Delaware. She is also the founder and president of the Global Ocean Forum. Her international ocean work has been recognized through a number of prestigious awards, including the Elizabeth Mann Borgese Meerespreis.

Cinzia Corinaldesi is an assistant professor at the Polytechnic University of Marche. Her teaching activity deals with applied marine ecology. Her research interests are on marine microbial ecology and biodiversity in the oceans. She is the author of 43 articles in ISI journals and several contributions to book chapters in the field of applied marine science and microbiology.

Ian Cresswell is the Science Director of CSIRO's Wealth from Oceans National Research Flagship, Australia's largest marine science programme. He is responsible for the quality of the Flagship's science, and the alignment of research activities for maximum benefit to the nation. Ian has a long history of working in environmental science and policy.

Roberto Danovaro is the President of the Stazione Zoologica Anton Dohrn and a full professor in marine biology (Polytechnic University Marche). He is author of more than 220 WoK articles on marine biodiversity and ecosystem functioning, and the impact of global change. His awards include the Medaille d'Or (French Oceanographic Society, 2010), the World Biology Prize (BioMedCentral, 2011), and the ENI Award (2013).

Dr **Kenneth Denman** is a chief scientist with Ocean Networks Canada, and an adjunct professor at the University of Victoria, British Columbia and the EC Canadian Centre for Climate Modelling and Analysis. He has been awarded several prizes, including the President's Prize of the Canadian Meteorological and Oceanographic Society and the T. R. Parsons Medal for excellence in ocean science.

Cristina Gambi is an adjunct professor at the Polytechnic University of Marche. Her research interest is on the relationship between biodiversity and ecosystem functioning in different deep-sea ecosystems. She is the author of 38 articles in ISI journals including two works cited by CORDIS and Faculty 1000 Biology.

Rolf Gradinger is a professor in biological oceanography and Associate Dean at the School of Fisheries and Ocean Sciences at the University of Alaska Fairbanks. His research has focused on the ecology of Arctic Sea ice and on the biodiversity of Arctic marine systems as part of the Census of Marine Life.

Henry P. Huntington works for the International Arctic Campaign of the Pew Charitable Trusts. His research focuses on human–environment interactions in the

Arctic, especially with indigenous peoples. He has been involved in several international assessments concerning the Arctic. He lives in Eagle River, Alaska, with his wife and two sons.

Dr **S. Salman Hussain** is a researcher in ecosystem economics and is the Coordinator of the UNEP TEEB project (The Economics of Ecosystems and Biodiversity), a position he took up in January 2014. His current research focus is on marine ecosystem economics.

Viviana Iavicoli is a senior researcher in international law and space law at the Institute for International Legal Studies (ISGI) of the National Research Council (CNR) in Rome. She is a board member of the Master in Space Institutions and Policies and lecturer in space law.

E. Gunnar B. Kullenberg was a professor at the University of Copenhagen and the University of Gothenburg, Executive Secretary IOC, and Executive Director of the International Ocean Institute. He has held visiting scholar positions at the United Nations Institute for Training and Research (UNITAR), the Ship-Ocean Foundation, and the Canada Centre for Inland Waters.

Eva Leu is a researcher at Akvaplan-niva AS in Oslo/Tromsø, Norway. She was a research scientist at the Norwegian Polar Institute and the Alfred-Wegener-Institute for Polar and Marine Research. Her main research interests are ecological consequences of climate change in Arctic marine ecosystems, with a special emphasis on microalgal ecophysiology and biochemistry. She holds a PhD in marine biology from the University of Oslo.

Aurora Mateos is a legal consultant and former Legal Adviser to the Intergovernmental Oceanographic Commission of the United Nations Educational, Scientific and Cultural Organization (UNESCO/IOC).

Professor **Rolph Payet** has a PhD in environmental science and is Minister for Environment and Energy in the Seychelles and the Pro-Chancellor of the University of Seychelles. He is a member of the Blue Ribbon Panel of the Global Partnership for Oceans and the Global Forum on Oceans, Coasts and Islands.

Charlotte Salpin is Legal Officer at the Division for Ocean Affairs and the Law of the Sea, Office of Legal Affairs of the United Nations. Previously, she worked at the United Nations Environment Programme. Her areas of expertise include marine biodiversity and marine scientific research.

Tullio Scovazzi is Professor of International Law, Faculty of Law, University of Milano-Bicocca, Italy. He occasionally attends, as legal expert for Italy, international secretariats, meetings, and negotiations on human rights, the marine environment, or cultural matters.

Dr **Tony Smith** leads the ecosystem based management (EBM) group in CSIRO's Wealth from Oceans Research Flagship. His research focus is on developing the scientific tools to support EBM, including risk assessment and modelling methods. He is a Member of the Order of Australia for services to marine science.

Dr **U. Rashid Sumaila** is a professor at the University of British Columbia`s Fisheries Centre. He specializes in bioeconomics, marine ecosystem valuation, and the analysis of global fisheries issues. Sumaila has published over 150 peer-reviewed journal articles, including in *Nature*, *Science*, and the *Journal of Environmental Economics and Management*.

Marjo Vierros is an adjunct senior fellow at the United Nations University Institute of Advanced Studies, where she coordinates the Global Marine Governance project. She is also a senior fellow with the Traditional Knowledge Initiative of the United Nations University. Her research interests include international ocean policy and community-based marine management.

Paul Wassmann is a professor in marine ecology at the University of Tromsø, Norway. He has worked extensively in the Barents Sea and the waters off Svalbard, in particular in the marginal ice zone, and from a system ecology viewpoint. He has led several Norwegian and European Arctic research projects and the ARCTOS network.

Francis Wiese is a marine ecologist and Science Director of the North Pacific Research Board in Anchorage, Alaska. His previous research focused on the impact of environmental and anthropogenic stressors on seabirds, sharks, marine mammals, and fish. In his current position he designs and implements integrated ecosystem research programmes.

Acronyms and abbreviations

ABE-LOS	Advisory Body of Experts on the Law of the Sea
ABNJ	Areas Beyond National Jurisdiction
ABS	Access and Benefit-Sharing
ACIA	Arctic Climate Impact Assessment
AFMA	Australian Fisheries Management Authority
Antigua Convention	Convention for the strengthening of the Inter-American Tropical Tuna Convention
APEC	Asia-Pacific Economic Cooperation
ATCM	Antarctic Treaty Consultative Meeting
Barcelona Convention	Convention for the protection of the marine environment and the coastal region of the Mediterranean
BAU	Business as usual
Cartagena Convention	Convention for the protection and development of the marine environment of the wider Caribbean region
CBA	Cost–benefit analysis
CBD	Convention on Biological Diversity
CBS	Convention on the conservation and management of pollock resources in the central Bering Sea
CCAMLR	Convention for the Conservation of Antarctic Marine Living Resources
CCN	Cloud condensation nuclei
CCSBT	Commission for the Conservation of Southern Bluefin Tuna
CEs	Choice experiments
CHM	Common heritage of mankind
CITES	Convention on International Trade in Endangered Species of wild fauna and flora
CLCS	Commission on the Limits of the Continental Shelf

CMS	Convention on Migratory Species of wild animals
CoC	International Code of Conduct for outer space activities
COFI	Committee on Fisheries of the Food and Agriculture Organization of the United Nations
COMEST	World Commission on the Ethics of Scientific Knowledge and Technology
COPUOS	Committee for Peaceful Uses of Outer Space
COSPAR	Committee on Space Research
CPR	Continuous plankton recorder
CPUCH	Convention on the Protection of the Underwater Cultural Heritage
CTBTO	Preparatory Commission for the Comprehensive Nuclear-Test-Ban Treaty Organization
CVM	Contingent valuation method
DHABs	Deep hypersaline anoxic basins
DIC	Dissolved inorganic carbon
DMS	Dimethyl sulfide
DSWC	Dense shelf water cascading
EAF	Ecosystem Approach to Fisheries
EBA	Ecosystem Based Approach
EBFM	Ecosystem Based Fisheries Management
EBM	Ecosystem Based Management
EBSAs	Ecologically and Biologically Significant Areas
EC	European Commission
EEZs	Exclusive Economic Zones
EIA	Environmental Impact Assessment
EPC	European Patent Convention
EPOCA	European Project on Ocean Acidification
FAO	Food and Agriculture Organization of the United Nations
GCOS	Global Climate Observing System
GCRMN	Global Coral Reef Monitoring Network
GEF	Global Environment Facility
GFCM	General Fisheries Commission for the Mediterranean
GHGs	Greenhouse gases
GLOSS	Global Sea Level Observing System
GMDSS	Global Maritime Distress Safety System
GOOS	Global Ocean Observing System
GPA	Global Programme of Action for the protection of the marine environment

GTOS	Global Terrestrial Observing System
GTS	Global Telecommunications System
HELCOM	Helsinki Commission
IATTC	Inter-American Tropical Tuna Commission
ICCAT	International Convention for the Conservation of Atlantic Tunas
ICG/CARIBE EWS	Intergovernmental Coordination Group for the Tsunami and Other Coastal Hazards Warning System for the Caribbean and adjacent regions
ICG/IOTWS	Intergovernmental Coordination Group for the Indian Ocean Tsunami Warning and Mitigation System
ICG/NEAMTWS	Intergovernmental Coordination Group for the Tsunami Early Warning and Mitigation System in the North-eastern Atlantic, the Mediterranean, and connected seas
ICG/PTWS	Intergovernmental Coordination Group for the Pacific Tsunami Warning and Mitigation System
ICJ	International Court of Justice
ICM	Integrated Coastal Management
ICP (or UNICPOLOS)	United Nations Open-ended Informal Consultative Process on Oceans and the Law of the Sea
ICSU	International Council for Science
ICZM	Integrated Coastal Zone Management
IDOE	International Decade of Ocean Exploration
IGBP	International Geosphere–Biosphere Programme
IGOS	Integrated Global Observing Strategy
IGY	International Geophysical Year
IHO	International Hydrographic Organization
IIOE	International Indian Ocean Expedition
IMCAM	Integrated Marine and Coastal Area Management
IMO	International Maritime Organization
IOC	Intergovernmental Oceanographic Commission
IODE	International Oceanographic Data and Information Exchange Programme
IOTC	Indian Ocean Tuna Commission
IPBES	Intergovernmental Platform on Biodiversity and Ecosystem Services
IPCC	Intergovernmental Panel on Climate Change
IPY	International Polar Year
ISBA (or ISA)	International Seabed Authority

ITIC	International Tsunami Information Center
ITLOS	International Tribunal for the Law of the Sea
ITPGRFA	International Treaty on Plant Genetic Resources for Food and Agriculture
ITSU	International Coordination Group for the Tsunami Warning System in the Pacific
ITU	International Telecommunications Union
IUCN	International Union for Conservation of Nature
IUU	Illegal, Unreported, and Unregulated fishing
IWRM	Integrated Water Resources Management
JNCC	Joint Nature Conservation Committee
JRC	Joint Research Centre
Lima Convention	Convention for the protection of the marine environment and coastal area of the South-east Pacific
LMEs	Large Marine Ecosystems
LOMAs	Large Oceans Management Areas
London Convention	Convention on the prevention of marine pollution by dumping of wastes and other matter
LOS Convention (or UNCLOS)	United Nations Convention on the Law of the Sea
MA	Millennium Ecosystem Assessment
MARPOL 73/78	International convention for the prevention of pollution from ships
MCZs	Marine Conservation Zones
MPAs	Marine Protected Areas
MSP	Marine Spatial Planning
MSR	Marine Scientific Research (in the context of the United Nations Convention on the Law of the Sea)
NADW	North Atlantic Deep Water
NAFO	Convention on future multilateral cooperation in the North-west Atlantic fisheries
NAO	North Atlantic Oscillation
NASCO	North Atlantic Salmon Conservation Organization
NEAF	Convention on future multilateral cooperation in the North-east Atlantic Fisheries
NOAA	National Oceanic and Atmospheric Administration
Noumea Convention	Convention for the protection of the natural resources and environment of the South Pacific region
NPAFC	North Pacific Anadromous Fisheries Commission
NPGO	North Pacific Gyre Oscillation

NTWC	National Tsunami Warning Centres
OBIS	Ocean Biogeographic Information System
OCB	US Ocean Carbon and Biogeochemistry Program
OECD	Organisation for Economic Co-operation and Development
OMZs	Oxygen Minimum Zones
OSPAR	Convention for the protection of the marine environment of the North-east Atlantic
OST	Treaty on Principles Governing the Activities of States in the Exploration and Use of Outer Space, including the Moon and Other Celestial Bodies (or Outer Space Treaty)
PDO	Pacific Decadal Oscillation
PEMSEA	Regional programme on building partnerships in environmental management for the seas of East Asia
PES	Payment for Ecosystem Services
PIC	Prior Informed Consent
POC	Particulate organic carbon
PSSAs	Particularly Sensitive Sea Areas
PTWS	Pacific Tsunami Warning System
RBM	River Basin Management
RFMOs/As	Regional Fisheries Management Organizations/ Arrangements
ROVs	Remotely Operated Vehicles
SACs	Special Areas of Conservation
SBSTTA	Subsidiary Body on Scientific, Technical and Technological Advice of the Convention on Biological Diversity
SDGs	Sustainable Development Goals
SEAFO	Convention on the conservation and management of fishery resources in the south-east Atlantic Ocean
SIDS	Small Island Developing States
SIOFA	Southern Indian Ocean Fisheries Agreement
SMTA	Standard Material Transfer Agreement of the International Treaty on Plant Genetic Resources for Food and Agriculture
SOLAS	International Convention for the Safety of Life at Sea
SPAMIs	Specially Protected Areas of Mediterranean Interest
SPAs	Special Protected Areas

SPRFMA	South Pacific Ocean Regional Fisheries Management Agreement
SST	Sea Surface Temperature
TEEB	The Economics of Ecosystems and Biodiversity
TEV	Total Economic Value
TOWS-WG	Working Group on Tsunamis and Other Hazards Related to Sea-Level Warning and Mitigation Systems
TRIPs Agreement	Trade Related Aspects of Intellectual Property Rights Agreement
TWS	Tsunami Warning System
UKOA	UK Ocean Acidification Research Programme
UNCLOS (or LOS Convention)	United Nations Convention on the Law of the Sea
UNCTAD	United Nations Conference on Trade and Development
UNEP	United Nations Environment Programme
UNESCO	United Nations Educational, Scientific and Cultural Organization
UNFCCC	United Nations Framework Convention on Climate Change
UNGA	United Nations General Assembly
UNICPOLOS (or ICP)	United Nations Open-ended Informal Consultative Process on Oceans and the Law of the Sea
UNU	United Nations University
VMEs	Vulnerable Marine Ecosystems
WCPFC	Convention on the conservation and management of the migratory fish stocks in the western and central Pacific Ocean
WHC	World Heritage Convention
WIPO	World Intellectual Property Organization
WMO	World Meteorological Organization
WTO	World Trade Organization
WTP	Willingness to pay

Foreword by Irina Bokova,
Director-General of UNESCO

Our planet is mainly ocean, our survival depends on the ocean, and our capability to thrive in the future will depend on a healthy ocean.

We are the stewards of the world's ocean, and it is our moral obligation and responsibility to pass on a healthy ocean to future generations and to ensure that ocean benefits can be enjoyed by humanity as a whole in an equitable manner.

The ocean is essential to life and well-being. We need blue carbon to help store CO_2 emissions. We need blue economies to benefit from the enormous and diverse opportunities offered by the marine environment – from tourism, transportation, fishing and recreation, to communication, scientific research, as well as the commercial applications of scientific findings in health and industry. We need the ocean also to support peaceful exchanges between countries which share transboundary waters and resources, and as a basis for scientific cooperation and global trade.

In 2000, the world agreed on a set of visionary Millennium Development Goals, inspired by the lessons of the Declaration of Principles and Agenda 21, the Plan of Action of the United Nations Conference on Environment and Development, held in Rio de Janeiro in 1992. Much progress has been achieved across the world, but this is insufficient and uneven. As countries accelerate towards the deadline of 2015, we must shape a new global sustainable development agenda, with the ocean at its heart.

The stakes are high and time is against us. The climate system has already been significantly altered by human activities. During Earth's history, the ocean has ensured the planet's resilience against climate variability – it is still assisting us today, but there is a mismatch between the pace of human-driven carbon emissions and the ability of the ocean to respond to alterations of the climate system. The alarm bell is ringing – we must take immediate steps to allow the world ocean to continue acting as a life-support system.

Everywhere, there are signs that we are crossing the boundaries of the planet. This is evident in the depletion of the capacity of the ocean to provide food and fibre in the form of fish proteins and other proteins, as well as the decreased water quality and related impacts on seaweed farming. We see this in the increased acidity of the ocean, which leads to vulnerability and the eventual disappearance of coral species on which the livelihood systems of many coastal peoples and populations depend. The rise in sea level is having a direct impact on small island systems and coastal areas, as well as on human settlements all around the world. These are just a few signals of deep distress.

This context is a call to action – to protect and use sustainably the ocean areas beyond national jurisdiction and to cherish them through collective stewardship and also greater scientific research. We need to take steps today to ensure that a healthy and productive ocean can continue assisting people and economies to thrive around the globe.

Ocean science is essential for this. In 1960, states created the Intergovernmental Oceanographic Commission, entrusting UNESCO with the responsibility of hosting the Commission and assisting in its mission in relation to ocean observation and forecasting. The role of the Commission has included risk reduction, marine sciences and capacity-building, and it is extending to informing policy and reinforcing the science–policy interface. To respond to current challenges, I am convinced that we need a strong, dynamic, and effective Intergovernmental Oceanographic Commission, to help states to strengthen the sustainability of the ocean and to derive shared benefits from healthy and resilient marine ecosystems.

This is also why it is important to respect the spirit of the 1982 United Nations Convention on the Law of the Sea. This landmark convention is designed to provide a significant contribution to fostering peace, justice, and equitable development for all peoples, especially for developing countries most in need.

This book is an important contribution to UNESCO's commitment to support and the sharing of knowledge to ensure a healthy and productive world ocean for the benefit of all. I wish to thank all of the authors for their engagement with these tasks. This matters for every society, today and tomorrow, in the run-up to 2015 and well beyond.

Irina Bokova

Foreword by Thomas E. Lovejoy

Elizabeth Kolbert, in *Field Notes from a Catastrophe*, paints a vivid picture of an oceanographic sample of sea butterflies or pteropods (tiny snails with 'feet modified like little flapping wings to keep the animals from sinking in the water column'). Held overnight, by the morning the CO_2 exhaled by the tiny animals had rendered the seawater in the container sufficiently acid that it etched their shells. This was one of the first inklings of ocean acidification, now recognized as a major consequence of global change, and an indication of the sensitivity of the marine environment.

Picture if you will, a coral reef submerged in a pleasant tropical sea. It is a colourful blend of multiple stakeholders: the coral itself, the aquatic organisms that feed from it, their waste that adds in time to the cycle of life, and the local human communities which have relied on the reef's productivity from time immemorial. It lies in delicate balance, where the alteration of one of these aspects leads to the collapse of the system. Even a change in the composition of the water itself can bring the whole system down.

This picture is an appropriate metaphor for the complex environmental and ecological change that characterizes the oceans today. The future of the planet's oceans involves diverse constituencies, from indigenous fishermen to UN lawyers, global tourists, and developing world policy makers. If each of these actors were asked how they saw the ocean, their answers would be quite different. The chapters in this book illuminate our knowledge of the oceans today, while reflecting the aspirations and expectation of these diverse constituencies.

Although 'global warming' is but one element of 'global environmental change', the warming of the oceans is nonetheless important in its own right. The warming is pervasive and goes 2000 metres deep and yet deeper. There will be dramatic effects on moisture flux, nutrient supply, and climate variability. Thermal expansion of water, in addition to melting ice on land and glacier retreat worldwide, will cause dramatic sea-level rise. About 600 million people live within 10 metres of sea level.

There will be drastic effects on ocean biodiversity as well. It appears that this has already caused a global fall of 6% in primary production. Were agricultural production to fall by a similar amount there would be instant alarm. This is similarly grave.

The impacts observed are the initial consequences of a 0.75 °C global increase. They pale in comparison with those that would ensue with a number of degrees Celcius increase, exacerbated by declining efficiency of carbon sinks and accelerating carbon–climate feedbacks.

Since 1950, the pH of the oceans has decreased by 0.1 pH unit (a logarithmic scale). This amounts to a 30% increase in acidity over a handful of decades. In contrast the oceans have been nurtured by the same 10,000 years of stable climate that brought forth human civilization.

In 2009 I had the honour of moderating the Expert Panel on Ocean Acidification in the New York UN Headquarters – an event that was organized by the UN Department of Economic and Social Affairs, the UN Division on Ocean Affairs and the Law of the Sea, and the UN foundation. Since 2009, the situation has, of course, only worsened and continues to do so with serious implications for the marine biota and for humanity.

The projected ocean acidification is not only higher than at any time in the last 120 million years, it is also increasing faster than anything documented over that time. It is accompanied by disruption of biogeochemical cycles, most notably that of nitrogen. Land run-off and nitrogen deposition from the air have the effect of nitrifying coastal and estuarine regions such as the Baltic Sea and the Gulf of Mexico. The consequent 'dead zones' have major deleterious effects on productivity and economic sustainability. Globally, they have doubled in number every decade for the past four decades.

Smaller and often forgotten organisms may seem trivial compared with such big issues. While pteropods may seem to be an evolutionary curiosity they constitute a critical base for food chains in northern oceans.

Mangroves in coastal and estuarine regions provide substantial ecosystem services. Their dense wall of biomass blocks storm surges and erosion of coasts, and blunts tsunamis. They are nurseries for fisheries, provide local communities with fishing hot-spots, and shelter countless organisms. They are also an important 'blue carbon' player in sequestering carbon. In spite of this, mangrove extent is expected to fall by 10 to 15% by 2100 (and has already fallen by 35% in the past 25 years, according to the Millennium Ecosystem Assessment).

Viewing Earth systems collectively, marine ecosystems can be shown to provide around two-thirds of ecosystem services globally. In addition to the obvious bounty of seafood and other resources, which account for trillions of dollars every year in the global economy, the role of the ocean in climate regulation, biogeochemical modulation, transport, and communications provides a large underpinning contribution.

This volume makes a very good case for determining the economic and non-economic value of biodiversity-dependent ecosystem services at multiple scales. Without question, the cost of inaction would be much higher than adopting a long-term approach of economic investment in ecosystem restoration and maintenance of existing ecosystem services.

One of the main findings of the Third Global Biodiversity Outlook (GBO3) (presented in Nagoya, Japan, in 2010 under the auspices of the Convention on Biological Diversity) was that continued overfishing would damage marine eco-systems to the point of collapse of fish populations. UN Secretary-General Ban Ki-Moon put it forthrightly in GBO3: '. . . conserving biodiversity cannot be an afterthought once other objectives are addressed – it is the foundation on which many of these objectives are built'. Restoration of ecosystems is essential to bring such services back to full force and value.

This is more than good science. It involves communication, international relations, politics, and law (e.g. national jurisdictions). For example, questions of national legal rights to genetic resources from the deep seabed and the high seas are essential to resolving possible subsequent public–private sector transfer. This newly recognized global gene pool is likely to re-open the debate on the law of the sea. Some fresh ideas about the evolving legal framework of the oceans are discussed with respect to biodiversity/bioprospecting.

Nonetheless, operational oceanography and international scientific cooperation on ocean-related matters should proceed and not wait for the legal and policy framework to be sorted out. If we can measure the ocean's influence on weather, a common good indeed, why can we not agree on measures to deal with other such common goods?

This international cooperation requires the best of interdisciplinary and syner-gistic science. Such a process has led to improvements in effective tsunami detection and preparation. Perhaps lessons might be learned from the legal frame-work drawn up for outer space or for Antarctica of the value to management of the high seas and deep seabed.

The oceans cry out to be managed in a holistic manner, such as applying an ecosystem approach that is inherent in the Convention on Biological Diversity. Perhaps there is some value or inspiration in emerging theory such as *Panarchy* (as applied to the study of the Arctic system in Chapter 5 of this book). Tools such as marine spatial planning show that some solutions to problems of the oceans already exist.

This volume highlights urgent ocean issues derived from established and developing science, and then presents some possible international legal and policy frameworks which offer sound solutions. The contents are relevant to policy makers, students, the private sector, and, indeed, any stakeholder. Examples range from local to subregional to global. It discusses the reciprocal nature of the effect of climate change on the oceans and the biodiversity therein.

Humanity tends to think locally as well as in the very short term. In fact we are disturbing global systems and the way the planet actually works. Why should near future cost–benefit outweigh that of the distant future when we are discussing such large and long-term problems? The good news is that there is great potential for better management of the oceans and a more complete oceans policy for the Ocean Planet.

Thomas E. Lovejoy

Preface

Humans are generally conscious of the value of the world ocean: no ocean, no life; a thriving ocean, a thriving humankind. Yet, when it comes to acting in a sustainable manner in our relationship with the ocean, we are often at a loss; it appears that the scale and pace of human action do not match the pace and scale of ocean processes well enough for the two to counterbalance and act synergistically. It appears that, from a lack of consciousness about what is right and what should be done, we prefer not to act. And it also appears that because of shortsightedness in terms of the human print on the ocean environment and life therein – and really 'shortsightedness' is the word, because the ocean is often out of sight – we comfortably (or less comfortably) hide behind our incapacity to see fully and to understand how increasingly modified and degraded the world ocean is, and we hope that it will recover by itself.

The world's oceans cover approximately 71% of the planet's surface, with an estimated volume of 1,335 billion cubic kilometres (Eakins and Sharman, 2010).[1] It is difficult to grasp the meaning of such immense figures. Intuitively, however, we all realize that such a large surface and volume must contribute substantially to the Earth's climatic balance. We also now know that life in the oceans is present all the way from the surface to its deepest areas. Evolutionary biology has demonstrated that the oceans have played a major role in the origin and shaping of life on Earth. Over historical time, humans have always depended greatly on the oceans for food, health, transportation, trade, and recreation – among other uses (Independent World Commission on the Oceans, 1998;[2] cf. also Chapter 2 of this book). In turn, a healthy ocean cannot be without proper management in light of the fact that over seven billion humans live on the planet today. In today's globalized economy and in an era characterized by global climate change and a global biodiversity crisis, the interdependence of humans' survival and ocean health is stronger than ever.

What are the key processes in the ocean that we cannot afford to alter without altering our own chances of a sustainable future for humankind? In which direction is human dynamics evolving and what are its consequences for the world ocean? And how can we match, as best we can, our knowledge (as well as our ignorance) with the tools – scientific, observational, management, policy, normative, educational – which we have developed in our long-standing relation with the ocean? These are the main questions which this book attempts to tackle. And by addressing itself to a mixed constituency of decision and policy makers, specializing students, and expert practitioners from multiple sectors, this book attempts to provide solutions to these questions in an integrated manner.

Building on the active efforts of the scientific and policy communities over the past 20 years aimed at solving issues related to the implementation of the law of the sea regime since the enactment of the Convention on the Law of the Sea, in 1994, this book describes the main known, emerging, and unresolved issues related to the oceans and the marine environment in general with which we are faced today and are likely to be faced with in the near future.

This book also attempts to provide options, supported by a thorough rationale, for a better science-based management of the oceans, for taking into account other relevant knowledge than scientific knowledge, and for enhancing the relevance and effectiveness of the legal and policy regimes governing the oceans. In fact, while the oceans are facing multiple problems, we already have available to us now a number of relevant scientific, policy, and participatory tools to implement solutions and, thus, we can render the oceans sustainable in this century.

The book targets graduate and postgraduate university students, policy makers, private sector operators, and other non-governmental stakeholders active in the area of ocean affairs. The intention of this book is to provide them with a comprehensive overview of ocean-related issues today and to point them towards approaches and tools needed to tackle these issues. The choice of targeting multiple audiences is justified by the assumption that the problems faced by the oceans today are complex because they are multifold and interconnected in nature. A comprehensive book in this area must speak to all constituencies concerned with ocean-related issues. The book also aims at building bridges between such diverse stakeholders.

In fact, the sustainable development of the oceans can only be achieved through the collective and coordinated action of multiple stakeholders, based on a common understanding of the problems and of the possible solutions available to us. Some of these solutions also point to possible opportunities, both in terms of the sustainability of the world's oceans as part of the life-support system on Earth, and of new areas for research and development, green growth, and proper governance.

Throughout the book, a major effort has been made to provide a variety of perspectives in disciplines and to connect, whenever appropriate and possible, these disciplines. It is also the intention of this book to present a balanced perspective between theory and practice, and the reader will be able to refer to the many practical examples and case studies around the world that are bridging multiple scales (global, regional, national, and sub-national) in relation to the questions raised in the book.

Overall, the book endeavours to provide a reference work and, therefore, major effort has been concentrated on synthesizing information available in the already published peer-reviewed literature and providing the reader with a comprehensive bibliography for her or his perusal. The book also attempts to bring added value to existing literature in the area of the oceans by bringing new evidence, while relying on a more narrative style, and using the opportunity to articulate at length complex notions, as opposed to the more technically complex prose style that characterizes scientific journals. Whereas, in this context, it is unavoidable that some chapters may use more technical language than others, an overall effort has been made to keep the language relatively accessible, and highly technical remarks have been presented in the chapter endnotes whenever possible.

Each chapter in the book addresses a set of specific issues and can therefore be read as a stand-alone contribution for those interested in those issues; for example, the impacts of human-induced climate change on the oceans; the various dimensions of the marine genetic resources debate; or, the adequacy of the current legal and policy regime for the oceans. Depending on the reader's interest and background, these parts can be read in isolation, in that they represent coherent separate units (albeit interlinked) within the book.

The chapters in the book present the following elements for study and reflection: an introduction to the main issues and opportunities facing the oceans today; the main human uses of the oceans and consideration of the current governance framework; perspectives on changes in the oceans and implications of such change, seen from a physical, and from a biological and ecological perspective; a specific case in point of the implications of these changes as exemplified by the emergence of a new ocean in the current generation's life span: the Arctic; implementing the ecosystem approach in the oceans; an introduction to the use of integrated economic valuation for marine ecosystems; institutions and governance for the oceans from the perspective of the role of international scientific cooperation; the main dimensions of the issue of deep-sea genetic resources – scientific and legal – as an example of a no longer emerging and yet major unresolved issue at the policy level; a reflection on whether the current regime on the law of the sea should be seen as the legal framework for all activities in the sea; possible analogies between problems and solutions of the law of the sea regime and the legal regime of outer space; and conclusions on possible ways forward so as to attain a more sustainable relationship of peoples with ocean areas and resources.

The editor's hope is that, ultimately, the contents of this book will assist in a concrete manner the process of decision-making related to ocean affairs at the international as well as multiple levels, with a view to agreeing on a bold agenda in relation to the contribution of the oceans to sustainable development.

Notes

1 Eakins, B.W. and Sharman, G.F. (2010). *Volumes of the World's Oceans from ETOPO1*. Boulder: NOAA National Geophysical Data Center.
2 Independent World Commission on the Oceans (1998). *The Ocean: Our Future*. Cambridge University Press, 248 pp.

Acknowledgements

The editor would like to thank all of the contributors to this book, who have made the realization of this project possible. They are all distinguished scholars, academics and practitioners, who carry out important work in order to elucidate how the ocean functions, and how best humans can benefit from it and also maintain its health.

His appreciation also goes to the many colleagues and friends who have provided critical comments which have enormously enhanced the quality of the book: Elisabeth Back Impallomeni, Elva Escobar, Valentina Germani, Kristina Gjerde, Lyle Glowka, Paul Oldham, Juan Luis Suáres de Vivero, and Unai Pascual. Thanks also go to Peter Bettembourg, William Ingram, and Emily Briant who contributed precious editing remarks on some of the book chapters.

The editor wishes to thank colleagues at Cambridge University Press and at UNESCO who have provided him with the needed support throughout the production of this book and thus helped to bring the whole project to a successful outcome.

Special thanks go to the Flemish government of Belgium, which has decided to purchase and make available a significant number of copies of this book to the targeted audiences in developing countries and at key international meetings related to ocean and marine affairs, with the aim of facilitating those dialogues, debates, and deliberations.

The editor wishes to take this opportunity of acknowledging the important roles as mentors and friends of Gunnar Kullenberg, Biliana Cicin-Sain, and Peter Bridgewater over his 20-year span of dealing with the complex problems of the oceans and the environment at international and intergovernmental levels.

The editor wishes to dedicate this book to Françoise Khoury, who has showed him how to apply the right judgement in life and has supported and encouraged him throughout all these years.

1

Issues regarding oceans and opportunities: an introduction to the book

SALVATORE ARICÒ

1.1 Recognizing change in the oceans and its implications

Global change manifests itself in relation to multiple systems and sectors, including in the oceans. It is critical to understand the main features of these changes – which are in a large part human induced – as well as their effects on ocean life and processes, human well-being, and sustainable development in general, in order to frame possible responses to issues concerning the oceans today.

Providing an overview of change in the oceans is a challenging task. Perspectives on current and future ocean conditions ought to be based on documented change and variations in the physical, chemical, and biological nature of the marine environment over basin-wide zones and decadal or longer time-scales.

The active scientific community has been pursuing research on and observations of the marine environment since the genesis of modern oceanography, at the end of the 19th century. The second half of the 20th century has seen the development of international scientific research cooperative programmes such as the International Indian Ocean Expedition (IIOE) (1957–onward).[1] We now rely on an increasingly operational Global Ocean Observing System (GOOS), which encompasses ocean observing systems at multiple scales (national, subregional, and regional). Systems to detect ocean hazards, such as tsunami events, are being made operational on an ocean basin basis (cf. Chapter 8); daily ocean forecast services such as weather forecasting systems and services are becoming a reality.

Oceans are changing before our eyes; the magnitude and pace of those changes deserve special attention, further study, reflection and – it appears – urgent action.

One such change is ocean warming. Storage and transport of heat in the ocean are central drivers of the Earth's energy budget, which affects climate variability, the storm regime, moisture flux, and which can trigger expansion of the oceans, which is a major contributor to rising sea levels. Increasing temperatures also have an impact on nutrient supply to primary producers and influence species invasion.

The increase in ocean heat content accounts for most of the increase in the Earth's heat levels, which is corroborated by model simulations on the evolution of sea surface temperatures. The observed warming has penetrated at least 2000 metres into the deep sea. Monitoring variability in ocean heat content is crucial in an era of global climate change.

Changes in ocean salinity at the global scale have also been observed. These changes are coupled with surface forcing of the hydrological cycle, and can lead to increased stratification, with related reduced nutrient supply to the euphotic zone. Of particular concern is the documented reduction of the Arctic ice cover since the late 1970s.

Ocean circulation is critical to correction of the global heat imbalance and in regulating the global hydrological cycle, and also in contributing towards ensuring the abiotic and biotic conditions necessary to the proper functioning of marine ecosystems. It is important to develop an overview of the complex dynamics of ocean circulation, as well as identifying examples of change in various parts of the world's ocean. These result in changes in the present climate of, for example, north-western Europe; these changes are further exacerbated by positive feedback mechanisms.

The biogeochemistry of the ocean is also changing as a consequence of atmospheric greenhouse gas levels. Prior to 1750, atmospheric CO_2 concentrations had been at 260–280 ppm for about 10,000 years; in 2011 they passed 390 ppm. Consequently, different ocean basins behave in different ways: those which can out-gas (emit CO_2) behave as net sinks but, in some other cases, large ocean basins may approach saturation. Changes in ocean biochemistry also need to be seen in relation to a less dynamic ocean circulation and increased density stratification due to surface warming; both of which are expected in a warmer climate, as these will slow down the vertical transport of carbon into the deep ocean.

The increase in CO_2 in the atmosphere and the passing of roughly a third of this increase into the world's ocean have resulted in ocean acidification, which carries potentially understated consequences for life in the oceans. The projected level of ocean acidity would be higher than anything experienced during the past 120 million years, and the ongoing acidification of the ocean is occurring faster than has ever been documented previously. Current changes in pH are likely to affect the structure and functioning of marine ecosystems through interference in shell formation due to the numerous taxa of benthic calcifiers (molluscs, echinoderms, crustaceans, bryozoans, serpulid polychaetes, foraminifera, sponges, and corals).

Biological production is also being affected by changes in the nutrient balance. This is mainly due to increased discharge of nitrate from the land due to fertilizer use and to nitrogen deposition from air. We need to be wary of the risks of eutrophication and the spreading of dead zones, where reduced levels of dissolved oxygen determine a condition known as 'hypoxia' (below 2 ml/l) or even 'anoxia'. The most affected areas are estuarine and coastal areas. Dead zones are characterized by a diversion of the energy flow towards microbial pathways, which has negative consequences for higher trophic levels in the ecosystem and for the proper functioning of the ecosystem as a whole. The observed extension of the oxygen minimum zones has consequences for the biogeochemical cycling of carbon, nitrogen, and many other important elements. Hypoxia and anoxia determine alterations in the composition of communities and have implications for ecosystem functioning. Marine organisms respond differently to hypoxia, and fish and crustaceans tend to be the most sensitive to reduced dissolved oxygen. The spreading of dead zones has to be seen in relation to coupled physical conditions in the water column (stratification, mixing, temperature, water exchange, circulation, and the air–sea interaction).

More generally, large-scale changes in ocean conditions combined with pollution, over-harvesting of fish stocks, and unsustainable fishing practices are determining the disruption of several ocean resources, which in turn has an impact on human well-being.

Rising sea level has also been observed and projected, which will affect the 600 million people who live near the ocean within 10 m of sea level. Global mean sea level has risen by 1.7 ± 0.2 mm/yr over the period 1900–2009 and by 1.9 ± 0.2 mm/yr since 1961. Several factors contribute to this phenomenon, including thermal expansion between sea level and 3000 m depth and the melting of glaciers and ice caps. We now have a comprehensive picture of the sea-level rise budget, and the related predictions confirm the gravity of the situation, in spite of the uncertainty of some of the projections. There is also a need to look at the projected rise in sea level, extreme storm surges, and waves from the standpoint of protecting life, property, and infrastructure in coastal regions.

High-latitude regions are particularly vulnerable to change. The Arctic is undergoing change which has the potential for significant consequences in relation to the global climate system, marine ecosystems, and human activities such as oil and gas exploration/exploitation and shipping. The salinity-driven stratification in the upper layers of high-latitude oceans is being altered as a consequence of an accelerated hydrological cycle and increased ice melting during the summer season. This has consequences for the food web and also in relation to invasion by species new to these areas. In the Arctic, documented change (decreases in the extent, thickness, and duration of the ice cover and the measured increase in ocean temperature in some parts of the Arctic Basin) is occurring relatively fast in comparison with other regions. Because of the interdependence between ocean basins, these changes have possible consequences at the global scale. A growing concern is that related to the Antarctic and Southern Ocean system, which is adapted to very stable conditions. In this area, however, observed and projected sea surface temperatures are too low to indicate whether significant ice melting is occurring, which calls for continued scientific observations of the Antarctic and Southern Ocean domain.

Future changes in ocean circulation and stratification are highly uncertain. The overall reaction of marine biological carbon cycling to a warm and high CO_2 world is not well understood, which is a cause for concern in the light of the ongoing debate about geo-engineering the planet.

Global change also has various consequences on ocean life, marine biodiversity, ecosystem functioning, and the services which ecosystems provide. These changes also concern deep-sea ecosystems which, contrary to the notion that they are extremely stable in terms of physico-chemical conditions, are also experiencing changes such as climate-driven temperature shifts, as a direct consequence of the prevailing surface climate conditions.

Climate-induced changes may differ strongly throughout the globe, especially along a latitudinal gradient (in general, warming appears more pronounced at the poles than at the equator). Changes in the global mean surface air temperature, not only over the past century, but also the unprecedented change in sea water temperatures recorded over the past 10–15 years, can affect the metabolic rates of marine organisms, population and community dynamics, and community structure and function. It is expected that change in the temperature of the global ocean will have consequences for the distribution of marine biodiversity throughout the globe.

The production of organic carbon in the surface ocean is also being affected, which has a consequence for food supply to deep-sea ecosystems. Change in the upper ocean temperature can affect the availability of nutrients for phytoplankton production and the subsequent flux of exported carbon to the deep-sea sediments. It appears that ocean warming has already caused a 6% decline in global ocean primary production. This tendency will continue through this century and will affect, in particular, the tropical ocean.

Long-term changes in plankton communities are likely to have an impact on commercial fish stocks, which in turn will have social and economic consequences. Changes in faunal abundance and composition, correlated to change in the productivity of the surface ocean, have been observed in the Pacific and the North Atlantic at greater than 4000 m depth. These changes, also expected in the equatorial abyss, result in dramatic effects on the functioning of deep-sea ecosystems due to the observed exponential relationship between biodiversity and ecosystem functioning in these ecosystems.

The combined effects of multiple stressors (environmental change – in particular, habitat destruction – associated with climatic change) on large ecological processes determine changes in larval dispersal and recruitment success, shifts in community structure and range extensions, and the establishment and spread of invasive species. The disruption of the connectedness among species may lead to a reformulation of species communities and to numerous extirpations and possibly the extinction of some species, and the loss of some species may have negative effects on ecosystem function. The role of alien species therefore should be assessed in a more integrated and dynamic context of shifting species' ranges and changing compositions and structures of communities.

Global change has an effect on multiple marine habitats. The response of marine systems to climate change will also depend on interactions with other human-induced changes in the marine environment, which will lead to different responses in different types of marine habitats and systems, from coastal water ecosystems (where current losses of seagrass ecosystems, mangrove forests, and coral reefs are expected to accelerate, with consequent reduction in services such as nutrient cycling, sediment stabilization, enhanced biodiversity and trophic transfers to adjacent habitats); to submarine canyon ecosystems and deep-sea basins (the former are being affected through the alteration of the seasonal transport of particulate organic matter associated with dense shelf water cascading; the latter by changes in water temperature due to climate change, with likely implications at the global scale).

For the first time in recent history, a new ocean is opening. The Arctic Ocean is highly variable and undergoing rapid change: there is evidence that within a few decades the Arctic will likely be an ice-free ocean during the summer season. Currently observed changes may reflect reorganization of the Arctic system. This may include new characteristics of production, diversity, function, distribution, and abundance of organisms. These changes are believed to be the beginning of a much more complex process than simply a northward shift in current species distributions.

Global change at the level of ocean regions and basins may entail non-uniform and non-linear responses in biological patterns. In the Arctic, such responses may affect the microbial and micro-phytoplankton levels, the size and/or concentration of primary producers, bloom regimes, the efficiency of the pelagic food web, migratory patterns and abundance of species aggregations and consequent impacts on commercial fisheries, and also cause the impoverishment of benthic communities due to reduced primary production as a result of early spring ice retreat. Moreover, the effects of these changes can also be cumulative in nature.

Exploring, understanding, adapting, and coping with an ocean which is changing before our eyes constitutes a unique opportunity for humanity, and our ability to understand and predict change will allow us to adapt and minimize the risks inherent in surprise.

The above-summarized changes in the oceans are documented, described, and analysed in detail in Chapters 3, 4, and 5 of the book.

1.2 Realizing that global change is largely determined by humans

We live in a planet the biophysical processes of which are increasingly impacted upon by human activity. Human uses of ocean areas have impacts at multiple scales.

Fisheries and aquaculture, shipping, recreation, energy, water extraction, military activities, underwater cables, scientific research, and mining represent the main uses. These all have potential and actual impacts on the marine environment, in the form of physical and chemical alteration of habitat features, biodiversity erosion and loss, climate change, ocean acidification, unsustainable depletion of species stocks, alteration of dispersal patterns, as well as social impacts such as disruption of local livelihoods due to the collapse of local natural resources, competition between industrial versus locally based fishing and aquaculture activities, and unsustainable tourism plans and initiatives.

Therefore there appears to be a clear need to reconcile human uses of ocean areas and resources therein with the imperative of preserving the functional integrity of the world ocean so as to ensure healthy and productive oceans for current and future generations. The first step in this direction is to fully realize the value of the oceans.

Marine ecosystems provide around two-thirds of the global aggregate of ecosystem services through the provision of seafood and other natural resources, worth trillions of dollars per annum, regulation of the Earth's climate, and the modulation of global biogeochemical cycles, water quality maintenance, opportunities for renewable energy production from the sea, trade-related services, and cultural and aesthetic benefits. There is a need to capture the broader change in social welfare which has derived from management intervention related to the marine environment.

Change in ecosystems can lead to welfare benefits even in the absence of a market price, hence the need for valuation methodologies in economics to value non-marketed benefits deriving from nature. The presumption that a cost or benefit arising in the near future affects our welfare more than that same cost or benefit arising in the distant future, also known as 'discounting', tends to privilege decision-making towards immediate benefits and delayed costs, which in turn impacts on sustainability and on inter-generational equity.

Accounting for marine ecosystem services such as those provided by marine protected areas can contribute to enhancing the socio-political enabling framework for such protection measures of the marine environment and provide a level playing field for the capacity of various stakeholders to affect decisions.

Valuation of various marine ecosystem services, including from the standpoint of monetized versus qualitative benefits, is therefore necessary. This is, however, rendered challenging, not only by considerations related to the need to develop tailored and sometimes novel methodologies specific to the marine environment, but also by bias in preference methods and limitations inherent in the application of stated preference techniques in the terrestrial domain to the marine domain. In fact, values derived in primary studies are context-specific in ecological and socio-cultural terms, and there is a clear paucity of data for open ocean environments.

Main human uses and impacts of ocean areas are described in Chapter 2, while the contribution of economic assessments to informing societal choices in relation to the marine environment is treated in Chapter 7 of the book.

While economic valuation does not provide all of the answers to informing decision- and policy-making, it can support the conservation outcome of the particular set of management actions envisaged. And, because the risks involved in commoditization of nature are real and there is a requirement to take into account not only 'credence attributes' of marine habitats (related to perceptions) but 'sensory' or 'experience' attributes as well (this is particularly true for the open and deep ocean environments), there is a need for economic valuation in the future to focus on both valuable as well as under-researched ecosystem services from the oceans (cf. Chapter 7).

1.3 A holistic and responsible approach to ocean management

The progressive realization of human impact on the oceans, coupled with the recognition that we should maintain healthy and resilient oceans rather than merely continuing to consume our natural capital, calls for the application of an ecosystem approach to the management of oceans and their resources.

The ecosystem approach has arisen largely due to the current crisis faced by biodiversity and natural resources, to which single species and single sector management approaches cannot respond adequately. In general terms, an ecosystem approach can assist in dealing with multiple impacts of human activities while at the same time maximizing long-term economic, social, and cultural benefits, and in mobilizing a wide range of stakeholders by defining use and conservation priorities.

Many different ecosystem approaches are available to us, including traditional and indigenous approaches to the use of the oceans and the resources therein. Analysing and comparing these approaches allow us to derive certain common principles which apply, as reflected in the United Nations Convention on the Law of the Sea (UNCLOS, or LOS Convention) and its 1995 Agreement for the Implementation of the Provisions of the United Nations Convention on the Law of the Sea of 10 December 1982, relating to the conservation and management of straddling fish stocks and highly migratory fish stocks (the UN Fish Stocks Agreement).

Case studies and lessons learned on the application of the ecosystem approach to the marine environment show that the CBD (Convention on Biological Diversity) Principles on the Ecosystem Approach are currently implemented through a variety of tools and approaches, including national and regional bioregional planning processes, marine protected areas, rehabilitation of ecosystems, fostering sustainable use of biodiversity, and participatory and bottom-up activities promoted by local stakeholders.

Currently, integrated marine and coastal area management is expanding to encompass the bioregional or large marine ecosystem scales, in an effort to also promote the application of the ecosystem approach to the management of marine areas beyond national jurisdiction. It is in this context that current efforts by the CBD related to the identification of 'ecologically and biologically significant areas' and the efforts under the Food and Agriculture Organization of the United Nations (FAO) to identify 'vulnerable marine ecosystems' should be considered.

The FAO ecosystem approach to fisheries is also to be recognized; this approach entails significant implications for the operation of Regional Fisheries Management Organizations (RFMOs – cf. Chapter 2), and the principles guiding its application appear to be largely consistent with those for the ecosystem approach developed in the context of the CBD. The FAO approach is also being increasingly tested and pursued in large marine ecosystems in various regions of the world, although current efforts have been focusing on the planning process related to an ecosystem approach to fisheries rather than on its actual implementation.

Indigenous holistic approaches to marine management can also be consistent and greatly contribute to the implementation of the ecosystem approach. Relevant case studies from different regions of the world demonstrate that, for example, in the Pacific region, communities that depend on resources from the sea for their subsistence tend to have an integrated view of peoples and nature, reflecting the perceived connection between all living things and their environments. In this regard, governments can learn significantly from these communities in their efforts towards fully embracing and applying an ecosystem approach while pursuing an integrated approach to ocean management.

Experiences and lessons learned in implementing the ecosystem approach are described in detail in Chapter 6.

While the policy relevance of applying the ecosystem approach is widely recognized, the application suffers from the frequent lack of availability of adequate analytical and scientific tools necessary for its implementation. Lack of an understanding of what the ecosystem approach entails may play a critical role in hampering its application. Hence it is important to present lessons derived from the successful implementation of the ecosystem approach, rather than its theoretical aspects. Relevant case studies must illustrate not only environmental and economic considerations, but also social and cultural factors affecting the management of ocean areas and resources, including from the perspective of how contemporary sector-based management approaches may have undermined the important contribution of indigenous and local knowledge systems to ocean management.

1.4 Towards an effective international governance of the oceans

An important step towards an effective governance of the oceans is the design and implementation of collaborative programmes involving multiple nations, especially in relation to transboundary ocean areas, shared resources, as well as areas beyond national jurisdiction. In this regard, a significant contribution can be provided by successful international scientific cooperation programmes.

Marine science and technology have an important role as regards contributing to good ocean governance through the provision to decision makers of information on marine science and technology. Transfer of technology and appropriate policies to ensure the production and dissemination of scientific information are also essential conditions to maximize the contribution of marine science and technology to a better governance of the oceans.

UNCLOS contains extensive provisions related to marine scientific research (MSR) and to development and transfer of marine technology. The Convention calls on competent international organizations to establish general criteria and guidelines to assist states in ascertaining the nature and implications of MSR. Since the adoption of UNCLOS, major technological developments have taken place in relation to the study and monitoring of the oceans, such as the use of satellites, aircraft, ships of opportunity, autonomous vessels, buoys, and floats. Hence additional guidance on how to deal with such developments can be provided under the above-mentioned criteria and guidelines.

The provisions of UNCLOS related to MSR do not apply to the collection of meteorological information in the marine environment under the World Meteorological Organization (WMO). These activities are routine observations for the collection of data which were recognized by WMO member states to be of common interest to all countries and to have universal significance, and collection of meteorological data for weather forecasts and warnings is recognized under the International Convention for the Safety of Life at Sea (SOLAS). The United Nations Framework Convention on Climate Change (UNFCCC) also calls on the parties to support and further develop international and intergovernmental programmes and networks or organizations involved in research, data collection, and systematic observation, including promoting access to, and the exchange of, data and analyses obtained from areas beyond national jurisdiction.

The development of tsunami early warning and mitigation systems can be seen from the perspective of the emergence of new regional regimes dealing with marine scientific research, especially with regard to the institutional arrangements put in place for such systems, which include governance mechanisms, operational standards and requirements, and separate agreements for the provision of tsunami watch services. These arrangements are based on commitment to principles in the policies of competent international organizations to establish general criteria and guidelines to assist states in ascertaining the nature and implications of marine scientific research (the Intergovernmental Oceanographic Commission (IOC), in the case of tsunami warning systems), rather than on compliance with binding agreements for their development and implementation, nor on modalities for their enforcement. An analysis of the evolution in the dynamics of the intergovernmental regional coordination groups established for tsunami early warning shows the adaptive nature of these governance mechanisms which reflects change in key concepts and principles, change in the group of leading actors, and expansion in the functional scope of the early warning systems.

The international Argo Project (the set-up and systematic running of networks of facilities for the continuous monitoring of ocean parameters such as temperature, salinity, and velocity of the upper 2000 m of the ocean) stresses the need for participating countries to agree on some form of international legal regulation related to the employment of thousands of voluntary observing ships and ships of opportunity, tide gauges, surface drifters, subsurface drifters, moored buoys, and profiling floats that may drift into national exclusive economic zones (EEZs). The Argo Project provides a clear example of a very successful programme aimed at collecting, organizing, and making available data on a free, unrestricted, and real-time basis which – as in the case of the data generated through the tsunami early warning system – are critical to the protection of life and property (information and data collected through the International Argo Project contribute to the development of, for example, sea-level rise projections). The project is intended to operate in a way which is consistent with UNCLOS. Yet, different interpretations exist with regard to whether it should be seen as an activity characterized as operational oceanography, not governed by the UNCLOS provisions related to MSR, or as part of the latter. A central feature of the Argo Project is that participating states are divided into implementers (i.e. deployers) and coastal states, whereby the floats belonging to the implementer may operate on the high seas or the territorial sea or the EEZ of the implementer or any other coastal state, while coastal states play a passive role by letting the implementer collect data from waters under their jurisdiction.

Activities falling under operational oceanography are among the most important developments in oceanography over the last forty years. These programmes are not regulated by UNCLOS or other legal instruments per se, and yet they demonstrate how states can collaborate internationally in MSR and how this has evolved since the adoption of UNCLOS. The competent international organizations in MSR are well positioned to move MSR programmes forward through international cooperation, and to ensure that developing countries can participate in them on an equitable basis, through adequate financial support and technical assistance.

The contribution of international scientific research to a better governance of the oceans is illustrated in detail through the two above-mentioned case studies in Chapter 8.

The many human uses of ocean areas and resources, which are perpetuated by multiple stakeholders at multiple levels, also call for a coherent, nested formal governance framework for ocean management.

Among the many international legal instruments relevant to the oceans are the UNCLOS and its implementing agreements, the CBD, the FAO instruments on fishing, relevant provisions under biodiversity-related conventions other than the CBD, and in relation to whaling, the IMO (International Maritime Organization) instruments on shipping, prevention of pollution and contamination, and on the designation of special areas, provisions related to the preservation of underwater cultural heritage, regional conventions and action plans, and instruments underpinning the work of the RFMOs. Chapter 2 presents a comprehensive review of such governance regimes.

Gaps both in the global as well as regional regimes have been documented. These include the limited reflection of modern governance principles (e.g. the ecosystem approach) and of tools (e.g. environmental and strategic impact assessment procedures); lack of compliance and enforcement mechanisms; lack of coverage of emerging issues such as bioprospecting and ocean geo-engineering, and of issues not benefiting from specific attention in the context of existing instruments, such as impacts of laying of cables and pipelines, of oil and gas exploration, and exploitation activities on extended continental shelves; and lack of coverage with respect to specific fish stocks at risk. Moreover, regional conventions related to the marine environment as well as the scope of RFMOs cover very limited portions of areas beyond national jurisdiction.

1.5 The adequacy of the current regime on the law of the sea (in the light of emerging and unresolved issues)

In a concise yet illuminating way, Treves (2010) illustrates the complex framework of the law of the sea and its institutions and the development of the law of the sea since the adoption of UNCLOS, achievements, and challenges for the future. Treves demonstrates that the adoption of the LOS Convention, its entry into force in 1994, and the now 165 associated ratifications and accessions have led to the radical transformation of the law of the sea from mainly customary law-based into a mainly treaty-based branch of international law. Customary international law continues to exist, but the relationship between customary and treaty rules has been deeply transformed.

In fact, in the early days of the law of the sea, the dominant (and prudent) view was that the correspondence of any general statement with customary law had to be assessed on a case-by-case basis. Today, with the exception of rules concerning the establishment and functioning of institutions, the rules set out in the LOS Convention correspond with customary international law; while customary law continues to play a role in relation to the international law of the sea, the widely recognized role of the LOS Convention as the 'constitution of the oceans' de facto imposes utmost caution in determining the emergence of customary rules in the context of the LOS Convention.

Treves (2010) reviews the institutions established in compliance with the rules of the LOS Convention, which are essential in making the Convention operational through the development of substantive rules and their updating. These are the Meeting of the States Parties to the Convention; the International Seabed Authority (ISA, or ISBA); the International Tribunal for the Law of the Sea (ITLOS); and the Commission on the Limits of the Continental Shelf (CLCS). For each of these, shortcomings and the challenges and opportunities that they face can be identified. Clarifications have been made regarding the relationship of the LOS Convention with organizations which existed before its entry into force, as well as in relation with generally accepted international rules and standards, such as laws and regulations of the coastal state concerning design, construction, manning or equipment of foreign ships exercising innocent passage, and international rules and standards setting minimum requirements with which domestic laws and regulations must comply, such as national laws, regulations, and measures concerning dumping.

The LOS Convention has brought the law of the sea under the jurisdiction of international courts and tribunals, with the notable exception of disputes related to fisheries, marine scientific research, delimitation, military activities, and enforcement activity in the exclusive economic zone. The settlement of disputes is entrusted to a plurality of adjudicating bodies: the ITLOS, the International Court of Justice (ICJ), and arbitration tribunals. Treves (2010) refers to the 'deterrence' effect of compulsory settlement, in that the various disputes have not appeared before courts and tribunals, but have rather been settled by the parties. Agreements concerning the law of the sea such as the 1995 UN Fish Stocks Agreement and the 2007 Wrecks Removal Convention also adopt the dispute-settlement provisions of the LOS Convention.

Treves (2010) explains the difficulties involved in amending and revising the LOS Convention because the Convention was set up to ensure the *stability* of its rules more than their *adaptability*. This makes it difficult to adapt to new problems of the oceans. Tools for change do exist outside the LOS Convention's framework. The 1995 Fish Stocks Agreement mentions the purpose of implementing certain provisions of the LOS Convention, while others such as the UNESCO Convention on the Protection of the Underwater Cultural Heritage (CPUCH) of 2001 do not. However, all of the agreements that fall outside the framework of the LOS Convention contain provisions which recognize the particular role of the LOS Convention, notably for the settlement of disputes. Other forums are relied upon to discuss issues emerging and unresolved issues related to the oceans, namely the UN Open-ended Informal Consultative Process on Oceans and the Law of the Sea (UNICPOLOS), the negotiations leading to the adoption of the yearly resolutions on the oceans and on fisheries under the UN General Assembly and, more recently, the Ad Hoc Open-ended Informal Working Group to study issues relating to the conservation and sustainable use of marine biological diversity beyond areas of national jurisdiction (cf. Chapters 9 and 10).

Treves (2010) advocates the notion of 'the primacy of the LOS Convention' when dealing with such emerging issues, as well as in relation to agreements that fall outside the LOS Convention framework. The practice being that few states dare to oppose the LOS Convention or even specific provisions of it. However, the LOS Convention does not have answers to all of the problems of the oceans. This applies, in particular, to subjects that are treated too summarily or are not con-sidered at all, as exemplified by the issues of straddling and highly migratory fish stocks and of underwater cultural heritage. Another example is that of issues related to the legal regime of genetic resources in the seabed beyond the limits of national jurisdiction, for which a specific process (the Ad Hoc Working Group) has been set up, which may lead to a process comparable with that which brought about the Third UN Conference on the Law of the Sea (cf. Chapters 9 and 10). Another category is that of issues which fall outside but also impinge on the LOS Convention, such as international environmental law, international human rights law, international trade law, the law of international security including terrorism and migration by sea, and developments in regional integration and cooperation and illegal immigration by sea.

Treves (2010) concludes with the remark that while most of the problems that fall in the first category can probably be dealt with through the built-in flexibility of the LOS Convention, the complex nature of the questions in the second category and the heterogeneous character of the rules involved make their solution difficult and challenge the rather precise and coherent framework provided by the LOS Convention. The author suggests the possibility that approaches which would overcome the built-in limitation of the notion of high seas may be required to deal with such questions as the ecosystem approach (cf. Chapter 6), concerns linked to climate change and its effect on the oceans (cf. Chapters 3, 4, and 5) and the preservation of marine biodiversity (cf. Chapters 4, 5, 6, 9, 10, and 12).

The approach based on the primacy of the LOS Convention is being increasingly challenged. Scovazzi (2010; cf. also Chapter 10 of this book) tackles the challenging question of whether UNCLOS provides the legal framework for dealing with the case of bioprospecting of marine genetic resources, with particular reference to areas beyond national jurisdiction.

Scovazzi (2010) also refers to the need to see the general aim of sharing benefits as a basic objective of the LOS Convention which contributes to 'the realization of a just and equitable international economic order which takes into account the interests and needs of mankind as a whole and, in particular the special interests and needs of developing countries, whether coastal or land-locked' (UNCLOS preamble). New cooperative approaches to accessing, studying, and using marine genetic resources from areas beyond national jurisdiction would fulfil the general objective of sharing the benefits of human activities in the sea. A possible source of inspiration and solutions is provided by relevant provisions of the Nagoya Protocol on access to genetic resources and the fair and equitable sharing of benefits arising from their utilization under the CBD, albeit the Protocol does not apply to areas beyond national jurisdiction per se.

Scovazzi (2010) also challenges the assumption that UNCLOS can provide the legal framework for all activities taking place in the sea (see Chapter 10). Chapter 10 recalls that changes in the original UNCLOS regime have been integrated into the Convention itself through the process referred to as 'evolution by integration'; and that, where the LOS Convention regime is insufficient, 'evolution by further codification' (adoption of a new instrument) may be required. New rules addressing marine genetic resources beyond national jurisdiction could therefore constitute a third UNCLOS implementation agreement.

1.6 'Hot waters': developments related to marine genetic resources

Treves (2010) states that '[t]he complexity of the problems involved, that straddle many aspects of the law of the sea and of the law of biodiversity, and the emergence of polarised positions that recall those that dominated the stage when the regime of polymetallic nodules began to be discussed in the 1960s and 1970s, makes this the most "fashionable" law-of-the-sea problem of the present.'

Genetic resources are genetic material of actual or potential value and are derived as defined under the CBD, the Nagoya Protocol, as well as in the context of normal practices of research and development. These resources provide great potential for the discovery of compounds with novel properties. This potential is also significant in the light of the fact that a large proportion of marine species are yet to be discovered. The application areas of these discoveries range over the development of new drugs to bioremediation and the enhancement of mariculture practices, as well as improving the efficiency of industrial processes. As such, marine genetic resources are of interest to science, governments, and the private sector.

In fact, the current debate on marine genetic resources, particularly those based on samples collected from areas beyond national jurisdiction, exemplifies well the dynamic nature of ocean science and policy and the need to bridge new findings of scientific research and observations (these resources were unknown at the time that the text of UNCLOS was being negotiated) with the legal and policy regimes for the oceans.

Discussions are ongoing regarding the legal regime applicable to marine genetic resources from areas beyond national jurisdiction, as the current lack of clarity may prevent both investors and governments from utilizing these resources in a sustainable manner, socially, economically, and environmentally. Exploration of the deep sea and the seabed is in its infancy, due to technological constraints and the high costs entailed by deep-sea research. Patents based on these resources have been filed by both public and private entities, mostly based on organisms collected in areas within national jurisdiction, and the trend in the number of such patents is that of steady growth. Source organisms including most of the marine taxa, including microorganisms from deep-sea ecosystems such as hydrothermal vents, and the related patents (which have been filed both by entities from developed as well as some technologically more advanced developing countries) are in some cases based on samples collected in marine areas beyond national jurisdiction in a wide range of geographical locations. These patents are still few in number but clearly indicate an interest in the commercial potential of deep-sea genetic resources.

There is a lack of information and sometimes also of transparency about the source of the samples (including whether they originate from areas within or beyond national jurisdiction), disclosure of information related to the development of the product in question, and how many patents have resulted in related commercial applications. Most of the technology (macro and laboratory) needed to access the deep-sea environment and to study the properties of the organisms collected is sophisticated and expensive and therefore tends to belong to a very limited number of countries. Increasingly cheap identification methods for marine genetic resources and practices in open science and the digitization of ensuing results have made information on genetic resources more accessible.

There is a need to consider the environmental impact of scientific research on marine resources, including the need to discriminate between the impact of research and development operations related to applications involving continuous harvesting, such as in the case of krill extracts, and of operations involving limited samples such as in the case of microbial resources from deep-sea ecosystems. Consideration should also be given also to the vulnerability of sensitive and pristine habitats and species which are rare or which present a limited distribution and, as described in Chapter 4, are already experiencing adverse impacts of climate change. Hence efforts to adopt a precautionary approach, such as the application of voluntary codes of conduct, are needed and commendable, although stronger protection measures could be necessary. These impacts may be reduced thanks to our increased capacity to produce synthetic derivatives of the compounds identified and to rely on microbial cultures. There is a need to identify, adopt, and promote practical measures to further capacity-building among countries, including in relation to facilitating access to samples and technology transfer.

The question of genetic resources from the deep seabed has triggered a process which may lead to the development of a new, dedicated implementing agreement under UNCLOS (cf. Chapters 9 and 10). Divergent views exist between those who consider the CBD to be competent in dealing with these resources and those who insist on the non-applicability of CBD provisions to deal with issues in areas beyond national jurisdiction. There is a possibility that future negotiation of a multilateral benefit-sharing mechanism under the Nagoya Protocol could assist the process of determining a benefit-sharing arrangement in relation to marine genetic resources in areas beyond national jurisdiction. A further difficulty exists in the challenge to combine legal boundaries established under UNCLOS with the biological and ecological features of marine life. Classification of the latter, in particular, in the open ocean and deep sea, is in its infancy (UNESCO, 2009).

In addition to the main legal and policy aspects related to marine genetic resources, a review of the international instruments that govern activities concerning these resources must also include intellectual property rights instruments, which involve both incentives for and obstacles to dealing with marine genetic resources. Since the inception of biotechnology patenting in the 1980s, debates have taken place on the impact of these patents on research and innovation; moreover, the scope of patentable subject matter varies depending on national law. A difficulty lies in the possibility that under relevant provisions of the World Trade Organization (WTO), member countries may be allowed to take measures excluding the granting of patents or the enjoyment of patent rights for inventions based on marine genetic resources from areas beyond national jurisdiction.

There is a general broadening and strengthening of patent protection versus its social benefits in many technological areas. Under the current patent regime and practice there is a risk that newly discovered marine genetic resources may be 'locked up' by patent monopolies. Consideration of the relationship between

UNCLOS and intellectual property instruments is important, as well as the relationship between UNCLOS and the Nagoya Protocol.

Because of difficulties related to the legal interpretation of the matter, some countries are willing to focus on practical measures to deal with these resources, including the promotion of MSR and development of codes of conduct; mechanisms for cooperation, sharing of information and knowledge resulting from research on marine genetic resources, including by increasing the participation of researchers from developing countries in relevant research projects; discussion of practical options for benefit-sharing, including options for facilitating access to samples; and consideration of relevant intellectual property aspects.

Chapter 9 provides a comprehensive review of scientific, technical, technological, legal, and policy aspects of marine genetic resources, with a focus on those to be found in areas beyond national jurisdiction.

1.7 Learning by analogy: the legal regime of outer space and the law of the sea regime

Inspiration for possible solutions to policy and legal gaps in relation to the law of the sea regime, derived from other regimes, are also explored in this book. Chapter 11 focuses on an analogy between problems and solutions of the legal regime of outer space and those of the law of the sea, with the idea that we can learn from such an analysis while tackling emerging and unresolved issues related to the oceans.

Outer space is defined as the area of space beyond the limits of national jurisdictions. The legal regime of space is often associated with other common space regimes such as the high seas, the deep seabed, and Antarctica.

Principles governing the use of outer space had already been anticipated in the late 1920s. During the 1950s, possible analogies with both the law of the sea and Antarctica were considered from the perspective of their possible implications with regard to the future legal regime for space law. Analogies with air law were rejected, due to the ensuing fragmentation of cosmic space into areas subject to different sovereignties, which would have implied that the use of outer space would be subject to the permission of the national state or states concerned. The historical political context was also a factor in determining that issues related to space law should be dealt with by the United Nations within the framework of peaceful use. The dominant presumption in those years was establishment of a free common area not subject to national jurisdiction, thus creating the enabling conditions for the freedom to use outer space.

In 1961, the UN General Assembly decided that activities carried out in outer space and celestial bodies should be in compliance with international law, including the Charter of the United Nations, and proclaimed the freedoms of exploration and use for peaceful purposes in outer space and celestial bodies which are not suitable for national appropriation. Principles governing space law were codified in the form of the Treaty on Principles Governing the Activities of States in the Exploration and Use of Outer Space, including the Moon and Other Celestial Bodies (the Outer Space Treaty) in 1967.

The Treaty proclaims the freedom for all states to explore and use outer space and stipulates that these activities should be carried out for the benefit and in the interests of all of mankind; outer space, including the Moon and celestial bodies, is not subject to national appropriation by any means; international law is applicable to outer space activities, which shall be peaceful; and states shall avoid harmful contamination in carrying out space activities. Analogies can be found with fundamental principles already developed in the law of the high seas, such as the applicability of international law to space activities, the non-appropriation principle, the freedom of exploration and use, and the respect of the environment, which constituted the basis for defining the regime of *res communis omnium* in the context of space law.

The historical evolution of the concept of the common heritage of humankind in the context of the law of the sea, which saw the concept introduced in the framework of the 1970 Declaration on Principles Governing the Sea-bed and Ocean Floor, and the Subsoil Thereof, beyond the Limits of National Jurisdiction, and the evolution of the concept in the context of space law, and specifically the 1979 Agreement Governing the Activities of States on the Moon and Other Celestial Bodies, which was concluded prior to the adoption of the final text of UNCLOS in 1982, deserve comparison. Several provisions under the Antarctic Treaty were also used as an inspiration to develop similar provisions under the Agreement Governing the Activities of States on the Moon and Other Celestial Bodies, including peaceful cooperation for scientific research and environmental protection.

The Outer Space Treaty includes a '*common benefit clause*' which should be considered as a limitation on the absolute freedom of space activities since it establishes that the exploration and scientific research '*shall be carried out for the benefit and in the interests of all countries, irrespective of their degree of economic or scientific development, and shall be the province of all mankind*'. This is explained in the light that while the law of the high seas is essentially based on old customs, space law was at the time *lege ferenda* (new law). These provisions should be intended in the sense that the spacefaring nations should help those states lacking the capacity to participate in space activities and analysed in the light of profit-oriented sector dynamics, such as that of data generated by the remote sensing activities of states. In this context, of relevance and importance are the efforts of the Working Group on Ethics of Outer Space, set up by the UNESCO World Commission on the Ethics of Scientific Knowledge and Technology (COMEST).

As to the 1979 Moon Agreement, the debate on the exploitation of lunar resources is characterized by profound divergences with respect to application of the common heritage of humankind approach, according to which only the owner of the common resources (mankind) is entitled to authorize the appropriation of a part of them to individual users. This debate encompasses discussions over private property rights, license of use, and the negotiation of a possible new Moon treaty and presents analogies with the regulation of resources of the Area.

Considerations on the sustainability of space activities are also presented in the chapter. In this regard, it is critical to remove obstacles that hamper the implementation of space treaties and to further elucidate pressing issues such as an environmental approach to space activities, in particular, with regard to space debris. The editor notes a similarity in the current debate on the law of the sea in relation to the need to remove obstacles to implementation and to clarify further how best to deal with issues such as the environmental impact of human activities in marine areas beyond national jurisdiction.

While different branches of international law can inform and influence each other, this does not necessarily imply an automatic acquisition in a given branch of previous experience gained and solutions found in another given branch. A general analogy remains, however, due to the fact that these issues belong to the categories of problems of areas beyond national jurisdiction, and scholars acquainted with the two regimes will continue benefiting from reciprocal stimulation.

1.8 Preliminary concluding remarks

The chapters in this book together present considerations on future approaches and practical steps in order to achieve a better management of the oceans. The book draws on facts and lessons learned in relation to the specific issues dealt with, which reflect main debates and critical steps in the international ocean agenda within the United Nations. Purposely, the conclusions drawn are not anticipated in this introductory chapter, as they represent the sum of the findings and reflections in the book as a whole. These are presented in the book's concluding chapter.

References

Scovazzi (2010). Is the UN Convention on the Law of the Sea the legal framework for all activities in the sea? The case of bioprospecting. In: *Law, Technology and Science for Oceans in Globalisation – IUU Fishing, Oil Pollution, Bioprospecting, Outer Continental Shelf.* D. Vidas (ed.). Leiden/Boston: Martinus Nijhoff Publishers.

Treves, T. (2010). The development of the law of the sea since the adoption of the UN Convention on the Law of the Sea: Achievements and challenges for the future. In: *Law, Technology and Science for Oceans in Globalisation – IUU Fishing, Oil Pollution, Bioprospecting, Outer Continental Shelf*. D. Vidas (ed.). Leiden/Boston: Martinus Nijhoff Publishers.

United Nations Educational, Scientific and Cultural Organization (UNESCO) (2009). *Global Open Oceans and Deep Seabed (GOODS) – Biogeographic Classification*. IOC Technical Series No. 84. Paris: UNESCO-IOC, 88 pp.

Note

1 The planning phase of the IIOE started under the auspices of the Scientific Committee on Oceanic Research (SCOR). The operational phase of the expedition began in 1962, after IIOE was transferred on to the Intergovernmental Oceanographic Commission (IOC) of UNESCO. Currently, a follow-up expedition is being planned to mark the fiftieth anniversary of the IIOE (cf. http://iocperth.org/IOCPerth/images/stories/IIOE-2_prospectus_final.pdf).

2

Main human uses of ocean areas and resources, impacts, and multiple scales of governance

MARJO VIERROS, U. RASHID SUMAILA, AND ROLPH A. PAYET

2.1 Introduction

The global ocean provides humankind with vital ecosystem goods and services that include the regulation of the Earth's climate, as well as provision of food and other goods, recreation, and spiritual values. The ocean is not only important for the Earth's economy, but also its environmental balance and survival (Noone *et al.*, 2013).

Human uses of the ocean include fishing (food), shipping, scientific research, the use of genetic resources, mining, underwater cables, energy, water, and recreation. While all those involved in these uses can be considered as ocean stakeholders, the concept of stakeholder is broader than just direct use. Stakeholders can include groups affected by management decisions; groups concerned by management decisions; groups dependent on the resources to be managed; groups with claims over the area of resources; groups with activities that impact on the area or resources; and groups with, for example, special seasonal or geographic interests (Vierros *et al.*, 2006). Ultimately, however, the entire population of the Earth depends in one way or another on the ocean for their survival, due to the climate regulating and oxygen providing services of these areas, and can thus be considered to be stakeholders.

Preserving and maintaining the services provided by the ocean will require integrated, ecosystem-based management approaches and governance structures at both global and local levels, which will take into account both direct human uses and conservation needs, as well as global benefits. Many habitats and species in the ocean are highly threatened by human activities, and the current governance regime is not sufficient and in many cases too fragmented to provide for effective management and protection of multiple and emerging threats.

2.2 Value and use of ocean areas, and environmental impacts of use

The goods and services provided by the ocean range from climate regulation to food, and recreational and spiritual value. For example, an analysis undertaken by The Economics of Ecosystems and Biodiversity (TEEB) project, found the value of coral reefs to humankind to be between US$130,000 and $1.2 million per hectare, per year (Diversitas, 2009). These calculations take into account the services provided by coral reefs in relation to food, raw materials, ornamental resources, climate regulation, moderation of extreme events, waste treatment, water purification, biological control, cultural services (including tourism), and maintenance of genetic diversity. Similarly, the services seagrasses provide in the form of nutrient cycling are valued at an estimated $1.9 trillion per year, while their support for commercial fisheries is estimated to be worth as much as $3500 per hectare per year (Waycott *et al.*, 2009).

The deep sea also provides important ecosystem services, but these are less well understood. Recent scientific research has shown that life in the deep sea plays a fundamental role in global biogeochemical cycles, including nutrient regeneration and production of oxygen, as well as the maintenance of the Earth's climate through the global carbon cycle (Armstrong *et al.*, 2010; Riser and Johnson, 2008). An estimated 50% of the carbon in the atmosphere that becomes bound or 'sequestered' in natural systems is cycled into the seas and oceans. Oceans not only represent the largest long-term sink for carbon but they also store and redistribute CO_2. Some 93% of the Earth's CO_2 is stored and cycled through the oceans (Armstrong *et al.*, 2010).

Fisheries provide an example of direct use of the oceans, and highlight their importance in provisioning food to humankind. According to the FAO, fish provide more than 2.6 billion people with at least 20% of their animal protein intake. This figure includes protein from the total of over 1000 species harvested from the world's capture fisheries. Since the 1960s, fishing fleets have shifted to fishing further offshore and in deeper waters to meet global demand (Morato *et al.*, 2006; Cochonat *et al.*, 2007).

Marine and inland fisheries and aquaculture together supplied the world with about 148 million tonnes of fish in 2010, with a landed value of $217.5 billion (FAO, 2012; Sumaila *et al.*, 2012). Using the average global multiplier for the fishing sector of 3, provided in Dyck and Sumaila (2010), these sectors created economic impacts of nearly $660 billion that year.

Overall, fish provided more than 2.9 billion people with at least 15% of their average per capita animal protein intake. Thus fish stocks are among the planet's most important renewable resources and have been supporting humanity for millennia. Fish support human well-being through employment in fishing, processing, and retail services (Pontecorvo *et al.*, 1980; FAO, 2010; Dyck and Sumaila, 2010; Teh and Sumaila, 2013), as well as providing food security for people, particularly in developing countries (Zeller *et al.*, 2006).

Shipping is another major use of the oceans, and is the principal means of transportation for goods around the world, thus making it central to world trade and economy. According to the UNCTAD Review of Maritime Transport (UNCTAD, 2010), more than 90% of international trade in goods is carried by sea, and an even higher percentage of developing-country trade is carried in ships. While prospects for shipping remain uncertain, and are tied to the world economy as a whole, there is likely to be a recovery in the shipping sector following the 2009 recession. In 2009, total goods loaded amounted to 7.8 billion tons, down from 8.2 tons recorded in 2008 (UNCTAD, 2010). According to the World Shipping Council (www.worldshipping.org/about-the-industry/global-trade), liner ships, which include container ships, transported about US$4 trillion worth of goods annually during 2009–2010, representing almost 60% of all seaborne trade. Ship-based tourism also continues to expand, and cruise ships generated an estimated $18 billion a year in passenger expenditure in 2010 (Brida and Zapata, 2010).

The bottom of the ocean hosts an extensive network of undersea communication cables. Since its establishment in the telegraph era, the network has expanded around the globe, and is now an integral part of modern society (Carter *et al.*, 2009). It is estimated that submarine cables carry in excess of US$10 trillion in transactions per day and carry over 97% of intercontinental data traffic due to their lower cost in comparison with satellite communication (APEC, 2012).

Scientific research is another common use of the ocean, with more work being undertaken in the easier-to-access nearshore areas. In the deep sea, scientific research is an expensive undertaking that requires not only sophisticated research vessels, but also specialized equipment, such as remotely operated vehicles (ROVs) and submersibles. While there is no systematic information on tracking research vessels worldwide, some publicly available databases are seeking to address this information gap. The website www.sailwx.info, hosted by a consortium of several research and operational oceanographic and meteorological facilities, allows live tracking of ships worldwide. While the database can be used to link to physical information about the ship, it does not provide information about the type of research activities that are being undertaken.

A related ocean use is biological prospecting, which is discussed in greater detail in Chapter 9. It should be noted, though, that biological prospecting is made feasible by the great potential for discovery in the ocean, with its diversity of habitats, species, and genera. The species diversity alone is estimated to be at around 5 to 10 million, which represents a huge genetic reservoir with great potential for discovery of products useful to humankind, such as medicines and enzymes.

Deep-sea habitats are the largest reservoirs of biomass and non-renewable resources (e.g. gas hydrates and minerals) (Danovaro *et al.*, 2008). Potential mineral resources in the deep sea include manganese nodules, cobalt-rich crusts, polymetallic sulphides, and phosphorites (Roberts *et al.*, 2005; van den Hove and Moreau, 2007), and there is also some speculation about future mining of frozen methane gas in the deep ocean (Glover and Smith, 2003), although this proposition is not without hazards, which include climate change impacts. It is likely that there will be increasing interest in extracting these deep-sea resources as technologies improve and minerals and other resources available on land are depleted.

Exploitation of mineral resources in the deep seabed, in particular polymetallic sulphide deposits associated with hydrothermal vent systems, is now closer to becoming a reality. The International Seabed Authority has entered into 15-year contracts for exploration for polymetallic nodules, polymetallic sulphides, and cobalt-rich crusts in the deep seabed with 15 contractors, pursuant to Resolution II of the Third UN Conference on the Law of the Sea. The first of these contracts was signed in 2001, with others following over recent years. The contracts allow the contractors to explore in specified parts of the deep oceans outside national jurisdiction. Under ISA regulations, each contractor has the exclusive right to explore an initial area of up to 150,000 square kilometres. Twelve of these contracts are for exploration for polymetallic nodules in the Clarion Clipperton Fracture Zone, with two contracts for exploration for polymetallic sulphides in the South West Indian Ridge and the Mid Atlantic Ridge, and one contract for exploration for cobalt-rich crusts in the Western Pacific Ocean (ISA, 2014; van Dover, 2011).

Many deep water seep environments are sites of significant reservoirs of petroleum and natural gas. In the Gulf of Mexico, offshore exploitation of oil and gas in proximity with seep communities has been occurring for decades. Reduced sea ice in the Arctic is likely to provide expanding opportunities for oil extraction, although such efforts could be hampered initially by movements of sea ice in some areas. It is believed that the Arctic seabed may contain substantial oil fields, the exploitation of which would bring with it environmental concerns about spills and their cleanup in remote, hard-to-reach environments (ACIA, 2005). Additionally, international debates as to which nations can claim sovereignty or ownership over the waters of the Arctic and their resources are likely to surface.

A burgeoning carbon economy provides incentive for commercial interests to experiment with schemes for ocean fertilization and deep-sea carbon sequestration. Ocean fertilization refers to the process of stimulating phytoplankton growth in the surface ocean through the addition of nutrients, in particular iron. These nutrients are thought to limit phytoplankton growth in many areas of the ocean, and it has been argued that the phytoplankton blooms resulting from fertilization will help remove CO_2 and mitigate climate change. As a result, ocean fertilization is being actively promoted by several commercial companies as a means of offsetting CO_2 emissions. A related climate change mitigation technique is ocean carbon sequestration, which refers to the direct forcing of liquefied CO_2 into the ocean in order to store and isolate it. This technique is still experimental.

Ocean energy from offshore wind farms, wave-energy generating devices, and ocean thermal vents offer potential for addressing climate change mitigation. Whilst technology development to fully exploit ocean energy is still costly and in an experimental stage, the International Energy Association estimates that total global installed capacity for ocean energy could be as high as 210 GW by 2050, as prices and access to ocean energy technologies are expected to fall.

Water generation from the ocean through a desalination process is now widespread in many water-scarce regions but is becoming a mainstay in coastal areas, with huge deficits in freshwater and groundwater pollution. The expansion of installed capacity for desalinated water has grown from about 8000 m^3/d (in 1970) to about 32 Mm^3/d (by 2001) worldwide, and this is expected to grow exponentially as climate change and other human activities reduce access to freshwater resources (Wangnick, 2002).

Aquaculture has expanded dramatically in coastal areas due to increasing demand and reduced supply of wild fish stocks. There are technological and financial challenges in expanding aquaculture beyond coastal areas and exclusive economic zones, but with the appropriate financial incentives and technologies it is possible for aquaculture in the high seas to become a reality. This may include mobile cage operations such as the 'ocean drifter', consisting of manned or autonomous cages capable of low-speed self-propulsion, operating in ocean gyres (Merrie *et al.*, 2014).

2.3 Impacts of human use of ocean areas

The marine environment, even in the deep sea, is no longer pristine and untouched, and pressures on ecosystems and species are increasing (van den Hove and Moreau, 2007). According to the Global Coral Reef Monitoring Network (GCRMN), the world has effectively lost 19% of its original area of coral reefs; 15% are seriously under threat of loss within the next 10–20 years; and an additional 20% in 20–40 years (Wilkinson, 2008). Seagrasses have been disappearing at a rate of 110 $km^2 \ yr^{-1}$ since 1980, and 29% of the known areal extent has disappeared since seagrass areas were initially recorded in 1879 (Waycott *et al.*, 2009). Oyster reefs have declined by more than 90% from historic levels in 70% of bays and 63% of the world's marine ecoregions. Most of these declines are due to direct human exploitation, habitat degradation, and clearance for development, and pollution.

The deep sea has also been impacted by human activities, in particular fisheries. Many deep water habitats, such as cold-water coral reefs and seamounts show impacts from bottom fishing activities. The biological resources of seamounts have been the target of intensive exploitation, resulting in overfishing and major crashes in stocks on some (Clark and Koslow, 2007), along with large impacts on the benthic communities of many studied seamounts, caused mainly by bottom fishing, particularly demersal trawling (Collie *et al.*, 2000; Koslow *et al.*, 2001; Waller *et al.*, 2007; Watson and Morato, 2004). Similarly, many cold-water coral reefs have been damaged by bottom fishing activities, but the extent of this damage has not been quantified (Hourigan, 2008).

With advances in technology, human capacity to reach remote areas has increased, leading to growing threats that include unsustainable fishing and shipping activities, pollution, ocean dumping, and oil, gas, and mineral exploration. In addition, climate change and ocean acidification increasingly threaten oceans and polar areas.

2.3.1 Fishing

The recent rapid increase in fishing effort (Watson *et al.*, 2013) has been accompanied by increasing use of technologically advanced fishing vessels, including bottom trawlers. The combined effects of the volume of fishing and the fishing gear applied have resulted in a number of environmental impacts including (i) overfishing of fish stocks (Pauly *et al.*, 2002); (ii) destruction of fish habitats (Sainsbury *et al.*, 1993); (iii) the fishing down of marine food webs (Pauly *et al.*, 1998); (iv) ecological disruption; and (v) by-catch problems (Alverson *et al.*, 1994).

Overfishing may cause ecological disruption and impact the health of the oceans because, when commercially valuable species are overexploited, there are negative ecosystem effects through impacts on non-targeted species. For instance, when top predators such as large shark species are fished out they trigger trophic effects in the shark food chain, which in turn leads to increasing numbers of species, such as rays, that are prey for large sharks. The result of this is declining stocks of smaller fish and shellfish that are eaten by these species, which then impacts the survival of marine mammals (e.g. Hansen, 1997) and the breeding success of seabirds (e.g., Anker-Nilssen *et al.*, 1997). It is suggested that fishing may even eliminate trophic groups or keystone species thereby altering the overall community structure of an ecosystem (Botsford *et al.*, 1997). Fishing usually results in unintentional killing of untargeted marine life, which can have significant effects on marine ecosystems, e.g. impacts on fish community structure by altering predator–prey relationships (e.g. Mehl, 1991).

2.3.2 Shipping

The environmental impacts of shipping include the spread of invasive alien species, pollution, and potential oil spills. With 90% of world trade carried by sea, the global network of merchant ships provides one of the most important modes of transportation for the spread of invasive species. Two major pathways for marine bioinvasion are discharged ballast water and hull fouling. Invasive species have caused species extinctions and damage to ecosystems and livelihoods, health, and economics in coastal areas throughout the world (Kaluza *et al.*, 2010). In the United States alone, the financial loss related to biological invasions is estimated at $120 billion per year (Pimental *et al.*, 2005).

Cruise ships have notable environmental impacts, particularly with regard to waste and sewage. It has been estimated that a 3000 passenger ship produces 15,000 to 30,000 gallons of blackwater (sewage, wastewater from toilets and medical facilities) and 90,000 to 225,000 gallons of greywater (wastewater from sinks, showers, galleys, laundry, and cleaning activities) each day (Brida and Zapata, 2010). It is also anticipated that cruise ship traffic in the Arctic will increase with the melting of sea ice (Huntington, 2009).

The likely increase of ship traffic in the Arctic has raised a number of urgent concerns, given the possibility that by 2015 the Arctic Ocean will be ice free for a short period in the summer. This would mean the disappearance of multi-year sea ice, as no sea ice would survive the summer melt season (Arctic Council, 2009). Ship traffic diverting from current routes to new routes through the Arctic is projected to reach 2% of global traffic by 2030 and 5% in 2050. In comparison, shipping volumes through the Suez and Panama canals currently account for about 4% and 8% of global trade volume, respectively.

The most significant threat from ships to the Arctic environment is the release of oil through accidental or illegal discharge. Other potential impacts include ship strikes on marine mammals, the introduction of invasive alien species, disruption of migratory patterns of marine mammals, and anthropogenic noise from shipping activity. Longer seasons for Arctic navigation may result in increased interactions between migratory species and ships (Arctic Council, 2009).

Growing Arctic ship traffic will also bring with it air pollution that has the potential to accelerate climate change in the world's northern reaches. Researchers estimate that engine exhaust particles could increase warming by between 17 and 78%. Most of this warming is likely to originate from the release of black carbon, or soot, from ships' diesel engines (Corbett *et al.*, 2010). Black carbon is a short-lived climate forcing pollutant, which is especially effective in accelerating the melting of ice and snow. Other ship emissions may also have unintended consequences for the Arctic environment and the cultures and well-being of Arctic populations, in particular indigenous residents (Arctic Council, 2009). Worldwide, ships release an estimated 1.2 million to 1.6 million tonnes of tiny airborne particles each year, which are high in carbon, sulfur, and nitrogen oxides. These particles have been linked to premature deaths worldwide, and are believed to cause heart and lung failure. Recent research indicates that shipping-related particulate matter emissions are responsible for approximately 60,000 cardiopulmonary and lung cancer deaths annually, with most deaths occurring near coastlines in Europe, East Asia, and South Asia (Corbett *et al.*, 2007).

2.3.3 Potential threats from other uses: scientific research, biological prospecting, mining, geo-engineering, and underwater cables

Underwater cables are thought to have no or minimal impact on the deep seabed, provided that they are sited to avoid sensitive habitats. However, their installation and repair may have some impact on the benthos as they can be subject to damage both from natural causes such as earthquakes and human causes such as deep water trawling (cf. www.suboptic.org/uploads/Economic%20Impact%20of%20Submarine%20Cable%20Disruptions.pdf). Installation requires ploughing to prepare the surface and jetting to bury the already installed cable, resulting in a 5–8 cm wide disturbance to the seabed. Repair may entail lifting of the cable and its reinstallation. Entanglements of whales in underwater cables have been reported in the past, but have now ceased with the transition from telegraph to coaxial cabling, and, subsequently, to fibre-optic systems (Carter *et al.*, 2009).

There have only been very minor documented impacts from scientific research on ecosystems and species in the global commons. Scientific research may entail physical disturbance or disruption, or the introduction of light into an ecosystem that is naturally deprived. Some evidence of disturbance caused by scientific research already exists. For example, the use of floodlights on manned submersibles may have irretrievably damaged the eyes of decapod shrimps that dominate the fauna at vents on the Mid-Atlantic Ridge (Herring *et al.*, 1999). Waste from Antarctic research stations has had an impact on the surrounding environment, resulting in locally elevated pollution levels, including sewage, hydrocarbons, and heavy metals, and resulting in changes to the structure of the adjacent benthic communities (Lenihan, 1992; Conlan *et al.*, 2004).

Actual or potential impacts from biological prospecting are similar to those from scientific research, given the close connection between the two activities. While there is little documentation about environmental impacts of this activity, they are thought to be relatively minimal at the early biodiscovery stages of collection, where the size of samples collected is small. If a given species has shown biotechnology potential, repeated collection may require larger quantities, raising the likelihood of environmental impact. However, synthetic manufacture in a laboratory of the chemical of interest generally eliminates the need for repeated collection (ATCM, 2009). Environmental impact remains a concern if the target organism is rare, has a restricted distribution, and/or the collection is focused on a particular population (Hunt and Vincent, 2006), or if the organism is already suffering from other environmental pressures, such as climate change. Also, anthropogenic pressures (such as helicopter landings, lights from submersibles, effects of camping or skidoos, etc.) can have an impact on pristine environments (ATCM, 2009; Herring *et al.*, 1999).

The extraction of polymetallic sulfide deposits will be relying on new technologies and methods; their impact is as yet unknown. It is expected that the drifting particles produced by deep-sea sulfide mining have the potential to smother, clog, and contaminate nearby vent communities. Organisms surviving these perturbations would be subject to a radical change in habitat conditions with hard substrata being replaced by soft particles settling from the mining plume. Mining could also potentially alter hydrologic patterns that supply vent communities with essential nutrients and hot water. A further problem may arise during dewatering of ores on mining platforms, resulting in discharge of highly nutrient enriched deep water into oligotrophic surface waters, which can drift to nearby shelf areas. Because most invertebrate diversity at vents is found in rare species, habitat destruction by mining can be potentially devastating to local and regional populations (Secretariat of the Convention on Biological Diversity, 2008). Other potential sites for mining include manganese nodule fields on abyssal plains and cobalt-rich crusts on the sides of seamounts. The environmental impacts of such mining are unknown, but would involve disturbance to benthic ecosystems and communities (Koslow, 2007).

Effects of offshore exploitation of oil and gas on seep ecosystem function and biodiversity are not well documented. Exploratory drilling and the installation and operation of production platforms will produce localized and widely spaced disturbances. Depletion of subsurface oil and gas reservoirs may eventually affect the energy supply to seep communities, but this remains to be investigated. A more widespread impact may come from the exploitation of subsurface gas hydrate deposits. These reserves of methane ice occupy significant volumes within the seabed of continental margins worldwide. Recent global estimates of gas hydrate reserves greatly surpass total known world petroleum reserves (Juniper, 2001). Although exploitation of subsea gas hydrates is probably many decades away, their extraction could involve large-scale disturbance of the seabed and consequent effects on seep communities (Juniper, 2001).

In addition to being inefficient, ocean fertilization has the potential to exacerbate chemical change in the oceans. Such chemical changes may have a range of important biological consequences, including potentially negative impacts. For example, the decay of excess phytoplankton uses oxygen, and may result in an undersaturation of oxygen in the water column. This would have substantial impacts on mid-water and deep-sea organisms and biodiversity. It is also likely that iron fertilization may lead to increases in production of N_2O, another green-house gas, reducing the overall efficiency of using iron fertilization to reduce climate impacts (Secretariat of the Convention on Biological Diversity, 2009).

For carbon sequestration, available techniques include direct injection of CO_2 into deep seawater, the storage of CO_2 as a liquid or a hydrate on the seafloor, and the injection of CO_2 into geological formations below the seafloor (Schubert *et al.*, 2006; Davies *et al.*, 2007; IPCC, 2005). The first two options are controversial because they are unlikely to offer a permanent solution because of oceanic circulation and eventual exchanges with the atmosphere. Today, only the third option, injection into sub-seabed geological formations, is allowed under the 2006 amendment of the London Convention on the Prevention of Marine Pollution by Dumping of Wastes and Other Matter, and there are significant research efforts in this direction (Schubert *et al.*, 2006). Depending on the volumes injected, measurable change in ocean chemistry could be expected and marine organisms near the injection site may be harmed (Secretariat of the Convention on Biological Diversity, 2009). The risk of CO_2 leakage from the deep subsurface and potential effects on deep-sea ecosystems and biodiversity are unknown, and will need to be assessed before operations proceed (Inagaki *et al.*, 2006).

2.3.4 Climate change and ocean acidification

The impacts of climate change and ocean acidification are likely to cause enormous changes to biodiversity and food webs in the oceans. Given the serious nature of these changes, they are discussed in detail in Chapters 3 and 4 of this book.

2.4 Governance

The international legal regime in the oceans is made up of a number of global and regional legal instruments. On the global level, the United Nations Convention on the Law of the Sea (UNCLOS) provides the comprehensive legal framework for all activities in the oceans. Other agreements are specific to certain topics, such as the Convention on Biological Diversity (CBD) on the conservation and sustainable use of biodiversity; sectoral activities (such as the instruments developed in the context of the FAO on fishing and the instruments adopted in the context of IMO on shipping); or relate to specific species, such as the Convention on Migratory Species of Wild Animals (CMS) and the Convention on International Trade in Endangered Species of Wild Fauna and Flora (CITES), which both cover marine species. On the regional level, a combination of regional seas conventions and action plans and regional fisheries management organizations and arrangements (RFMO/As) provide for the protection and preservation of the marine environment and fisheries regulation respectively.

2.4.1 Global legal instruments

The United Nations Convention on the Law of the Sea (UNCLOS)

The United Nations Convention on the Law of the Sea (UNCLOS) was adopted by the Third United Nations Conference on the Law of the Sea on 10 December 1982 and entered into force on 16 November 1994. UNCLOS is generally considered to reflect customary law. UNCLOS establishes a framework for ocean governance, specifying the rights of states within the various maritime zones but also their duties, including in relation to the sustainable management of marine living resources (Articles 116–119) and the protection and preservation of the marine environment (Articles 192–212). UNCLOS lays down a comprehensive legal regime for the world's oceans and seas, establishing rules governing all uses of the oceans and ocean resources (Kimball, 2005; Treves, 2010). The UNCLOS instrument is discussed in more detail in Chapters 9, 10, and 11 of this book.

In the 30 years since UNCLOS was adopted, the ocean has changed significantly, and, as this chapter has shown, the impacts of human activities are now felt in the deepest and most remote parts of the ocean (Gjerde, 2012). These changes will continue in the future, with climate change and acidification, as well as new ocean uses, accelerating and cumulatively impacting biodiversity. It is recognized that UNCLOS was a product of its time (Tladi, 2011), and this leaves some significant gaps and weaknesses that undermine not only the protection and preservation of the marine environment, but also the potential role of the ocean for sustainable development for all (Gjerde, 2012).

In this respect, it should be noted that the UN General Assembly (UNGA), as the global policy framework providing comprehensive and cross-sectoral guidance on oceans and the law of the sea matters, continues to keep abreast of, and pass resolutions on, topical and emerging oceans issues. The UNGA adopts two resolutions annually, on oceans and the law of the sea, and on sustainable fisheries. It has established a number of working groups to assist it in its work, including the Open-ended Informal Consultative Process on Oceans and the Law of the Sea (the Consultative Process); the ad hoc open-ended informal Working Group to study issues related to the conservation and sustainable use of marine biological diversity beyond areas of national jurisdiction (the UN Working Group); and the Regular Process for Global Reporting and Assessment of the State of the Marine Environment, including Socio-economic Aspects (Regular Process) Working Group.

The UN Working Group was established by the UNGA in 2004, and presently a number of governments are calling for an implementing agreement under UNCLOS for the conservation and sustainable use of marine biodiversity beyond national jurisdiction. This would address a package of five issues: (1) area-based management measures, including marine protected areas (MPAs); (2) environmental impact assessments; (3) marine genetic resources, including questions related to sharing of benefits; (4) capacity-building; and (5) technology transfer (Gjerde, 2012).

The UN Fish Stocks Agreement

The UN Fish Stocks Agreement is an implementing agreement of UNCLOS, which was adopted on 4 August 1995 by the United Nations Conference on Straddling Fish Stocks and Highly Migratory Fish Stocks (24 July–4 August), and entered into force on 11 December 2001. It is to be interpreted and applied in the context of, and consistently with, UNCLOS (cf. Chapter 10).

The UN Fish Stocks Agreement applies to straddling fish stocks[1] and highly migratory fish stocks. The objective of the Agreement is to ensure the long-term conservation and sustainable use of these stocks. It elaborates on the relevant provisions of UNCLOS by requiring fisheries management to be based on the precautionary and ecosystem approaches, and provides for monitoring, control, and surveillance. It also includes detailed provisions on enforcement and international cooperation, especially at the regional level (Kimball, 2005). To ensure effective implementation of the Convention a requirement was included for the establishment of regional and subregional fisheries management organizations (RFMOs) (UNEP, 2010).

The Convention on Biological Diversity (CBD)

The Convention on Biological Diversity (CBD) was adopted in Rio de Janeiro in 1992. The objectives of the CBD are the conservation of biodiversity, the sustainable use of its components, and the equitable sharing of benefits derived from genetic resources.

Article 22 of the CBD obliges its parties to implement the CBD consistently with the rights and obligations of states under the law of the sea. In areas beyond national jurisdiction, the scope of the CBD is limited to 'processes and activities carried out under a Party's jurisdiction or control, which may have adverse impacts on biodiversity' (Article 4b). The CBD does not apply to the components of biodiversity in areas beyond national jurisdiction, as it would within national jurisdiction. As a result, the CBD highlights the need for cooperation among parties in respect of areas beyond national jurisdiction for the conservation and sustainable use of biodiversity, either directly or through competent international organizations (Article 5).

The CBD contains provisions for the precautionary approach, the ecosystem approach, protected areas (Article 8), and monitoring and identification (Article 7). In particular, the ecosystem approach is seen as the primary framework for action under the Convention, and the fifth meeting of the Conference of the Parties endorsed a description, operational guidance, and principles of the ecosystem approach (decision V/6) (Please refer to Chapter 6 for more information on the ecosystem approach and its application to the marine environment). The CBD also expressly mandates the establishment of protected areas and recognizes that the conservation of biological diversity is a common concern of humankind and an integral part of the development process.

The CBD has a programme of work relating to marine and coastal biodiversity, and is undertaking work to identify Ecologically and Biologically Significant Areas (EBSAs) in the world's oceans. The CBD has also undertaken work, inter alia, on marine protected areas, biodiversity impacts of fishing, ocean noise, and environmental impact assessment in the oceans.

The International Convention for the Regulation of Whaling

The International Convention for the Regulation of Whaling was adopted in 1946, and aims to ensure the proper and effective conservation and development of whale stocks. It applies to factory ships, land stations, and whale catches under the jurisdiction of the parties to the Convention and to all waters in which whaling is carried out. The Convention established an International Whaling Commission, composed of member states, to organize scientific studies and investigations and to collect, analyse, and disseminate data. The Commission's main task is to review and revise as necessary the measures laid down in the Convention. It can fix the limits of open and closed waters, designate sanctuary areas, prescribe seasons, catch, and size limits for each species of whale, as well as prohibit types and methods of fishing (UNEP, 2010).

The Commission has established two large-scale high seas sanctuaries where commercial whaling is prohibited. The first of these is the Indian Ocean Sanctuary, and the second the Southern Ocean Sanctuary. While there is no commercial whaling in either of these areas, the taking of whales for the purpose of scientific research is permitted under conditions specified in the Convention (Article VIII).

The Convention on the Conservation of Migratory Species
of Wild Animals (CMS)

The Convention on the Conservation of Migratory Species of Wild Animals (CMS or Bonn Convention) aims to protect terrestrial, marine, and avian migratory species throughout their range. Parties agree to take, individually or in cooperation, appropriate and necessary steps to conserve migratory species and their habitat. For species in danger of extinction throughout all or significant portions of their range (Appendix I), the range states must take immediate action to protect them. For species in an unfavourable conservation status (Appendix II), range states are urged to conclude binding agreements on the full range of threats in order to improve their status.

Range is defined as all areas of land or water that a migratory species inhabits, stays in temporarily, crosses, or overflies at any time on its normal migration route. In areas beyond national jurisdiction, range states are defined as including all states whose vessels are taking the species beyond the limits of national jurisdiction, and they are required to prohibit taking of Appendix I species. Range states should also control activities undertaken within national jurisdiction that may endanger species beyond national jurisdiction (Kimball, 2005).

The Convention on International Trade in Endangered Species
of Wild Fauna and Flora (CITES)

The Convention on International Trade in Endangered Species of Wild Fauna and Flora (CITES) aims to ensure that the international trade in specimens of wild animals and plants does not threaten their survival. The import and export of species covered by CITES has to be approved by the national authorities of the member states in accordance with the rules and regulations laid down by the Convention.

Species are listed in three appendices resulting in different levels and types of protection. Among the marine listings (in Appendix II) are many species of cetaceans, marine turtles, seahorses, corals, and commercial marine fishing species such as basking sharks, whale sharks, the great white shark, and the humphead wrasse. However, a proposal to list bluefin tuna in Appendix I (which would have meant a complete ban on trade) failed in 2010.

Regarding marine areas beyond the limits of national jurisdiction, the CITES provisions on 'introduction from the sea' cover transportation into a state of any species taken beyond national jurisdiction. For any species included in Appendix I or II, this requires the prior grant of a certificate from the management authority of the state of introduction. This restriction does not apply to species included in Appendix II if they are taken in conformity with the relevant convention by flag ships of a state party to both. There is still work in progress within CITES and its bodies to fully clarify the term 'introduction from the sea'.

The Convention on the Protection of the Underwater Cultural Heritage (CPUCH)

The UNESCO Convention on the Protection of the Underwater Cultural Heritage was adopted in November 2001. The Convention entered into force on 2 January 2009. The Convention protects cultural heritage both within and beyond national jurisdiction and gives preference to preserving such heritage on site. Thus, parties to the Convention have a responsibility to protect cultural heritage found in the area. The Convention defines underwater cultural heritage as all traces of human existence having a cultural, historical, or archaeological character which has been partially or totally under water for at least one hundred years. Shipwrecks and other historical or cultural objects can attract the settlement of species, and thus protective measures taken under the Convention may have the added benefit of protecting the associated biodiversity (UNEP, 2010).

International Maritime Organization (IMO) instruments

International rules and regulations concerning maritime safety, the efficiency of navigation, and the prevention and control of marine pollution from ships have been developed under the auspices of the International Maritime Organization (IMO). IMO is considered to be the competent international body under UNCLOS to establish special protective measures in defined areas where shipping presents a risk. These apply uniformly to all ships and include routing and discharge restrictions and reporting requirements.

International Convention for the Prevention of Pollution from Ships (MARPOL 73/78)

Discharges from ships, both accidental and intentional, are regulated by the International Convention for the Prevention of Pollution from Ships, 1973, as modified by the Protocol of 1978 relating thereto (MARPOL 73/78). MARPOL 73/78 regulates vessel design, equipment, and operational discharges from all ships. It also provides for the designation of special areas where more stringent discharge rules apply, including in respect of oil, noxious liquid substances, and refuse from ships. Special areas are defined as areas where, for technical reasons relating to their oceanographic and ecological condition and to their sea traffic, the adoption of special mandatory methods for the prevention of sea pollution is required (Kimball, 2005; UNEP 2010).

Particularly Sensitive Sea Areas (PSSAs)

In addition to the special areas described above, the IMO has adopted a resolution providing for the designation of Particularly Sensitive Sea Areas (PSSAs).

According to the IMO, a PSSA is 'a comprehensive management tool at the international level that provides a mechanism for reviewing an area that is vulnerable to damage by international shipping and determines the most appropriate ways to address that vulnerability'.

Procedures for identification of PSSAs and the adoption of associated protective measures have been set forth under IMO PSSA guidelines, the latest version of which was adopted in December 2005 (Resolution A. 982 (24)). The guidelines define a PSSA as 'an area that needs special protection through action by IMO because of its significance for recognized ecological, socio-economic, or scientific attributes where such attributes may be vulnerable to damage by international shipping activities'. Although the currently existing PSSAs are located within national jurisdiction, they could also be designated in areas beyond national jurisdiction.

Convention on the Prevention of Marine Pollution by Dumping of Wastes and Other Matter 1972 (London Convention)

The London Convention and the 1996 Protocol thereto aim to control and prevent all sources of marine pollution caused by the deliberate disposal of wastes or other substances at sea. The Convention differentiates between matter whose dumping is prohibited (listed in Annex I) and those which require a permit. Issuance of a permit requires consideration of various factors including the characteristics of the proposed dumping site. Under the London Convention, states with common interests in protecting the marine environment in a given geographical area are to enter into regional agreements. Parties must also cooperate in the development of procedures for the effective application of the Convention on the high seas, including procedures for reporting dumping by vessels or aircraft (UNEP, 2010).

The London Convention has been important in considering the emerging issue of ocean fertilization. In 2008, the parties to this Convention adopted a resolution on the regulation of ocean fertilization, which stated that given the present state of knowledge, ocean fertilization activities other than legitimate scientific research should not be allowed. Parties further agreed that scientific research proposals should be assessed on a case-by-case basis. Similar text can also been found in recent (2008 and 2010) decisions of the Conference of the Parties to the CBD.

Ballast water and sediments

The International Convention for the Control and Management of Ship's Ballast Water and Sediments (2004, not yet in force) aims to prevent, minimize, and ultimately eliminate the transfer of harmful aquatic organisms and pathogens due to ballast water exchange. The Convention requires ships to conduct ballast water exchanges at least 200 nautical miles from the nearest land and in waters deeper than 200 metres, wherever possible (Regulation B-4, Annex).

It should be noted that the ratification of this Convention has been slow, and despite its importance in blocking (or at least greatly reducing) one of the major vectors for introduction of invasive alien species, it has yet to enter into force.

2.4.2 Gaps in the global legal regime

A comprehensive legal analysis by IUCN (Gjerde *et al.*, 2008) identified the following regulatory gaps in the global legal regime with a particular focus on marine areas beyond national jurisdiction.

- Modern conservation principles (such as the ecosystem approach and precautionary approach) and tools (such as marine protected areas and EIAs and SEAs) are not consistently incorporated and/or applied in all relevant existing instruments.
- Lack of specific requirements for EIAs, monitoring, and reporting area-based measures and other modern conservation tools to the full range of ocean-based human activities in areas beyond national jurisdiction.
- Lack of rules or a process to coordinate regulation of interactions between activities occurring in the high seas water column and on the extended continental shelf of coastal states.
- Lack of effective compliance and enforcement mechanisms.

In addition to these general gaps, there are a number of activities that do not have detailed international rules and standards. These activities include:

- bioprospecting
- laying of cables and pipelines
- construction of various types of installations
- unregulated fisheries such as some discrete high seas fish stocks or sharks
- grey-water discharges from shipping
- oil and gas exploration and exploitation activities on extended continental shelves
- military activities (recognizing that government ships on non-commercial service benefit from sovereign immunity)
- new and emerging activities such as climate change mitigation techniques and potential construction and operation of floating energy and aquaculture facilities
- ocean geo-engineering.

In addition, there is no regime that would assess the cumulative impacts over time and across different sectors.

2.4.3 The regional legal regime

The regional agreements considered here are regional seas conventions and action plans and the instruments adopted by regional fisheries management organizations and arrangements. Both elaborate and supplement the UNCLOS regime in their respective regions, and provide for special areas where a higher level of protection can be established. They incorporate ecosystem approaches to various degrees, but there is a growing trend to move beyond single species management and towards ecosystem-based management, and management of seascapes. The text below looks at the degree to which this has been achieved, and the types of modern management approaches and tools that are employed.

Regional seas conventions

There are currently 18 regional seas agreements and programmes, 13 of which have been established under the auspices of the United Nations Environment Programme (UNEP). Some agreements, such as those in the North-East Atlantic and the Antarctic, pre-date the establishment of UNEP. Most regional seas have adopted binding framework conventions, while others have non-binding action plans as a basis for their cooperation. Several have protocols relating to specially protected areas and wildlife. Only six of these regional seas conventions explicitly cover areas beyond national jurisdiction. These are the Convention for the Protection of the Marine Environment of the North-East Atlantic (OSPAR Convention), the Convention for the Protection of the Marine Environment and the Coastal Region of the Mediterranean (Barcelona Convention), the Convention for the Protection of the Natural Resources and Environment of the South Pacific Region (Noumea Convention), the Antarctic Treaty, the Convention for the Protection and Development of the Marine Environment of the Wider Caribbean Region (Cartagena Convention), and the Convention for the Protection of the Marine Environment and Coastal Area of the South-East Pacific (Lima Convention). In most cases, the area beyond national jurisdiction covered by these conventions is relatively small.

Table 2.1 (compiled based on Gjerde *et al.*, 2008, UNEP, 2010, and convention websites) summarizes the coverage and provisions of these regional seas conventions.

As can be seen from the table, the regional seas conventions (with the exception of the OSPAR and Barcelona Conventions and the Antarctic Treaty) only cover very limited areas beyond national jurisdiction. The map in Figure 2.1 illustrates this point. In addition, while these conventions generally contain provision for conservation tools such as marine protected areas and species protection measures, as well as for the control of pollution, not all of them provide comprehensively for

Table 2.1 *Coverage and provisions of regional seas conventions.*

Agreement	Coverage	Mandate and provisions	Conservation approaches and tools	Ongoing efforts in areas beyond national jurisdiction
OSPAR Convention (North-East Atlantic)	The 'OSPAR Maritime Area' extends from the shores of the contracting parties to substantial adjacent areas beyond national jurisdiction	– Promote concerted action to prevent and eliminate marine pollution – Achieve sustainable management of the maritime area, to meet the needs of present and future generations – Does not cover fishing and shipping	– Ecosystem approach, network of marine protected areas	– Establishment of marine protected areas beyond national jurisdiction – Biogeographic classification
Barcelona Convention (The Mediterranean)	Most Mediterranean countries have not declared EEZs, thus a marine area beyond national jurisdiction generally starts at the 12 nm territorial sea limit	To reduce pollution in the Mediterranean Sea and protect and improve the marine environment of the area, thereby contributing to its sustainable development	Protocol concerning specially protected areas and biological diversity; has led to the establishment of Specially Protected Areas of Mediterranean Interest (SPAMIs)	– Establishment of marine protected areas beyond national jurisdiction – Biogeographic classification
Noumea Convention (South Pacific)	Only covers small enclaves beyond national jurisdiction, which are enclosed by national EEZs (so-called 'donut holes'). Does not cover other high seas areas in the South Pacific	To ensure that resource development is in harmony with maintenance of the unique environmental quality of the region and the evolving principles of sustained resource management	– Protocols on dumping and cooperation in combating oil pollution – While it does not provide for any protocol on protected areas, the Convention provides for the establishment of specially protected areas and protection of wild flora and fauna	None

Table 2.1 (cont.)

Agreement	Coverage	Mandate and provisions	Conservation approaches and tools	Ongoing efforts in areas beyond national jurisdiction
The Antarctic Treaty	Covers continents and surrounding seas south of latitude 60 degrees South (note this excludes the Antarctic Convergence area, and is thus different from the area covered by CCAMLR)	To ensure that Antarctica shall continue forever: – to be used for peaceful purposes – to facilitate scientific research – to promote international cooperation in scientific investigations	The Protocol for Environmental Protection (Madrid Protocol) establishes EIA procedures as well as Antarctic specially protected areas and Antarctic specially managed areas	Protected area designation, species protection and management, EIA
Cartagena Convention (Wider Caribbean Region)	Applies only to small high seas enclaves in the Gulf of Mexico and the Caribbean, as well as areas where parties have not yet declared EEZs	– To prevent, reduce, and control pollution of the convention areas – To ensure sound environmental management, using for this purpose the best practicable means at their disposal and in accordance with their capabilities	– Three protocols: on land-based sources of pollution, specially protected areas and wildlife (SPAW), and in combating oil spills – Establishment of protected area networks under SPAW protocol	None
Lima Convention (South-East Pacific)	Applies to high seas, but only insofar as pollution in the high seas may affect the 200 nm maritime area	To prevent, reduce, and control pollution of the marine environment and coastal area of the South-East Pacific and to ensure appropriate environmental management of natural resources	The Protocol for the Conservation and Management of Protected Marine and Coastal Areas of the South-East Pacific provides for establishment of MPAs, including extended continental shelves	None

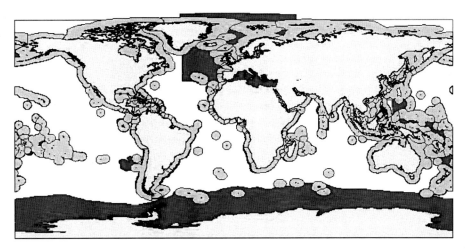

Figure 2.1 Area beyond national jurisdiction (in red) covered by Regional Seas Conventions. (*Map is for information only and expresses no opinion on boundaries.*) A black and white version of this figure will appear in some formats. For the colour version, please refer to the plate section.

application of the ecosystem approach, nor monitoring and assessment. Only the Antarctic Treaty has provisions for environmental impact assessment. None of the conventions address all human uses, with bioprospecting and new and emerging uses some of the most common gaps.

Regional fisheries management organizations and arrangements

Regional fisheries management conventions are generally administered by regional fisheries management organizations (RFMOs). While there are some 30 regional fishery bodies, some of which have been established under the FAO Convention and some independently by states, there are approximately 15 RFMOs with full responsibility for agreement on binding conservation and management measures. Most cover only areas beyond national jurisdiction, although three cover only areas within national jurisdiction (Kimball, 2005).

The scope of each RFMO's conservation responsibilities varies in accordance with the associated convention. Some have competence over most or all marine living resources, while others manage only a particular species. Some are mandated to develop measures based on ecosystem and precautionary approaches, while others manage a target fishery resource without consideration of ecosystem effects. In the former category there tend to be the more recently concluded agreements, while agreements pre-dating UNCLOS do not often include these concepts. In fact, recent research has shown that newer RFMOs generally conform better with the newer trends, particularly those pertaining to conservation measures that have evolved significantly over the years (Cullis-Suzuki and Pauly, 2010).

In response to concerns about declining fisheries and biodiversity in the oceans, there have been recent efforts within the international community to strengthen the conservation and management regimes of RFMOs, and to improve their performance in accordance with the demands of international fishery instruments. The UN Fish Stocks Review Conference agreed, in May 2006, that RFMOs should undergo performance reviews on an urgent basis, including independent evaluation, ensuring that results were publicly available. The December 2006 UN General Assembly Resolution on Sustainable Fisheries also called upon countries to develop and apply best practice guidelines for RFMOs, and to undertake performance reviews of RFMOs, based on transparent criteria. As a result, many RFMOs are taking steps to strengthen governance through implementing the ecosystem approach to fisheries and are adopting the precautionary approach.

Table 2.2 (compiled based on Gjerde *et al.*, 2008, UNEP, 2010, and convention websites) summarizes the coverage and provisions of regional fisheries management conventions, including the extent to which they incorporate ecosystem and precautionary approaches. FAO advisory bodies with no regulatory power are not included in the table.

2.4.4 Geographic gaps in the regional legal regime

It is evident from Table 2.2, and from the analysis by Gjerde *et al.* (2008), that there are large geographical gaps in the coverage of regional fisheries management conventions. Even when regional seas conventions are considered in addition to the fisheries agreements, large swathes of the global commons are left uncovered. One of the largest regulatory gaps is the Arctic area. The Arctic has no regional legally binding instrument for fisheries management or biodiversity conservation, except for the OSPAR Convention in the North-East Atlantic. OSPAR, however, only covers a portion of the Arctic. The Arctic Council was established as a high-level intergovernmental forum to provide a means for promoting cooperation, coordination, and interaction among the Arctic states, with the involvement of the Arctic indigenous communities and other Arctic inhabitants on common Arctic issues, in particular issues of sustainable development and environmental protection in the Arctic. While the Arctic Council has adopted a number of political measures to address various issues affecting the Arctic, has been successful in achieving cooperation, and has undertaken substantive work on topics such as Arctic biodiversity and climate change impacts, it has no regulatory competence.

Table 2.2 *Coverage and provisions of Regional Fisheries Management Conventions, including the extent to which they incorporate ecosystem and precautionary approaches.*

Regional fisheries management convention	Mandate and provisions	Conservation approaches and tools	Ongoing efforts for management of habitats and non-target species
The Convention on Future Multilateral Cooperation in North-East Atlantic Fisheries (NEAF)	To ensure the long-term conservation and optimum utilization of the fishery resources in the convention area, providing sustainable economic, environmental, and social benefits	– Covers resources of fish, molluscs, crustaceans and including sedentary species, excluding those covered by other international agreements – Ecosystem approach newly incorporated	– Closure of areas from bottom fishing – By-catch measures
North Atlantic Salmon Conservation Organization (NASCO)	Conservation, restoration, enhancement, and rational management of salmon stocks. taking into account the best scientific evidence available	– Only covers salmon stocks (prohibits fishing of these on the high seas of the convention area) – Ecosystem and precautionary approaches incorporated	
International Convention for the Conservation of Atlantic Tunas (ICCAT)	To maintain populations of tuna and tuna-like fishes found in the Atlantic Ocean at levels which permit the maximum sustainable catch for food and other purposes	– Covers 30 tuna and tuna-like fishes – No reference to ecosystem or precautionary approach, and no direct mandate to safeguard marine ecosystems	None, except requirement to collect data on shark by-catch
Convention on the Future of Multilateral Cooperation in the Northwest Atlantic Fisheries (NAFO)	Optimum utilization, rational management, and conservation of fishery resources of NAFO convention area	– Covers straddling fish stocks and discrete high seas fish stocks (not salmon, tuna, marlins, and whales) –No direct mandate to safeguard marine ecosystems	– Bottom fishing closures in areas of seamounts, coral, and sponge habitat – Some by-catch measures

Table 2.2 (cont.)

Regional fisheries management convention	Mandate and provisions	Conservation approaches and tools	Ongoing efforts for management of habitats and non-target species
Convention on the Conservation and Management of Fishery Resources in the Southeast Atlantic Ocean (SEAFO)	To ensure the long-term conservation and sustainable use of fishery resources on the high seas, other than highly migratory stocks, taking into account other living marine resources and the protection of the marine environment	– Ecosystem approach incorporated – Does not regulate highly migratory stocks (those not addressed by ICCAT are unregulated in this area)	– Closure of areas from fishing activity – Regulation of seabird and other by-catch
Southern Indian Ocean Fisheries Agreement (SIOFA)	To ensure the long-term conservation and sustainable use of fishery resources other than tuna that fall outside national jurisdictions	– Refers to the precautionary approach, ecosystem approach, and duty to protect biodiversity in the marine environment	Measures not yet agreed upon. Only current ecosystem conservation measure relates to voluntary seamount protected areas
Agreement for the establishment of Indian Ocean Tuna Commission (IOTC Convention)	To promote cooperation with a view to ensuring the conservation and optimum utilization of stocks and encouraging sustainable development of fisheries based on such stocks	– Only covers tuna and tuna-like species – No provisions for ecosystem or precautionary approach	None, except some measures relating to shark, seabird, and turtle by-catch
Agreement for the Establishment of the General Fisheries Council/Commission for the Mediterranean (GFCM)	To promote development, conservation, and management of living marine resources	While the ecosystem approach is incorporated, its implementation has been slow	– Trawl ban on all areas deeper than 1000 m. Also shallower area closures from bottom fishing – By-catch measures
Central Bering Sea Convention (CBS Convention)	Conservation, management, and optimum utilization of pollock resources in the convention area	– Only covers pollock stocks – Ecosystem approach not incorporated	None

North Pacific Anadromous Fisheries Commission (NPAFC)	To promote the conservation of anadromous fish stocks in the convention area	Only covers seven species of salmon	Only with regard to minimizing by-catch and scientific research
Convention on the Conservation and Management of the Migratory Fish Stocks in the Western and Central Pacific Ocean (WCPFC)	To ensure, through effective management, the long-term conservation and sustainable use of highly migratory fish stocks in accordance with UNCLOS and UNFSA	– The ecosystem approach has been incorporated – Does not apply to discrete or straddling fish stocks	Only some by-catch mitigation measures
Convention for the Establishment of an Inter-American Tropical Tuna Commission (IATTC) and Convention for the Strengthening of the Inter-American Tropical Tuna Convention (Antigua Convention) (not yet in force)	The objective of the IATTC Convention is to ensure long-term conservation and sustainable use of tuna and other species taken by tuna-fishing vessels in the Eastern Pacific Ocean, in accordance with relevant rules of international law	– New convention addresses an expanded number of tuna species, as well as by-catch – New convention has not yet entered into force	Measures to prevent fishing from adversely impacting non-target species adopted in 2010
South Pacific Ocean Regional Fisheries Management Agreement (SPRFMA)	Through the application of the precautionary approach and ecosystem approach to fisheries management, to ensure the long-term conservation and sustainable use of fishery resources, and in so doing, safeguard the marine ecosystems in which these resources occur (entered into force in 2012)	The ecosystem and precautionary approaches included	Interim measures to protect vulnerable marine ecosystems and ban on expanding fishing areas until assessment completed

Table 2.2 (cont.)

Regional fisheries management convention	Mandate and provisions	Conservation approaches and tools	Ongoing efforts for management of habitats and non-target species
Commission for the Conservation of Southern Bluefin Tuna (CCSBT)	To ensure, through appropriate management, conservation and optimum utilization of southern bluefin tuna	– Only applies to southern bluefin tuna – The wider impacts of fishing on living marine resources and marine ecosystem not considered	None, except limited seabird by-catch measures
Convention for the Conservation of Antarctic Marine Living Resources (CCAMLR)	To ensure the conservation, including rational use, of Antarctic living marine resources	– Covers all species and ecosystems within convention area – The ecosystem approach and precautionary approach well developed	– Bioregional classification and proposed network of MPAs – Different types of fishing closures, from species-specific to full fisheries closures – Area-wide gillnet and trawl ban – Seabird and other by-catch measures

Other geographic gaps exist in:

- The Atlantic Ocean, where the only regional environmental instruments are in the North-East Atlantic (OSPAR Convention) and in the Caribbean (Cartagena Convention). The latter only covers small areas beyond national jurisdiction. Similarly, legally binding fisheries management instruments (except for tuna and tuna-like species) are missing in the East Central, West Central and South-West Atlantic;
- The Indian Ocean, where no legally binding instrument exists for biodiversity conservation. In addition, there is no legally binding instrument for fisheries management in the Northern Indian Ocean, except for tuna and tuna-like species;
- The Pacific Ocean, where there is no legally binding instrument for biodiversity conservation beyond national jurisdiction, except for small high seas enclaves covered by the Noumea Convention. In addition, there are no legally binding fisheries management instruments in the Central and North-Eastern Pacific (except for tuna), and the Western Pacific for fisheries resources other than highly migratory fish stocks.

The map in Figure 2.2 demonstrates the geographic gaps in the coverage of regional agreements (both regional fisheries and regional seas agreements).

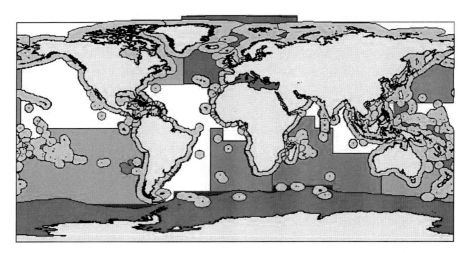

Figure 2.2 Geographical coverage of regional fisheries management organizations (RFMOs) and regional seas conventions. White areas are gaps in coverage. Note that the map only includes RFMOs covering multiple species and incorporating some form of ecosystem approaches. (*Map is for information only and expresses no opinion on boundaries.*) A black and white version of this figure will appear in some formats. For the colour version, please refer to the plate section.

2.4.5 Gaps in application of modern conservation approaches and tools

Even in areas covered by regional agreements, this coverage does not necessarily translate to comprehensive ecosystem-based management of all biodiversity. Many agreements only cover specific species, rather than all fisheries and biodiversity. Similarly, many agreements (particularly older agreements and those covering only one or a few species) do not yet incorporate modern conservation principles, such as the ecosystem approach and precautionary approach, and do not comprehensively take into account the impact of fishing activities on the environment, or on non-target species (i.e. by-catch). While many RFMOs have now identified and protected vulnerable marine ecosystems on the deep seabed, these efforts are far from comprehensive. Of the regional seas agreements, only OSPAR and the Antarctic Treaty have made progress in identification and declaration of marine protected areas in the global commons. These two agreements are also the only ones with environmental impact assessment provisions.

An additional challenge for the international community is to bring agreements into force once negotiations have been concluded. Of the fisheries agreements, the Antigua Convention and SPRFMA are not yet in force.

2.5 Conclusions

The above discussion demonstrates that the international legal regime covering the oceans is complex and fractured. Many of the older agreements are reflections of their time, and do not fully consider the impact of human activities on ecosystems and non-target species. Modern conservation principles, such as the ecosystem and precautionary approach, and tools such as MPAs and EIA, are not comprehensively incorporated. There also remain geographical gaps in the regional regime, which leave large parts of the global commons without a regional agreement.

The future will bring a number of unforeseen challenges that will include a warming climate and its impacts. Related to this may be the emergence of new technologies (ocean fertilization, carbon sequestration, floating wind farms, floating and seabed nuclear power stations) that are not yet fully addressed by existing legal regimes. In addition, the likely future increase in shipping and human use in the Arctic is not yet fully regulated.

All international treaty regimes and agreements can be undermined by parties and non-parties alike. Where implementation and enforcement of obligations is left to states parties, there is a risk of non-implementation or non-compliance and the development of an 'implementation gap' between parties who take their obligations seriously and those who do not. There is also the ever-present risk of activities by non-parties undermining the efforts of the parties (Rayfuse, 2008). Thus, even the best legal regime can, in the end, be voided if enforcement provisions are not sufficient.

References

Alverson, D. L., Freeber, M. H., Pope, J. G. *et al.* (1994). *A Global Assessment of Fisheries Bycatch and Discards*. FAO Fisheries Technical Papers No. 339, Rome: Food and Agriculture Organization of the United Nations, 233 pp.

Anker-Nilssen, T., Barret, R. T., and Krasnov, J. V. (1997). Long- and short-term responses of seabirds in the Norwegian and Barents Seas to changes in stocks of prey fish. In: *International Symposium on the Role of Forage Fishes in Marine Ecosystems, Anchorage, Alaska, USA, 13–16 November 1996*. University of Alaska Sea Grant College Program, AK-SG-97–01, pp. 683–698.

Antarctic Treaty Consultative Meeting (ATCM) (2009). *The Antarctic Biological Prospecting Database*. WP001 for ATCM 17, Baltimore, USA.

Arctic Climate Impact Assessment (ACIA) (2005). Cambridge University Press, 1042 pp. Available at www.acia.uaf.edu, last accessed on 8 June 2014.

Arctic Council (2009). *Arctic Marine Shipping Assessment 2009 Report*. Available at www.arctic.noaa.gov/detect/documents/AMSA_2009_Report_2nd_print.pdf, last accessed on 8 June 2014.

Armstrong, C. W., Foley, N., and Tinch, R. *et al.* (2010). Ecosystem Goods and Services of the Deep Sea. A HERMIONE Project report. Available at http://median-web.eu/IMG/pdf/ecosystem_goods_and_services.pdf, last accessed on 8 June, 2014.

Asia-Pacific Economic Cooperation (APEC) (2012). Economic Impact of Submarine Cable Disruptions. Available at www.suboptic.org/uploads/Economic%20Impact%20of%20Submarine%20Cable%20Disruptions.pdf, last accessed on 8 June 2014.

Botsford, L. W., Castilla, J. C., and Peterson, C. H. (1997). The management of fisheries and marine ecosystems. *Science* **277**: 509–515.

Brida, J. G. and Zapata, S. (2010). Cruise tourism: Economic, socio-cultural and environmental impacts. *Int. J. Leisure and Tourism Marketing* **1**(3): 205–226.

Carter L., Burnett D., Drew S. *et al.* (2009). Submarine Cables and the Oceans: Connecting the World. Cambridge: UNEP-WCMC Biodiversity Series No. 3-1. ICPC/UNEP/UNEP-WCMC.

Clark, M. R. and Koslow, J. A. (2007). Impacts of fisheries on seamounts. In: *Seamounts: Ecology, Fisheries, and Conservation*. Pitcher, T. J., Morato, T., Hart, P. J. B., Clark, M. R., Haggan, N., and Santos, R. S. (eds.). Blackwell Fisheries and Aquatic Resources Series 12. Oxford: Blackwell Publishing, pp. 413–441.

Cochonat, C., Durr, S., Gunn, V. K. *et al.* (2007). The deep sea frontier: Science challenges for a sustainable future. Luxembourg, Office for Official Publications of the European Communities, 53 pp.

Collie, J. S., Escanero, G. A., and Valentine P. C. (2000). Photographic evaluation of the impacts of bottom fishing on benthic epifauna. *ICES Journal of Marine Science* **57**: 987–1001.

Conlan, K. E, Kim, S. L, Lenihan, H. S, and Oliver, J. S. (2004). Benthic changes over ten years of organic enrichment by McMurdo Station, Antarctica. *Mar. Poll. Bull.* **49**: 43–60.

Corbett, J. J, Lack, D. A., Winebrake, J. J. *et al.* (2010). Arctic shipping emissions inventories and future scenarios. *Atmospheric Chemistry and Physics,* **10**(19): 9689, doi: 10.5194/acp-10-9689-2010.

Corbett, J. J., Winebrake, J. J., Green, E. H. *et al.* (2007). Mortality from ship emissions: A global assessment. *Environ. Sci. Technol.* **41**(24): 8512–8518.

Cullis-Suzuki, S. and Pauly, D. (2010). Failing the high sea: A global evaluation of regional fisheries management organizations. *Marine Policy* **34**(5): 1036–1042.

Danovaro, R., Dell'Anno, A., Corinaldesi, C. *et al.* (2008). Major viral impact on the functioning of benthic deep-sea ecosystems. *Nature* **454**: 1084–1087.

Davies, A. J., Roberts, J. M., and Hall-Spencer, J. M. (2007). Preserving deep-sea natural heritage: Emerging issues in offshore conservation and management. *Biological Conservation* **138**(3–4): 299–312.

Diversitas (2009). *What Are Coral Reef Services Worth? $130,000 to $1.2 Million Per Hectare, Per Year.* ScienceDaily, 28 October 2009, available at www.sciencedaily. com/releases/2009/10/091016093913.htm, last accessed on 8 June 2014.

Dyck, A. J. and Sumaila, U. R. (2010). Economic impact of ocean fish populations in the global fishery. *Journal of Bioeconomics* **12**(3): 227–243.

FAO (2010). *The State of World Fisheries and Aquaculture 2010.* Rome, Italy: Food and Agriculture Organization of the United Nations.

FAO (2012). *The State of World Fisheries and Aquaculture 2012.* Rome, Italy: Food and Agriculture Organization of the United Nations.

Gjerde, K. M. (2008). *Regulatory and Governance Gaps in the International Regime for the Conservation and Sustainable Use of Marine Biodiversity in Areas beyond National Jurisdiction.* IUCN Marine Series No. 1. Available at http://data.iucn.org/ dbtw-wpd/edocs/EPLP-MS-1.pdf, last accessed on 16 June 2014.

Gjerde, K. M (2012). Challenges to protecting the marine environment beyond national jurisdiction. *The International Journal of Marine and Coastal Law* **27**: 839–847.

Glover, A. G. and Smith, C. R. (2003). The deep-sea floor ecosystem: Current status and prospects of anthropogenic change by the year 2025. *Environmental Conservation* **30**(3): 219–241.

Hansen, D. J. (1997). Shrimp fishery and capelin decline may influence decline of harbour seal (*Phoca vitulina*) and northern sea lion (*Eumetopias jubatus*) in Western Gulf of Alaska. In: *International Symposium on the Role of Forage Fishes in Marine Ecosystems, Anchorage, Alaska, USA, 13–16 November 1996*, University of Alaska Sea Grant College Program, AK-SG-97–01: pp. 197–207.

Herring, P., Gaten, E., and Shelton, P. M. J. (1999). Are vent shrimps blinded by science? *Nature* **398**: 116.

Hourigan, T. F. (2008). The status of cold-water coral communities of the world: A brief update. In: *Status of Coral Reefs of the World 2008.* Wilkinson, C. (ed.). Townsville, Australia: Global Coral Reef Monitoring Network and Reef and Rainforest Research Centre.

Hunt, B. and Vincent, A. C. (2006). Scale and sustainability of marine bioprospecting for pharmaceuticals. *Ambio* **35**(2): 57–64.

Huntington, H. P. (2009). A preliminary assessment of threats to Arctic marine mammals and their conservation in the coming decades. *Marine Policy* **3**(1): 77–82.

Inagaki, F., Kuypers, M. M., Tsunogai, U. *et al.* (2006). Microbial community in a sediment-hosted CO_2 lake of the southern Okinawa Trough hydrothermal system. *PNAS* **103**(38): 13899–13900.

IPCC (2005). *IPCC Special Report on Carbon Dioxide Capture and Storage.* Prepared by Working Group III of the Intergovernmental Panel on Climate Change (Metz, B., Davidson, O., de Coninck, H. C., Loos, M., and Meyer, L. A. (eds.)). Cambridge University Press, United Kingdom and New York, NY, USA, 442 pp.

ISA (2014). International Seabed Authority website, www.isa.org.jm/en/home, last accessed on 8 June 2014.

Juniper, K. S. (2001). Background paper on deep-sea hydrothermal vents. In: *Managing Risks to Biodiversity and the Environment on the High Sea, Including Tools Such as Marine Protected Areas: Scientific Requirements and Legal Aspects*. Thiel, H. and Koslow, A. (eds.). Proceedings of the Expert Workshop held at the International Academy for Nature Conservation, Isle of Vilm, Germany, 27 February–4 March 2001. German Federal Agency for Nature Conservation. Available at www.bfn.de/fileadmin/MDB/documents/proceed1.pdf, last accessed on 16 June 2014.

Kaluza, P., Kölzsch, A., Gastner M. T. *et al.* (2010). The complex network of global cargo ship movements. *J. R. Soc. Interface* **7**(48): 1093–1103.

Kimball, L. (2005). *The International Legal Regime of the High Seas and the Seabed beyond the Limits of National Jurisdiction and Options for Cooperation for the Establishment of Marine Protected Areas (MPAs) in Marine Areas beyond the Limits of National Jurisdiction.* Secretariat of the Convention on Biological Diversity, Montreal, Technical Series No. 19, 64 pp.

Koslow, J. A., Gowlett-Holmes, K., Lowry, J. K *et al.* (2001). Seamount benthic macro-fauna off southern Tasmania: Community structure and impacts of trawling. *Marine Ecology Progress Series*, **213**: 111–125.

Koslow, T. (2007). *The Silent Deep: The Discovery, Ecology, and Conservation of the Deep Sea.* Sydney, Australia: UNSW Press.

Lenihan, H. S. (1992). Benthic marine pollution around McMurdo Station, Antarctica: A summary of findings. *Marine Pollution Bulletin* **25**: 318–323.

Mehl, S. (1991). The northeast Arctic cod stock's place in the Barents Sea ecosystem in the 1980s: An overview. In *ProMare*. Sakshaug, E., Hopkins, C. C. E., and Øritsland, N. A. (eds). Symposium on Polar Marine Ecology, Trondheim, Norway, *Polar Research* **10**.

Merrie, A., Dunn, D. C., Metian, C., *et al.* (2014). An ocean of surprises – Trends in human use, unexpected dynamics and governance challenges in areas beyond national jurisdiction. *Global Environmental Change* **27**: 19–31.

Morato, T., Cheung, W. W. L. and Pitcher, T. J. (2006). Vulnerability of seamount fish to fishing: Fuzzy analysis of life-history attributes. *Journal of Fish Biology* **67**: 1–13.

Noone, K. J., Sumaila, U. R., and Diaz, R. J. (eds.) (2013). *Managing Ocean Environments in a Changing Climate: Sustainability and Economic Perspectives.* London, UK: Elsevier, 359 pp.

Pauly, D., Christensen, V., Dalsgaard, J. *et al.* (1998). Fishing down marine food webs. *Science* **279**: 860–863.

Pauly, D., Christensen, V., Guénette, S. *et al.* (2002). Towards sustainability in world fisheries. *Nature* **418**(6898): 689–695.

Pimental, D., Zuniga, R., and Morrison, D. (2005). Update on the environmental costs associated with alien invasive species in the United States. *Ecol. Econ.* **52**: 274–288.

Pontecorvo, G., Wilkinson, M., and Anderson, R. *et al.* (1980). Contribution of the ocean sector to the United States economy. *Science* **208**(4447): 1000–1006.

Rayfuse, R. (2008). Protecting marine biodiversity in polar areas beyond national jurisdiction. *RECIEL* **17**(1): 3–13.

Riser, S. C. and Johnson, K. S. (2008.) Net production of oxygen in the subtropical ocean. *Nature* **451**: 323–326.

Roberts, S., Aguilar, R., Warrenchuk, J. *et al.* (2005). *Deep Sea Life: On the Edge of the Abyss.* New York: Oceana.

Sainsbury, K. J., Campbell, R. A., and Whitelaw, A. W. (1993). Effects of trawling on the marine habitat on the north west shelf of Australia and implications for sustainable fisheries management. In: Hancock, D.A. (ed.,) *Australian Society for Fish Biology*

Workshop, Victor Harbor, South Australia, 12–13 August 1992. Canberra: Australian Government Publishing Service, pp. 137–145.

Schubert, R., Schellnhuber, H.-J., Buchmann, N. *et al.* (2006). *The Future Oceans – Warming Up, Rising High, Turning Sour.* Special Report of the German Advisory Council on Global Change. Berlin.

Secretariat of the Convention on Biological Diversity (2008). *Synthesis and Review of the Best Available Scientific Studies on Priority Areas for Biodiversity Conservation in Marine Areas beyond the Limits of National Jurisdiction.* CBD Technical Series No. 37: 63 pp.

Secretariat of the Convention on Biological Diversity (2009). *Scientific Synthesis of the Impacts of Ocean Fertilization on Marine Diversity.* CBD Technical Series No. 45: 53 pp.

Sumaila, U. R., Cheung, W. W. L., and Dyck, A. *et al.* (2012). Benefits of rebuilding global marine fisheries outweigh costs. *PLoS ONE* **7**(7), e40542, doi: 10.1371/journal.pone.0040542.

Teh, L. C. L. and Sumaila, U. R. (2013). Contribution of marine fisheries to worldwide employment. *Fish and Fisheries* **14**(1): 77–88.

Tladi, D. (2011). Ocean Governance: A Fragmented Regulatory Regime. In: *Oceans: The New Frontier.* Jacquest, P., Pachuari, R. K., and Tubiana, L. (eds.). Delhi: TERI Press, pp. 99–110.

Treves, T. (2010). The development of the law of the sea since the adoption of the UN Convention on the Law of the Sea: Achievements and challenges for the future. In: *Law, Technology and Science for Oceans in Globalisation – IUU Fishing, Oil Pollution, Bioprospecting, Outer Continental Shelf.* Vidas, D. (ed.). Leiden/Boston: Martinus Nijhoff Publishers.

UNCTAD (2010). *Review of Maritime Transport 2010.* Geneva, Switzerland: United Nations Conference on Trade and Development. Available at www.unctad.org/en/docs/rmt2010_en.pdf, last accessed on 16 June 2014.

UNEP (2010). High seas MPAs: Regional Approaches and Experiences. UNEP (DEPI)/RS.12 /INF.6.RS. 12th Global Meeting of the Regional Seas Conventions and Action Plans. Bergen, Norway, 20–22 September, 2010.

van den Hove, S. and Moreau, V. (2007). Deep-Sea Biodiversity and Ecosystems: A Scoping Report on their Socio-economy, Management and Governance. Nairobi: United Nations Environment Programme.

Van Dover, C. L. (2010). Mining seafloor massive sulphides and biodiversity: What is at risk? *ICES Journal of Marine Science*, doi: 10.1093/icesjms/fsq086.

Vierros, M., Aricò, S., and Douvere, F. (2006). *Implementing the Ecosystem Approach in Open Ocean and Deep Sea Environments.* United Nations University, Institute of Advanced Studies. Available at http://unu.edu/publications/policy-briefs/implementing-the-ecosystem-approach-in-open-ocean-and-deep-sea-environments.html, last accessed on 8 June 2014.

Waller, R., Watling, L., Auster, P. *et al.* (2007). Anthropogenic impacts on the Corner Rise Seamounts, North-West Atlantic Ocean. *Journal of the Marine Biological Association of the United Kingdom* **87**(5): 1075–1076.

Wangnick, K. (2002). *2002 IDA Worldwide Desalting Plants Inventory.* Produced by Wangnick Consulting for the International Desalination Association. Gnarrenburg, Germany.

Watson, R. and Morato, T. (2004). Exploitation patterns in seamount fisheries: A preliminary analysis. In: *Seamounts: Biodiversity and Fisheries. Fisheries Centre Research Report* **12**(5). Morato, T. and Pauly, D. (eds.)

Watson, R. A., Cheung, W. W. L., Anticamara, J. A. *et al.* (2013). Global marine yield halved as fishing intensity redoubles. *Fish and Fisheries* **14**(4): 493–503.

Waycott, M., Duarte, C. M., Carruthers, T. J. B. *et al.* (2009). Accelerating loss of seagrasses across the globe threatens coastal ecosystems. *PNAS* **106**(30): 12377–12381.

Wilkinson, C. (ed.) (2008). *Status of Coral Reefs of the World 2008*. Townsville, Australia: Global Coral Reef Monitoring Network and Reef and Rainforest Research Centre: 296 pp.

Zeller, D., Booth, S., and Pauly, D. (2006). Fisheries contributions to GDP: Underestimating small-scale fisheries in the Pacific. *Marine Resource Economics* **21**(4): 355–374.

Note

1 Straddling stocks are fish stocks that migrate through, or occur in, more than one exclusive economic zone.

3

Physical and chemical changes in the ocean over basin-wide zones and decadal or longer time-scales: perspectives on current and future conditions

GUNNAR KULLENBERG, EDDY CARMACK, AND KENNETH DENMAN

3.1 Introduction

Comprehensive international research programmes and reviews have firmly established that human activities are substantially impacting the total environment, including the ocean, and are triggering global and regional changes in the physical, chemical, and biological conditions which affect ecosystems and human societies. These programmes include the International Geosphere–Biosphere Programme (IGBP), the Large Marine Ecosystem Programme (LME), activities of the Intergovernmental Oceanographic Commission (IOC) and World Heritage Convention of UNESCO, as well as integrated assessments of the Intergovernmental Panel on Climate Change (IPCC) and the United Nations Environment Programme (Kullenberg, 2010; Clarke, 2010).

As a consequence of the growing needs of the increasing human population, diverse pressures on the marine environment have increased markedly over past decades. These include offshore oil and gas exploration and exploitation (e.g. into deeper and more hazardous areas of the Arctic Basin), expansion of fisheries into new areas (e.g. the Southern Ocean krill fisheries), transportation and shipping (e.g. across previously ice-covered areas of the Arctic Basin), offshore extraction of renewable energy, aquaculture production, enhanced use of the coastal zone for urbanization, land reclamation, infrastructure installations, ports and transportation, and recreation and tourism. Taken together these factors constitute a global change parallel to the population growth since the 1950s, with large consequences for our physical environment. However, what of all these factors constitutes *significant* change? Carmack and McLaughlin (2011) conclude that 'changes in the *physical* environment are considered significant when they affect the biosphere, including humans'. This definition agrees in principle with the IGBP and LME approach. Observations and model simulations show that impacts of climate change on the ocean include the redistribution of oceanic water mass boundaries and habitats, and identify the need for time-series observations over a wide range of ocean zones that will permit science-informed policy decisions.

In the present chapter we briefly summarize documented changes and variations in the physical, chemical, and biological nature of the marine environment over basin-wide zones and decadal or longer time-scales, omitting any in-depth analysis of possible consequences for biology and the ecosystem. In this brief overview of changing ocean conditions, many aspects cannot be covered, but are all required to obtain an understanding of the processes and the concerns for global environmental change. Reference is made to mainstream overviews presented in the regular assessments of the IPCC with respect to climate change, and publications such as Field, Hempel, and Summerhayes (2002) for the marine environment and Robinson and Brink (1998) for regional and shelf seas; and Sherman, Aquarone, and Adams (2009) for LMEs.

3.2 Ocean warming

Ocean temperatures and air–sea fluxes play a key role in regulating physical and biological systems. Storage and transport of heat in the ocean are central drivers of the global energy budget (Church *et al.*, 2011; Levitus *et al.*, 2012). They influence slow climate variability such as El Niño (Zebiak, 1989) and the North Atlantic Oscillation (Curry and McCartney, 2001), and influence storm tracks and moisture flux across the oceans (Hoskins and Hodges, 2002). Expansion due to warming is a major contributor to sea-level rise (Johnson and Wijffels, 2011; Church *et al.*, 2011). Increasing temperatures contribute to upper layer stratification, and thus may alter nutrient supply to primary producers (Barber, 2007). Ocean temperatures also set thermal range limits for fish (Pörtner and Farrell, 2008) and conditions for species invasion (Occhipinti-Ambrogi, 2007).

Overall the global ocean has warmed significantly during the period of instrumental records (Levitus *et al.*, 2012). For example, the ocean mean temperature in the 0–2000 m layer increased by 0.09 °C over the period 1955–2010, and global mean sea surface temperature (SST) increased by 0.67 °C over the last century (Rayner *et al.*, 2006). Warming in most LMEs accelerated in the late 1970s/early 1980s. Approximately two thirds of the increase in ocean heat content during 1955–1998 occurred in the upper 700 m, with patterns of change having clearly defined basin-wide scales (Levitus *et al.*, 2012).

The increase in ocean heat content is much larger than in any other store of energy on Earth over the assessed periods 1961–2003 and 1993–2003 and accounts for approximately 93% of the increase in heat content of the Earth system (Levitus *et al.*, 2012). Ocean heat content variability is therefore a critical variable for detecting the effects of the observed increase in greenhouse gases (GHG), and for resolving the planet's overall heat and energy balance.

Satellite observations of LMEs show SST warming in 61 out of 64 domains, ranging from 0.08 °C (Patagonia) to 1.35 °C (Baltic Sea). Relatively rapid warming of 0.6 °C was found in middle and high-latitude LMEs over the past 25 years. Warming exceeding 0.96 °C over this period was observed in the North Sea, East China Sea, Sea of Japan/East Sea, Newfoundland/Labrador Shelf, and Black Sea LMEs (Sherman, Aquarone, and Adams, 2009).

Multi-model simulations aimed at elucidating the response of ocean ecosystems to climate warming (Sarmiento *et al.*, 2004) show an increase in SST everywhere in the global ocean. This analysis brings out changes in biome areas that may be expected to result from global warming, including a large reduction in the marginal sea ice biome and an increase in the permanently stratified subtropical gyre biome. The subpolar gyre biome expands while the seasonally stratified subtropical gyre biome contracts.

Recent scientific findings show, furthermore, that human actions are now reaching all observed ocean areas. They confirm that the ocean has absorbed 90% of the heat increase of the planet over the past 50 years. The ocean's heat content has increased about 20 times more than that of the atmosphere. The warming has penetrated at least 2000 metres into the deep sea. There have been interpretations of observations suggesting that the meridional over-turning in the North Atlantic has slowed down, associated with a related decrease of the formation of North Atlantic deep water (NADW), but evidence from sedimentary records is more consistent with an increased production of NADW during cyclic climate warming periods, rather than the reverse (Wefer and Berger, 2001). Altered deep water formation, the related overflow across the Denmark Strait, and the thermohaline circulation may then be coupled to a decrease in the surface layer density due to climate-induced sea ice melt and subsequent export of low salinity Arctic waters through the Fram Strait into the Greenland, Iceland, and Norwegian seas.

3.3 Salinity and fresh water

Ocean storage and transport of fresh water are critical elements of the global climate, including the water cycle (e.g. Schanze *et al.*, 2010) and anthropogenic climate change (e.g. Held and Soden, 2006). Understanding of the ocean's freshwater budget has increased greatly over the past decade due to the Argo Project, an array of profiling floats that measures temperature and salinity all year round in the upper 2 km of the ice-free global ocean (Roemmich and Gilson, 2009). Ocean salinity and hence freshwater content are changing on gyre and basin-wide scales, and observed changes can have major implications for water mass composition and circulation patterns.

Mixed layer salinity and evaporation minus precipitation (E−P) are well correlated over much of the ocean so that near-surface salinity acts as a proxy rain gauge (Schmitt, 2008; Yu, 2011). As such, surface forcing has led to an intensification of the global hydrological cycle over the past 50 years, shown by increasing salinity at the sea surface in areas dominated by evaporation and decreasing surface salinity in areas dominated by precipitation (Durack and Wijffels, 2010). These surface-forced changes also extend to subsurface levels. Changes of salinity are of global scale with similar patterns in the basins: subtropical waters show an increase; subpolar surface and intermediate waters have freshened in the Atlantic and the Pacific; the intermediate water at around 1000 m depth in the Southern Hemisphere has freshened in both the Atlantic and Pacific parts. In the Northern Hemisphere, a freshening has occurred in the North Pacific, with an attendant decrease in the oxygen content, indicating less ventilation than before; but in the North Atlantic the intermediate layer, at about 900–1200 m, has become saltier due to increased outflow from the Mediterranean, which is becoming saltier (cf. Curry *et al.*, 2003). A decrease in salinity has also been observed in northern seas.

In areas with increasing evaporation there are increasing salinity trends; in higher latitude areas there is generally freshening due to more precipitation and higher run-off there. The Atlantic is becoming saltier over much of the water column. The increasing difference in volume-averaged salinity between the Pacific and the Atlantic oceans suggest changes in freshwater transport between these ocean basins. The pattern of salinity changes is compatible with changes in the hydrological cycle, in particular with changes in precipitation and inferred larger water transports in the atmosphere from low to higher latitudes and from the Atlantic to the Pacific. The results of the analysis of model simulations by Sarmiento *et al.* (2004) agree with the response of the salinity distribution to an overall enhancement of the atmospheric hydrological cycle, due to the increased moisture bearing capacity of the warmer air. The subtropical region of high evaporation generally becomes saltier, while high-latitude areas of greater rainfall become fresher. The temperature and salinity changes result in an overall reduction of the surface water density, which can be expected to lead to increased vertical stratification and possibly reduced nutrient supply to the euphotic zone. The model simulations also show a reduction of the winter maximum surface mixed layer depth, especially at latitudes greater than 40°.

In the Arctic the ice cover shrank by 2.7%/decade between 1979 and 2009, with a decrease of 7.4%/decade in the summer minimum sea ice cover (Lemke *et al.*, 2007). However, summer minimum sea ice cover for the years 2007–2012 has shown to be the six lowest years of summer minimum sea ice cover observed; the record low summer sea ice cover in 2012 was only 51% of the average for the years 1979–2000 (US National Snow and Ice Data Center, 2012). Variability of salinity and temperature in areas of Icelandic and Faeroese waters has been observed over several decades in the associated shelf and slope regions (Hansen *et al.*, 1998). The annual mean SST north of Iceland in the century 1870–1970 shows a high at around 1870–1875, a rather persistent low over the period 1880–1915, followed by above average for 1925–1965. Temperature and salinity observations in late spring in the same region north of Iceland over the half-century 1950–2000 show deviations from the 1961–1980 mean at roughly decadal time-scales. The temperature is up to about 2 °C above average over the period 1950–1960, about 2 °C below in the 1960s, with the coldest period in 1965–1971, and subsequent oscillations of about +/− 1.5 °C, with positive values in the 1990s. Similar variability occurs in the salinity records. In both there are marked interannual variations. These observations also demonstrate the variability of the inflowing volume of Atlantic water north of Iceland. An appreciable mixture of polar water has been observed in the area since the mid 1960s.

3.4 Ocean circulation

Ocean circulation is key to correcting the global heat imbalance by transporting excess heat from the low to the high latitudes, in redressing the global hydrological cycle, and in establishing a marine ecology based on advection of abiotic and biotic properties. This section describes ocean surface current changes, transports derived from ocean surface currents, and features such as rings inferred from surface currents. Surface currents are obtained from *in situ* and satellite (altimetry and wind) observations. The changes of heat and salt content are linked to changes in ocean circulation. The salinity changes are consistent with an increase in the hydrological cycle over the oceans and will drive changes in ocean advection.

In the North Atlantic the warming has extended well below 1000 m, and is particularly pronounced under the Gulf Stream and the North Atlantic Current at 40 °N (Bindoff *et al.*, 2007; Levitus *et al.*, 2012). This is consistent with a predominantly positive phase (increasing trend) of the North Atlantic Oscillation (NAO) during past several decades.

The NAO follows the variation of the pressure difference between the Icelandic low and the Azores high. The climates of western Europe, north-western Asia, and the north-west coast of North America are closely linked with this pressure difference, thus reflected in the NAO index (e.g. Voituriez, 2003). When the NAO is high, strong westerlies dominate, bringing oceanic air towards western Europe. On the other hand a low NAO implies weak westerlies, favouring polar or continental air over north-western Europe. Observations from around the year

1850 provide for direct measurement of the NAO, and analysis of the growth rings of trees have made it possible to reconstruct the NAO since around 1700 (Voituriez, 2003). The time series show several periods of oscillation. However, the reasons for these oscillations have not yet been clarified.

Within the Arctic Ocean the ice coverage is decreasing faster than predicted by climate models, with consequences to both the wind-driven and thermohaline circulations. The change in size of the ice–snow cover areas provides an important positive feedback through the change of reflection of the incoming energy, a change in albedo. Decreasing ice–snow cover areas implies a decrease of reflection and an increase of energy absorption resulting in more ice–snow melting. Vice versa, when the ice–snow cover increases, this leads to increased reflection, supporting further snow cover and ice generation. Changes in ice cover within the Arctic Basin may also be related to ocean circulation. Part of the inflowing water of Atlantic origin can penetrate further north into the Arctic Basin at the Eurasian side, forming there an intermediate and deep water layer. Studies of historical data of this water mass, extending over one hundred years, show that since the 1970s this water has warmed significantly and migrated upwards closer to the sea ice (cf. Carmack and Melling, 2011). It was likewise confirmed that while moving anticlockwise around the basin, the water of Atlantic origin cooled. Taken together these findings suggest an increase of the heat flux from the layer with Atlantic water to the sea ice, leading to a thinning of the ice. This process could have preconditioned the ice cover for the dramatic decrease observed during the International Polar Year 2007 (Carmack and Melling, 2011).

In the Greenland, Iceland, and Norwegian seas the surface layer density is strongly dependent upon the salinity, which varies with the inflow of relatively high saline water from the south and inflow of low-salinity melt water from the Arctic Basin. High salinity combined with cooling, strong winds and evaporation in the winter season, together drive the vertical circulation, convection, and deep water formation in the northern North Atlantic; particularly in the Greenland Sea. However, the rate of renewal of the interior deep waters also depends upon the mixing in the interior of the ocean. There will always be some mixing due to the energy supplied by the tidal motion. The energy input from the wind, combined with other air–sea interaction processes, including breaking wind-generated waves, generates mixing in the top several hundred metres. The wave processes have a noticeable influence on the exchange of oxygen between the atmosphere and the ocean.

Time-series of observations during 1950–2000 of the atmospheric pressure difference between the Icelandic low and the Azores high, measured between Iceland and Lisbon (the NAO), the Gulf Stream intensity, and the water temperature in the Labrador Sea, show decadal variability in all (Latif and Meincke, 2001). The NAO indicates an increasing trend since the 1960s, the water temperature in the Labrador Sea shows a decreasing trend since the 1970s, and the Gulf Stream intensity shows an oscillation opposite to that of the temperature. This means that a decreasing temperature in the Labrador Sea may be associated with a strengthening of the Gulf Stream and vice versa. There appears to be some consistency with the oscillation of the NAO. However, how this ocean–atmosphere interaction works is still subject to scientific debate.

Increasing transport in the Gulf Stream and North Atlantic circulation may lead to a deepening of the thermocline, due to the geostrophic constraint, and a compression horizontally of the current system (Stommel, 1958). This would lead to a withdrawal or weakening of warm surface water from northern parts of the region and a cooling there. This in turn leads to an increased pressure and weakening of the westerlies. The northern, polar, wind system may then enhance the transport of cold water towards Iceland–Greenland. At the same time the decreasing westerlies would lead to a weakening of the wind-driven circulation with a corresponding rise of the thermocline, forcing the warm surface water radially outwards from the Sargasso Sea region. This in turn leads gradually to a change in the pressure difference and a strengthening of the westerlies. The system could in this way be sustainable.

A low NAO corresponds with weak westerlies. This may allow the relatively high saline water from the south to penetrate relatively further north than during a high NAO index, which implies strong westerlies. These will force the surface layer water to deviate towards the east and south, decreasing the penetration of this water northwards. Coupled ocean–atmosphere model simulations seem consistent with this conceptual interpretation. But how does this fit with the possible decrease of the meridional over-turning, also called the thermohaline circulation, referred to in Section 3.2? Warming periods may coincide with an increase of heat transfer towards the north and increasing NADW formation. This in turn leads to a gradual increase of the atmospheric pressure difference, an increasing NAO, as experienced since the 1960s, and strengthening of the westerlies, leading to a gradual decrease of the penetration northwards of the high saline water from the south. In parallel, the fronts between different air and water masses at middle and high latitudes have a tendency to shift southwards. This generates a southward displacement of the intertropical convergence zone, at the thermal equator, with an associated strengthening of the north-east trade wind. This can intensify the North Atlantic circulation gyre in mid-latitudes.

During the most recent ice age in the North Atlantic the over-turning, thermohaline circulation, also referred to as the global ocean conveyor belt, was partially shut down. The northern North Atlantic was much cooler than at present. The northward transport of heat and salt of the North Atlantic current (also referred to as the North Atlantic Drift), being the extension of the always existing Gulf

Stream, was much reduced. During the cooling period of ice formation and low deep water generation the salt export from the northern waters was decreasing, or was weak. The inflow of low-saline melt water from the Arctic Basin was also weak. These together implied a gradual increase of the surface layer salinity and density, decreasing the difference between the top layer and the deep water. In the polar, high-latitude regions the buoyancy provided by inputs from several sources also play an important role in ocean forcing, together with the large-scale wind forcing. The process may continue until a critical density difference is reached, allowing the deep water formation to start or increase, reactivating the conveyor belt. This allows for gradually increasing inflow from the south of relatively warm and saline water.

The NAO will then gradually increase due to the warming, strengthening the westerlies and weakening the northward penetration. In parallel, the density difference between surface layer and deep water increases due to the warming and inflow of low saline water from the Arctic Basin. The tidal energy for the interior mixing remains the same. The outlined conceptual cyclic process may thus continue.

The northward transfer of heat driven by the thermohaline circulation in the North Atlantic provides a warming of the present climate of north-western Europe by 5 to 10 °C (North and Duce, 2002). Data from ice cores as well as sedimentary records confirm that the thermohaline circulation is not stable (Wefer and Berger, 2001). Throughout the last ice age several flips in the circulation occurred, causing major climate change in this region, now referred to as Dansgaard–Oeschger events. Cold climate episodes could start with a temperature drop of over 5 °C over Greenland over a few decades or less, but persisting for centuries.

The problem now is that the increase of GHG in the atmosphere leads to enhanced melting of ice in the north, including in the Arctic Basin, generating a decrease of the surface layer density. With the energy available for surface layer and interior mixing being essentially the same as before, this implies a possible decrease of the deep water formation and an associated decrease of inflow from the south of warm and saline water. This, then, counteracts the warming in the north. The critical numbers in this conjecture are the density difference between surface and deep waters and the amount of energy available for the interior mixing. If this is only provided by the tidal motion it may be assumed to be essentially constant.

Strong variations of the North Atlantic circulation have been observed during past decades. These appear coupled with regular low frequency changes of the NAO, which shows an overall increase since the early 1960s. Sedimentary records suggest that the flow of warm high-salinity water presently reaching the northern North Atlantic was practically cut off during the last glacial maximum. Related oceanic and atmospheric fronts were then pushed southwards, and the North

Atlantic current was diverted towards Portugal. The formation of deep water in the North Atlantic was then reduced, leading to a related decrease in heat transfer to the north – a positive feedback.

Ocean circulation and climatic conditions are closely linked. The ocean has a much longer 'memory' than the atmosphere. Researchers at the Geological Survey of Denmark and Greenland and the Universities of Aarhus and Gothenburg (Kuijpers, personal communication 2010) have found a link between sea surface warming in the North Atlantic and cold winters in Europe. The periods of cold winters may be linked to the variability of SST in the North Atlantic called the Atlantic Multidecadal Oscillation. The multidecadal data indicate that we may be entering a similar climate regime to that in the 1940s, with noticeably colder winters than in previous decades. The multidecadal oscillation of the SST does fit with the low frequency changes in the NOA and variations of the North Atlantic circulation observed during the last couple of decades.

The observation that warm SST in the North Atlantic appears linked with cold winters in Europe seems to fit with the observation of opposite oscillation between the Labrador Sea and the Gulf Stream. An increasing temperature in the Labrador Sea corresponds with a weakening of the Gulf Stream and vice versa. A weakening of the Gulf Stream and the North Atlantic Gyre flow implies relatively cold climate and winters in Europe. During the cold period the NAO is low, with weak westerlies allowing the warm and salt water of the North Atlantic current to penetrate relatively further north than during periods of strong westerlies, coupled with a high NAO, which deviates the North Atlantic current towards the east and south-east. Thus it seems that a relatively warmer SST period would correspond with a relatively colder European period. Are the ocean and the continental temperatures out of phase? Can this be coupled with the strength of the westerlies? Sedimentary records seem to indicate that an increased production of NADW is consistent with warming periods. These records suggest that the oceanic and continental climate conditions are slightly out of phase, possibly particularly during transition periods between the climate cycles (Guiot *et al.*, 1989).

Multidecadal basin-wide oscillations are also well documented in other ocean regions, in particular in the context of searching for correlations with interannual to multidecadal fluctuations in fisheries yields. In the North Pacific Ocean, the Pacific Decadal Oscillation (PDO) correlates with SST and sea surface height anomalies and with large-scale shifts in North Pacific salmon production (Mantua *et al.*, 1997). A related index, the North Pacific Gyre Oscillation (NPGO), correlates with sea surface salinity, chlorophyll a pigments, oxygen, and nutrient concentrations, all indicators of planktonic foodweb productivity (Di Lorenzo *et al.*, 2008).

Similar oscillations have been recorded in several LMEs through studies over several decades. An example is provided by the Yellow Sea LME, where four SST regimes have been identified over the past 140 years: a warm period before 1900; a cold period during 1901–1944; a warm regime with cooling trend during 1945–1976; and a warm regime with a warming trend during 1977–2007. The SST regime shifts, fluctuations in herring abundance and rainfall in Eastern China show good correlations (Tang, 2009). These regime shifts in the Yellow Sea may be coupled with multidecadal changes in the Pacific Ocean and the related Pacific Decadal Oscillation. This has been well documented with good correspondence between global air temperature, the PDO, the Atmospheric Circulation Index and the Regime Indicator series over a century (Chavez *et al.*, 2003).

The tropical ocean plays a major role in the natural climate variability at interannual scales. In the tropical Pacific Ocean the El Niño–Southern Oscillation, ENSO phenomenon, represents the most pronounced climate variability process on Earth, with strong effects on the weather in many parts during El Niño occurrences. The reversed, cold period, La Niña, also has strong effects, with opposite characteristics. Initial model results had suggested that with rising SST the frequency of the El Niño would increase, with associated extreme weather events. It has been observed that high growth rates of atmospheric carbon dioxide generally correspond with El Niño, meaning relatively low ocean uptake or high degassing. But some episodes do not reflect this, as during the 1992–1993 El Niño, resulting from the volcanic eruption of Mt Pinatubo in mid 1991. A similar but not as pronounced phenomenon has recently been clarified in the form of the Indian Ocean Dipole. This is observed as SST anomalies. These are coupled with zonal winds, changing in direction from westerlies to easterlies when the SST is cool in the east and warm in the west. This can have a very strong influence on the south-west monsoon, and on related drought or precipitation conditions in parts of Africa and south-east Asia and Australia (Yamagata *et al.*, 2004).

3.5 Biogeochemistry and atmospheric greenhouse gas levels

The ocean has a key role in global biogeochemical cycles, and the documented changes discussed above can affect the marine biogeochemical cycles of, for example, carbon, oxygen, and nutrients. The ocean has always been a large reservoir for carbon dioxide, but following the Industrial Revolution the net uptake (to 2011) is estimated to be ~157 GtC,[1] which has led to a gradual acidification of the ocean. The fraction of carbon dioxide emissions from fossil fuel burning and cement production that has be taken up by the ocean has decreased from ~(48 ± 9)% during 1800–1994, to ~(34 ± 6)% during 1990–1999, to ~(30 ± 7)% during the decade 2002–2011.[2] Prior to 1750, atmospheric CO_2 concentrations were 260–280 ppm for about 10,000 years. During recent ice ages, peak levels were 180 ppm, during the warmest interglacial peaks, nearly 300 ppm, and during the Pliocene at about 3 million years ago, may have been around 400 ppm. In 2011 it passed 390 ppm (Le Quéré *et al.*, 2012). The estimated oceanic mean uptake per decade has been increasing: from 1.5 ± 0.5 GtC/yr in the 1960s, to 2.5 ± 0.5 GtC/yr for 2002–2011 (Le Quéré *et al.*, 2012). The tropical ocean is out-gassing CO_2. The extra-tropical Northern Hemisphere Ocean is a net sink and the Southern Ocean is a large sink, although there are (controversial) recent indications that it is nearing saturation (e.g. Le Quéré *et al.*, 2008). The highest latitudes appear to be essentially neutral. The warming of the ocean results in release of additional carbon dioxide due to reducing solubility – a positive feedback.

A more sluggish ocean circulation and increased density stratification due to surface warming, both expected in a warmer climate, will slow down the vertical transport of carbon, alkalinity, and nutrients as well as the replenishment of subsurface waters that have not been in contact with the atmosphere. This provides a positive feedback to the atmospheric GHG concentration. Changes in the ocean circulation can also affect the regional circulation of shelf and coastal seas, leading to either an increased export of nutrients plus carbon from shallow seas into the open ocean, or to increased upwelling of nutrients plus carbon onto the shelf areas and coastal seas. The physical bottleneck feedback mechanism dominates the biological one, resulting in an overall positive feedback to climate change. Continued development depends on the future ocean circulation for which model projections show a large range, meaning great uncertainty.

As of 1994 (Sabine *et al.*, 2004), over 50% of the anthropogenic carbon taken up by the ocean was confined to the top 400 m, and was undetectable in much of the deep ocean. It takes decades and centuries for the carbon to transfer into deep ocean waters. The deepest penetration is found in the North Atlantic and the Antarctic waters related to the deep water formations in these areas. At low latitudes, oxygen concentrations have been decreasing in the thermocline 100–1000 m since the 1960s (Stramma *et al.*, 2008).

As carbon dioxide enters the ocean from the atmosphere, it partitions into dissolved carbon dioxide gas, bicarbonate, and carbonate ions. The relative proportions of these three components of 'dissolved inorganic carbon' (DIC) are determined by the pH. As more anthropogenic carbon dioxide enters the ocean, the pH decreases (i.e. the ocean becomes more acidic), which increases the fraction of dissolved carbon dioxide and reduces the fraction of DIC existing as carbonate ions. This means that the partial pressure of carbon dioxide in the ocean increases, thus opposing the influx of carbon dioxide from the atmosphere relative to what it would be before the pH started to decrease. The second effect, reduced carbonate ion, means that the waters become less saturated with respect to carbonate ion, making it more difficult for organisms with calcium carbonate skeletal structures to form calcium carbonate and increase or maintain their skeletal structures.

Currently the surface layers of the ocean are saturated with calcite and aragonite, the most common forms of calcium carbonate in marine organisms, but under-saturated below the saturation horizon, starting at depths varying from about 200 m in parts of the high latitudes and the Indian Ocean, to 3500 m in the Atlantic Ocean. Calcium carbonate dissolves when it sinks below the saturation horizon. Because of the addition of anthropogenic carbon dioxide to the oceans, the saturation horizon has become shallower, especially at high latitudes (e.g. Feely *et al.*, 2004). Reduced alkalinity due to sea ice melt water may contribute to the decrease in carbonate saturation state, evidenced in part of the Arctic Basin (Carmack *et al.*, 2010), and predicted for the entire Arctic Ocean and large regions at high latitudes (e.g. Steinacher *et al.*, 2009; Denman *et al.*, 2011).

The increasing acidification of the oceans and the associated increase in dis-solved carbon dioxide and decrease in saturation state for the common forms of calcium carbonate are not due to climate change, but only to the increase of CO_2 in the atmosphere and the invasion of roughly a third of that increase into the world ocean. This is a near-certain development with potentially serious biological consequences. Recent observations indicate that surface ocean pH has decreased by ~0.1 units since the start of the Industrial Revolution, 250 years ago (Bindoff *et al.*, 2007; OCB-EPOCA-UKOA, 2010 and further update). As measured by the concentration of H^+ ions, ocean acidity has increased by 26%. A continued rise in the atmospheric carbon dioxide levels may lead to a further decrease in pH of 0.4 by 2100 (Meehl *et al.*, 2007), i.e. a further increase in H^+ ions of 150%. This level of ocean acidity would be higher than anything experienced during the past 120 million years. Acidification leads to a decrease of carbonate ion, which can lead to a decrease in biological production of corals and calcifying phytoplankton and zooplankton.

How extensive might these and other changes be that may result from the increasing acidification expected over the next century? The mass extinction in the oceans that occurred at the end of the Permian period, about 250 million years ago, has been described by Payne and Clapham (2012). They conclude that the Permian extinction (the Great Dying) was triggered by a lack of dissolved oxygen, an excess of carbon dioxide, enhanced ocean acidity, and warming, leading to a complete change in the dominant animals of the ocean. Corals, sea sponges, and shelled animals were devastated. No major group of marine invertebrates was spared. Substitutes included snails and bivalves like clams and scallops, which became dominating. This shift provided the foundation for present day marine ecology according to the researchers. What happened appears to have been a 'perturbation of the global carbon cycle'. This in turn may have been due to the largest volcanic event of the past 500 million years, which formed the Siberian Traps, and injected large amounts of carbon gas into the atmosphere.

The dissolution of carbonate from the ocean floor will increase. Ecological changes due to acidification can be severe for corals, for pelagic ecosystems, and for the marine food web at higher trophic levels. Acidification may lead to shifts in ocean ecosystem structure and dynamics, which may alter the biological production and export from the surface of organic carbon and calcium carbonate. Models suggest that the overall effect of carbon–climate interactions is a positive feedback, giving higher CO_2 than models not including coupling or interactions.

Acidification of the ocean has occurred throughout geological periods, possibly mainly due to volcanic activity. However, recent reviews, such as that of Zeebe (2012), conclude that none of the mass extinctions in the last 450 million years is a good analogue of what we might expect over the next century, primarily because the ongoing ocean acidification is occurring faster than any that is documented in the paleo-record. Past catastrophic-type activities have lasted thousands of years. Human activities have generated a serious acidification over decadal time-scales, and there is little indication that we are yet doing anything effective to reduce fossil fuel emissions.

3.6 Biological production and nutrient changes

The functioning of ocean ecosystems depends strongly upon climate conditions, including near-surface density stratification, ocean circulation, temperature, salinity, wind field, and ice cover (cf. Cermeño and Falkowski, 2009). In turn, ocean ecosystems affect the chemical composition of the atmosphere, including concentrations of CO_2, N_2O, O_2, dimethyl sulfide (DMS), and sulfate aerosol. Feedbacks are very complex because they involve physical conditions and responses. Warmer water could lead to higher photosynthetic uptake of CO_2, but stratification increase could lead to lower nutrient levels in the euphotic zone. Sulfate aerosol particles are responsible for globally averaged temperatures being lower than would be expected from the GHG levels alone. The aerosols scatter and absorb radiation and act as cloud condensation nuclei (CCN), reducing precipitation efficiency, likely to result in a reduction of the annual mean net radiation at the top of the atmosphere.

Large uncertainties about this remain. However, the influence of climate conditions on ocean productivity is well documented through long time-series of observations. These include the Continuous Plankton Recorder (CPR) in the North Atlantic, displaying close correlation between different phytoplankton species and the NAO (Beaugrand *et al.*, 2009).

Nutrient supply to the ocean has changed through increased nitrate release from the land due to use of fertilizers and nitrogen deposition from the air in polluted areas. The nitrogen cycle is 'accelerated' with an increase in NO_X and nitrite oxide emissions driven by human actions: increased fertilizer use, agriculture, and fossil fuel combustion. Dust deposition provides an important source of micronutrients like iron, zinc, and others which could change the biological production pattern, and enhance photosynthetic carbon fixation. A warmer climate may lead to a decrease in dust deposition. This would provide for a positive feedback, decreasing photosynthesis and the uptake of CO_2. Changes in plankton species composition and regional shifting of high production zones due to changing climate could lead to further feedbacks. An increase in plankton blooms is indicated for high northern latitudes.

There are variable signals with regard to changes in nutrient concentrations, with few studies reporting decadal changes. Changes have been observed in all ocean basin deep waters, but with no clear pattern. An inferred decrease in primary production of 6% from the early 1980s to the late 1990s is plausible, with a decrease in nutrient inputs from intermediate waters. This is consistent with decreased ventilation of these water layers, suggested by their observed decrease in oxygen content, which may also be linked to an increased rate of oxygen consumption through faster oxidation in warmer water. The uncertainties in these inferences are underlined by the indications from analysis of model simulations presented by Sarmiento *et al.* (2004). These suggest a global increase of ocean primary production in the range 0.7 to 8.1%, with large regional differences.

Shifts and trends in plankton biomass have been observed in the North Atlantic, the North Pacific, and the Southern Indian Ocean. Trends of decreasing chlorophyll a and primary production since 1978 are observed in LMEs (Sherman, Aquarone, and Adams, 2009). It should be noted that changes in ocean productivity have been a key to reconstructing the conditions during other climate periods, including winds and ocean currents, by means of isotope palaeontology.

While the data are insufficient to conclude whether observed changes are caused by natural variability or are due to trends, they do indicate large-scale changes in ocean conditions. According to reports from the IPCC, the ongoing changes in the ocean have very strong repercussions on basic human needs. Fish perish as climate change disrupts oceanic systems. Through the combined impacts of climate change, pollution, over-harvesting, damaging fishing practices, in particular bottom trawling, migration of species, infestations of exotic species, and changes in biodiversity, the major fishing grounds are all affected. A decline of 13% of cumulative catches in the 64 LMEs since the maximum in 1994 has been confirmed, with 80% of mean annual marine fisheries occurring within these LMEs. Decreasing trends have been established for chlorophyll a and primary production since 1998. Up to 75% of fishing grounds may be affected by changes in ocean circulation due to climate change interfering with established circulation, ventilation, over-turning, and mixing. Up to 80% of the primary fish catches are exploited beyond or close to their harvesting capacity. Increasing SSTs over the coming decades are expected to bleach and kill up to 80% of coral reefs, not to mention the additional effects of increasing ocean acidification. And the atmospheric concentration of carbon dioxide is still increasing.

3.7 Eutrophication, oxygen levels, and biodiversity

Ocean biological and ecosystem conditions are influenced by climate change, and are also influenced in several other direct ways by human actions. A comprehensive and integrated approach is required in order to obtain an understanding of ocean changes and their significance for our life-support systems. Nutrient enrichment of marine systems, in particular through excess use of nitrogen fertilization, has been well documented in many coastal areas and shelf seas. This has driven an exponentially increasing spread of dead zones with dissolved oxygen levels below 2 ml/l (hypoxia) since the 1960s, for instance in the Baltic Sea, the Kattegatt, Black Sea, Gulf of Mexico, and East China Sea (Diaz and Rosenberg, 2008). Natural nutrient enrichment through ocean circulation, coastal and equatorial upwelling supporting highly productive zones, can generate low or intermediate oxygen levels at varying depths in the water column. In these areas the ecosystem and benthos seem, however, to be adapted to low levels of dissolved oxygen. This is not the case in the estuarine and coastal areas subject to excess nutrient enrichment from land run-off. In these areas the low oxygen levels lead to mass mortality and major changes in community structure and biodiversity.

Observations made over several decades show a decline in oxygen levels since the 1940s onwards, and localized declines were documented in the Baltic Sea in the 1930s. However, it was only in the 1960s that hypoxia became more widely observed. Paleo-indicators suggest that hypoxia has not been a naturally recurring condition in many of the systems. Since the 1960s the number of dead zones has approximately doubled each decade, according to Diaz and Rosenberg (2008).

The Baltic Sea is an interesting case because of its importance for the surrounding countries, its physical, chemical, and biological characteristics, strong river inflows, topography with several basins separated by sills, stratification, and estuarine characteristics, with narrow and shallow connections to the ocean (e.g. Kullenberg, 1983).

The impact of climate change on the Baltic Sea ecosystem over the past 1000 years was recently elucidated by Kabel *et al.* (2012). They demonstrated the role of a surface water temperature change of about 2 °C between the present and the Little Ice Age (about 1350–1850) for oxygen concentrations in the Baltic deep waters. The importance of changing biogeochemical conditions triggered by temperature variations was highlighted by the increase in summer period cyanobacteria blooms. The variations in cyanobacteria blooms during recent decades are coupled with the increase in summer sea surface temperatures. The Baltic study showed correlation between warmer periods and increased total organic carbon in laminated sediments. It was estimated that about half of the carbon originated from cyanobacteria blooms. Sediment studies likewise showed high organic carbon content during the Medieval Climate Anomaly in around 950–1250, when the sea surface temperature was comparable with present levels, or about 2 °C higher than during the Little Ice Age. The results were corroborated by model runs using a three-dimensional circulation model with an integrated biogeochemical model. The model outputs showed a higher oxygen content in the bottom waters during the cold period than during the warm periods. It is noteworthy that similar results were obtained when the nutrient loads in the model were kept as high as at present. The results thus indicate a strong surface layer temperature influence on the spread of anoxic areas, probably in combination with changing primary production. In another study reasonable correlation was suggested between the oxygen content and a temperature increase in the Baltic deep and bottom waters at times during the 20th century, on the basis of observations (Kullenberg, 1970).

It is has not yet been explained how hypoxia affects the habitat requirements of different species or the resilience of an ecosystem. Habitat compression can occur in the water column and at the sediment–water interface, or in the sediment. Habitat compression and loss of fauna as a result of hypoxia have very significant effects on the ecosystem function and energetics. The energy flow can be diverted into microbial pathways, and as the benthos die off the microbial pathways quickly dominate the energy flow. This has strong negative consequences for higher trophic levels and growth of predators, with functioning of the ecosystem being strongly dependent upon the biodiversity. The importance of benthic fauna diversity for the functioning of deep-sea ecosystems with oxygen levels well over 2 ml/l, typically over 5 ml/l, has been demonstrated by Danovaro *et al.* (2008; cf. Chapter 4). The results suggest an exponential decline in deep-sea ecosystem functioning linked to loss of benthic biodiversity. They also indicate one possible consequence, a reversal of the energy flow in the ecosystem towards microbial levels. These results suggest that a reduction in functional biodiversity may be associated with an exponential decline in the ecosystem processes.

The significance of these findings may be highlighted by recalling the discovery of deep-sea vents in the 1970s and the subsequent documentation of rich and diverse communities of organisms in the deep sea. The discovery of ocean vents, where hot gases erupt from the ocean floor, changed our understanding of life. The gases support a surprisingly active life and ecosystem, which does not use light and photosynthesis, but depends on bacterial processes for organic production. Further discoveries in the 1980s documented life at cold methane seeps, which use methane as the energy source. Thus the deep-sea ecosystem has a rich diversity.

The 'surprise' discoveries referred to highlight the importance of avoiding 'surprise' disruptions of deep-sea ecosystems through lack of understanding and adequate management of human interferences.

The key factors influencing the degree of ecosystem degradation in hypoxia are duration and levels of dissolved oxygen concentration. Diaz and Rosenberg (2008) conclude that 'currently hypoxia and anoxia are among the most widespread deleterious anthropogenic influences on estuarine and marine environments and now rank with over-fishing, habitat loss and harmful algal blooms as major global environmental problems'. Management restrictions of nutrient inputs have improved conditions in some cases, for instance rivers in the eastern USA and in the UK, but have not improved conditions in the Chesapeake Bay in eastern USA. Reduction of nutrient inputs from fertilizing the land is required, at least to levels around those of the 1950s, thus before the time of the spread of eutrophication. This situation could be seen as in parallel with the need for a reduction in atmospheric greenhouse gas levels.

The development of dead zones is also strongly coupled with the physical conditions in the water column: the stratification, the mixing, temperature, water exchange, circulation, and the air–sea interaction. The potential effects of climate change can influence these conditions considerably, which may lead to further spreading of dead zones. For instance, a higher water temperature leads to an increase in the rate of oxidation of organic matter and thus of oxygen consumption, which may not be compensated for by an increased rate of oxygen supply. The changes already observed in the ocean related to global climate change highlight the need to evaluate this potential at global and regional levels, for instance through the LME Programme. The significance for our society is underlined by the importance of estuarine and coastal sea food to human development.

The inherent interdependences highlight the necessity for an integrated and comprehensive evaluation of ongoing developments. It is not sufficient to address just one problem, or one component of the whole system. A system-wide approach is required. This observation is certainly not new, but needs to be stressed. It is supported by confirmation that integrated fisheries management that limits the harvesting rate to appropriate levels can reverse the trend of declining fisheries. Highly migratory species subject to international fishing are threatened: for example, Atlantic bluefin tuna, and other fish such as cod, flounder, and sole caught in 'mixed fisheries'. It should be possible to address these problems through a proper implementation and enforcement of the 1995 UN Fish Stocks Agreement, which is also part of the Law of the Sea (cf. Chapters 6 and 10). Hilborn (1996) notes that the environmental cost of producing more food on land through greater use of fertilizers and antibiotics, leading to more chemical run-off and increasing dead zones in estuaries and coastal seas, is potentially much higher than would result from fewer fish in the ocean. However, the interdependence within the marine ecosystem must also be noted. An increasing proportion of global fisheries is made up of small fish like sardines, anchovies, and menhaden. These species are also very important for many larger species of fish, marine mammals, and sea-birds, highlighting the need to limit catches of small fish in order to avoid further destruction of the whole ecosystem.

3.8 Sea-level change

Rising sea level has serious implications for the estimated 600 million people who live near the ocean within 10 m of sea level (Nicholls *et al.*, 2011). Global mean sea level rose by 1.7 ± 0.2 mm/yr over the period 1900–2009 and has risen by 1.9 ± 0.2 mm/yr since 1961. Since the start of the satellite altimetry record in 1993, the rate of sea-level rise has been about 3 mm/yr, with the rate from the satellite record (which covers the open ocean) being about 0.04 mm/yr higher than the conventional record from coastal and island measurements, corrected for glacial isostatic adjustment (Church and White, 2011). During the period 1955–1998, thermal expansion (0–3000 m) accounted for about 0.4 mm/yr (Bindoff *et al.*, 2007), rising to about 0.8 mm/yr during the period 1972–2008 (Church *et al.*, 2011). The contribution from the melting of glaciers and ice caps was about 0.7 mm/yr during this latter period. For the first time, Church *et al.* (2011) have 'closed' the sea-level rise budget: for the period 1972–2008, the observed sea-level rise from tide gauges and altimeter data (2.1 ± 0.2 mm/yr) agrees with the sum of the contributions (1.8 ± 0.4 mm/yr).

Church and White (2011) estimate the global average sea-level rise during the period 1880–2009 to be 201 mm (= 20.1 cm). The 2007 IPCC projection for the 21st century spans 18–59 cm, depending on which scenario is used, but future rapid changes in ice sheet flow were excluded due to large uncertainty in the projections. For comparison, during the last warm period at about 125,000 years ago, polar regions were significantly warmer than at present (~3–5 °C) with an estimated sea level 7–9 m higher than today (Kopp *et al.*, 2009). Between the last glacial maximum 24,000 years BP and 4000 years BP, sea level rose ~120 m to the present level (Fig. 6.8 in Jansen *et al.*, 2007).

Recent measurements of ice sheet mass balance changes from satellite-borne gravity measurement (Shepherd *et al.*, 2012) allow improved projections of sea-level rise during the 21st century. For example, Vermeer and Rahmstorf (2009), using semi-empirical methods relating sea-level rise to temperature rise, estimate for the IPCC 2007 scenarios a sea-level rise of 75–190 cm for the period 1990–2100. Nicholls *et al.* (2011) review projections of sea-level rise to

2100 published since 2007, and conclude that for a warming of more than 4 °C 'a pragmatic estimate of sea-level rise by 2100' is between 0.5 and 2 m. From their risk analysis, region by region, they state that there is a 'real risk of the forced displacement of up to 187 million people over the century (up to 2.4% of global population)'. The most recent comparison of data and models shows that the rate of sea-level rise of the past few decades is greater than projected by the IPCC models. (Rahmstorf *et al.*, 2012).

With respect to long-term global change, changes in ocean volume are good indicators and can provide strong constraints on model simulations of the climate, while ocean volume and density changes provide the same on global ice mass budgets. Significant interannual and decadal variations of global sea level are, however, also found which are not related to ocean volume redistributions. Differences in evaporation and precipitation in the hydrological cycle, together with the global topography, generate sea-level differences between the major oceans. The net effect of the Northern Hemisphere hydrological cycle is a sea-level distribution with levels higher by 0.5–1 m in the North Pacific subtropical gyre compared with the North Atlantic, and a sub-Arctic gyre in the North Pacific with 0.2–0.8 m higher water levels than in the North Atlantic Arctic seas. Furthermore, the subtropical North Atlantic has a sea level higher by 0.3–0.5 m compared with the northern seas and the Eurasian Arctic Basin (Carmack and McLaughlin 2011). This difference supports the northward flow of water from the warm North Atlantic gyre into the Arctic Ocean. No pattern of the longer-term regional distribution of potential sea-level rise has emerged, although model projections display common features. These include a maximum in sea-level rise in the Arctic Ocean and a minimum in the Southern Ocean, south of the Antarctic Circumpolar Current (Church, 2006 and Gregory *et al.*, 2001, cited by Church).

Land motion corrections to tide gauges from models of global isostatic adjustments can contribute ± 2 mm/yr globally. Including global isostatic adjustments, the current best estimate of sea-level rise is given as 3.2 ± 0.4 mm/yr over the period 1993–2009 (Church and White, 2011). The balance of evidence suggests that there has been a significant change in the long-term rate of sea-level rise, with roughly half due to ocean temperature change, and other parts due to inflow of water from land, including ice melting.

Predictions of extreme storm surges and wave heights are more uncertain, owing to their dependence upon possible long-term trends and changes in atmospheric pressure fields and winds. It can be concluded that there is a need to take into account possible future increases in sea level, extreme storm surges, and waves in designs for protecting life, property, and infrastructure in coastal regions. The same applies to offshore infrastructures and from the economic and environmental standpoints.

3.9 High-latitude regions

The high-latitude oceans of both the Northern and Southern Hemispheres remain under-sampled and poorly explored, despite their strong roles in global climate and life-support systems. The interaction between air, water, and ice is of great importance for the heat balance of the ocean–atmosphere system and thus the climate. Wind forcing can be extreme, with large seasonal variations. Sea ice formation accompanies cooling over the shelves and extracts heat and fresh water from the water column. These processes make the polar ocean boundaries especially important in understanding global ocean circulation (Royer and Stabeno, 1998). Sediments of the polar seas hold keys to reconstruction of past ocean circulation and ice ages. But still, documentation of change in the polar regions remains sparse (Wassmann *et al.*, 2011). These explanations justify the efforts made during the International Polar Year (IPY) 2007–2008, the International Geophysical Year (IGY) 1957–1958, the International Southern Ocean Studies during the International Decade of Ocean Exploration (IDOE) 1970–1980, and the calls for continued and upgraded monitoring of these areas.

Global climate models and observations agree that warming will occur first and faster in high-latitude oceans (Loeng *et al.*, 2005, cited by Carmack and McLaughlin 2011). The IPY observations confirmed that the Arctic is changing faster than models predict. Models suggest that climate warming can lead to a contraction of the highly productive marginal sea ice biomes, by 42% in the Northern and 17% in the Southern Hemispheres (Sarmiento *et al.*, 2004). Change is currently occurring in the Arctic, with potentially major consequences for the global climate system, marine ecosystem, oil and gas exploration/exploitation, and shipping.

Carmack and McLaughlin (2011) present an in-depth review of physical and chemical observations in the sub-Arctic and Arctic seas around northern North America, obtained during the joint efforts of Canada's Three Oceans programme and the Joint Ocean Ice Study. Oceanic domains and linkages between the sub-Arctic and Arctic parts are analysed. Changes that are identified include: ocean warming, sea ice melting, upper layer freshening, altered stratification, increased acidification and, in the sub-Arctic Pacific, increased hypoxia.

An understanding of the changes in the Arctic Ocean requires consideration of its coupling to the global ocean and the sub-Arctic Pacific and Atlantic in particular. These areas are characterized by a salinity dominated stratification which influences all processes associated with horizontal and vertical fluxes (Carmack, 2007). Global ocean stratification is largely driven by the hydrological cycle, with evaporation in the warm oceans and condensation in the cold. Relatively colder and fresher waters from the Pacific flow northwards into the Arctic Ocean, through the Bering Strait, with a higher sea level on the Pacific side, overflowing the warmer and saltier Arctic water, with additional inflows from the continental rivers supporting the stratification required for the formation of sea ice. The combination of sea ice formation and fresh water run-off generates buoyancy changes. Thus the buoyancy forcing of the coastal circulation becomes very important in regulating vertical mixing. The fresh water loop is closed with the export of sea ice and low salinity waters through the Fram Strait and the Canadian Arctic Archipelago into the convective gyre regions of the Greenland, Iceland, Norwegian, Irminger, and Labrador Seas (Carmack and McLaughlin 2011).

Historical records of the storage of fresh water in the Arctic show strong multidecadal variability over the past century (Proshutinsky *et al.*, 2009). In some parts of the region the freshwater inputs may be changed by human activities, including river deviations, damming, and hydroelectric power generation. This may influence the coastal water circulation in combination with natural changes in the freshwater cycle. The significance may be reflected, although it cannot be proven, in the Great Salinity Anomaly in the North Atlantic during the 1980s, which changed the rate of deep water formation there for several years (Royer and Stabeno, 1998).

This phenomenon may be coupled with the strong negative anomalies of the NAO in the period 1955–1965. A weakening of the westerlies associated with the relatively high Icelandic pressure (low NAO), favoured northerly to easterly winds with advection of polar water towards the Iceland–Greenland region (e.g. Voituriez, 2003). The low density water, forming a vast lens not mixing easily with the underlying layers, was carried by prevailing currents along Greenland into the Labrador Sea and on into the North Atlantic (Voituriez, 2003).

The analysis of Carmack and McLaughlin (2011) concludes with a summary of the changes which have potential biological consequences. Ocean temperature has measurably increased in parts of the Arctic Basin. This is partly associated with local warming due to the decreased albedo, but also includes water of Atlantic origin. The salinity-driven stratification in the upper layers of high-latitude oceans is increasing due to the combined effects of an accelerated hydrological cycle and increasing ice melting during the summer season. A major result of this is a possible decrease in transfer of nutrients into the productive euphotic zone, resulting in a decrease in primary production. The freshening of the Arctic Ocean also leads to a change in phytoplankton species, with the smallest cells dominating the larger ones (Li *et al.*, 2009). This may have strong consequences for the food web if it persists. A 'tipping point' may be reached if the combined effects of warming and increased stratification support food webs that favour low energy predators like jellyfish.

The changes in stratification are different between the Eurasian and the North American Canadian parts of the Arctic (Carmack and Melling, 2011). On the

Eurasian side the stratification and vertical stability are decreased, possibly allowing an increased heat flow upwards from the intermediate warmer water of Atlantic origin. On the Canadian side the stratification is increased, associated with the inflowing low salinity water from the Pacific.

Decreasing ice cover extent, thickness, and duration are basic features of changes in the Arctic. The observations of the 2007–2008 IPY showed record breaking minimal sea ice extent (Carmack *et al.*, 2010). The changes include earlier melting and delayed freeze-up in some parts, for example, the Canada Basin. Consequences include increased ice motion in response to wind forcing and increased potential for transfer of nutrients to the shelf areas, when the ice retreats beyond the shelf break. An unexpected consequence of the increased sea ice melt is its effect on the carbonate chemistry and related decrease in pH and increase in acidification in parts of the Canada Basin. The low alkalinity sea ice melt water tends to decrease the saturation state of calcium carbonate (Yamamoto-Kawai *et al.*, 2009).

Changes in the atmospheric forcing lead to shifts in position of fronts separating Pacific and Atlantic waters, with associated alterations in the circulation in the Arctic Basin. This has potential implications for fish production, due to changes in nutrient supplies to the euphotic zone and changes in primary production. Changes in light conditions will also influence this. Major circulation alterations and changes in water column properties are driven by changes in advection, for instance transport of water of Atlantic origin northwards into the basin. These advective processes also influence the distribution of biological species including phytoplankton and zooplankton, and invasion of new species coupled with the regional warming. Invasion of new species is documented in parts of the Arctic (Carmack *et al.*, 2010).

Low levels of oxygen in poorly ventilated subsurface waters are well documented in some sub-Arctic regions, and concern for oxygen decrease and hypoxia in ocean waters around northern North America is increasing. However, the Arctic Ocean and bordering sub-Arctic Atlantic do not display widespread evidence of hypoxia (Carmack *et al.*, 2010).

In the Arctic, changes occur relatively fast in comparison with other regions, evidenced by observations. These also bring out the significance of the linkages between ocean basins and the interdependencies. Observations and modelling simulations show that changes in the Arctic will have an impact on the global thermohaline over-turning circulation (the great conveyor belt) and on properties of downstream ocean basins. Changes in the physical environment of the Arctic Ocean are affecting the marine ecosystem. However, quantification of these effects remains elusive, for instance, whether organic production will increase or decrease in conditions of reduced ice cover. This underlines the need to understand the mechanisms and causal links that drive a change, in addition to observing and describing it. However, the vulnerability of the Arctic to climate change and several different potential human impacts is a cause for increasing concern.

Concern for the Antarctic was displayed through the initiation, after IGY (1957–1958), of the Antarctic Treaty of December 1959. Its aim is 'in the interests of all mankind that Antarctica shall continue forever to be used exclusively for peaceful purposes and shall not become the object of international discord'. This can be understood as an effort to help preserve the pristine character and integrity of the ecology of the Antarctic and at least parts of the Southern Ocean surrounding the continent. Subsequent treaties, for instance, the Convention on the Conservation of Antarctic Marine Living Resources (CCAMLR) (cf. Chapters 2 and 11), underline the significance of such an aim. It is from the pristine Antarctic environment that unique information about past changes of climate has been secured. The historical construction of atmospheric carbon dioxide concentration and temperature from the Lake Vostok ice core in Antarctica demonstrate the strong correlation between the two climate signals going back 400,000 years (Petit *et al.*, 1999).

The Weddell Sea is a major breathing hole of the world ocean, where deep and bottom water formation occurs. The sea ice biosphere and primary production in the marginal ice zone are supporting the whole Southern Ocean ecosystem. In these areas there exists a highly diversified community at the interface between the ice and the water, productive the whole year around even though located under the ice cover. There is a rich occurrence of krill. The Antarctic and Southern Ocean ecosystem is adapted to very stable conditions. Its resilience and vulnerability to change are of concern. A number of international research programmes have, since IGY, endeavoured to elucidate the conditions in these sensitive areas. However, the Southern Ocean still remains largely under-sampled and unexplored.

Unfortunately the arrival of humans has initiated environmental change with considerable local impacts. Furthermore, the Antarctic continent and the Southern Ocean are now increasingly being exploited for various purposes, including tourism and adventure, research, and fisheries. These activities are undoubtedly leading to further changes in the environment. Southern Ocean krill fisheries have increased dramatically over recent years due to new technology. In the context of both global sea-level change and the conditions of the Southern Ocean ecosystem there is concern for potential loss of Antarctic ice, as is now happening in the Arctic and Greenland. However, present surface temperatures, as well as those projected for the century, are too low for significant melting of ice to occur (Church, 2006). An increase in snow fall, with resulting increase of water in the form of ice, has also been forecast, but has so far not been observed. The need to pursue efforts to obtain more time-series of observations from the Antarctic–Southern Ocean domain is obvious.

3.10 Uncertainties

Many uncertainties remain regarding ocean changes. Limitations in sampling, especially in the Southern Hemisphere, mean that decadal variations can only be evaluated with moderate confidence. Observed changes in the thermohaline circulation of the North Atlantic are inconclusive due to large decadal variations. Climate models suggest that the climate system responds to changes in that circulation and also that it might gradually decrease during this century due to warming and freshening of the North Atlantic surface layer. However, there is low confidence in the trends of thermohaline circulation and the global ocean fresh-water budget. Global average sea-level rise for the past 50 years is likely to be larger than explained by warming and ice melting, since it is not possible over this period to satisfactorily quantify the known processes causing sea-level rise. The longer term perspective of considerable sea-level rise and associated unstable coastal conditions can however not be doubted if the current situation continues.

Future changes in ocean circulation and stratification are highly uncertain. The overall reaction of marine biological carbon cycling to a warm and high CO_2 world is not well understood. Several small feedbacks could add up to one large feedback. The response of marine biota to ocean acidification is not clear, either with regard to individual physiology or ecosystem functioning. Potential impacts are serious, particularly for those species depending upon calcium carbonate, some of which may even become extinct in the current century. Finally, palaeontology, using different techniques, has elucidated many previous climate changes. However, although we can identify many positive feedbacks, we still do not know what first triggered the changes. Could they be related to varying heights of the continents, allowing more or less coverage of snow and ice? Great uncertainties and a lack of understanding of critical processes also lead to very grave concern about large geo-engineering projects under consideration to address the climate change problem.

There is an increasing need to maintain high productivity, biodiversity, and vulnerable habitats, particularly in coastal areas often subject to large land reclamation projects, in view of their important role in global material and ecosystem balances and their increasing vulnerability to many other linked processes. These include overfishing, aquaculture, and transportation.

The effects of warming are not well quantified, with several feedbacks (cf. Pörtner and Farrell, 2008). Ocean changes reflected in the hydrological cycle, the role of aerosols, and variations in cloudiness are likewise not well quantified. Changes in extreme weather events are documented, but it is not clear how the frequency and duration of such events are changing, even if it seems that they are coupled with climate variability and change.

The protection of biodiversity in international waters and on the seabed is not adequately covered by existing conventions, and the impact of changes in the biodiversity in these large zones is not known.

Human health is affected by organisms transmitting disease, increasing harmful algal blooms and red tides, which are coupled to changing ecological conditions. Changes or imbalances in the established biodiversity may trigger such occurrences and generation of large medusa. The consequences of change in the oceanic foodweb are significant but remain poorly quantified. However, recent progress in 'end-to-end' foodweb models (e.g. Fulton, 2010; Moloney *et al.*, 2011) provides a process-based modelling approach to make projections of the future state of oceanic ecosystems and living marine resources, when such models are embedded in IPCC-class models of a changing climate (Stock *et al.*, 2011).

3.11 Conclusions

Observations and certainties of change in the ocean confirm that planet Earth is under severe stress and that humanity is to blame for this situation and for having so far failed to take the required action. Climate change, nitrogen use, and loss of biodiversity are major concerns well reflected in the changing ocean. Several of these predicted changes are confirmed by observations of ocean basins over decadal time-scales. Recently a panel of marine scientists convened by the International Programme on the State of the Ocean and the IUCN concluded that marine 'degradation is now happening at a faster rate than predicted' (Rogers and Laffoley, 2011).

The impact of change threatens to alter the marine environment at a scale in time and space not experienced or documented before, with effects on most current issues, sectors, peoples, and ecosystems – the whole life-supporting system. This confirms the imperative of achieving integrated and sustainable management, already acknowledged at the Rio Conference in 1992. The necessity to take into account interactions between the ocean, the atmosphere, and the land, and the ecosystems, including human society, is obvious; without a healthy ocean, there is no healthy life on Earth. The ocean system must be treated as a whole, as spelled out in the UN Convention on the Law of the Sea (cf. Chapters 1, 10, and 11).

However, in spite of strong efforts to promote the development of ocean research in association with negotiations for the Law of the Sea, including through the International Decade of Ocean Exploration 1970–1980, and the discovery of deep-sea vents in the 1970s, the era of deep-sea exploration may only now come into being, stimulated by private expeditions such as those reaching the greatest depths of the ocean in March 2012, over 50 years after the first such effort, as well as through increased realization of the role played by the ocean in human society, globalization, and comprehensive human security (Chua, Kullenberg, and Bonga, 2008).

References

Barber, R. T. (2007). Picoplankton do some heavy lifting. *Science* **315**: 777–778. doi: 10.1126/science.1137438.

Beaugrand, G., Luczak, C., and Edwards, M. (2009). Rapid biogeographical plankton shifts in the North Atlantic Ocean. *Global Change Biology* **15**: 1790–1803.

Bindoff, N. L., Willebrand, J., Artale, V., *et al.* (2007). Observations: Oceanic climate change and sea level. In: *Climate Change 2007: The Physical Science Basis. Contribution of Working Group I to the Fourth Assessment Report of the Intergovernmental Panel on Climate Change.* Solomon, S., Qin, D., Manning, M., Chen, Z., Marquis, M., Averyt, K. B., Tignor, M., and Miller, H. L. (eds.). Cambridge: Cambridge University Press, pp. 385–428.

Carmack, E. C. (2007). The alpha/beta ocean distinction: A perspective on freshwater fluxes, convection, nutrients and productivity in high-latitude seas. *Deep Sea Research Part II: Tropical Studies Oceanography* **54**(23–26): 2574–2598.

Carmack, E. and McLaughlin, F. (2011). Towards recognition of physical and geochemical change in sub-Arctic and Arctic Seas. *Progress in Oceanography* **90**(1–4): 90–104.

Carmack, E. and Melling, H. (2011). Cryosphere: Warmth from the deep. *Nature Geoscience* **4**(1): 7–8.

Carmack, E. C., McLaughlin, F., Vagle, S., *et al.* (2010). Structures and property distributions in the three oceans surrounding Canada in 2007: A basis for a long-term ocean climate monitoring strategy. *Atmosphere–Ocean* **48**(4): 211–224.

Cermeño, P. and Falkowski, P. G. (2009). Controls on diatom biogeography in the ocean. *Science* **325**(5947): 1539–1541. doi: 10.1126/science.1174159.

Chavez, F. P., Ryan, J., Lluch-Cota, S., and Ñiquen, M. C. (2003). From anchovies to sardines and back: Multidecadal change in the Pacific Ocean. *Science* **299**: 217–221.

Chua, T.-E., Kullenberg, G., and Bonga, D. (eds.) (2008). *Securing the Oceans: Essays on Ocean Governance – Global and Regional Perspectives.* Quezon City: GEF-UNDP-IMO Regional Programme on Building Partnerships in Environmental Management for the Seas of East Asia (PEMSEA) and the Nippon Foundation, 770 pp.

Church, J. (2006). Global sea levels: Past, present and future. In: *IOC Annual Report No. 13*, 8–16. Paris: UNESCO.

Church, J. A. and White, N. J. (2011). Sea-level rise from the late 19th to the early 21st century. *Surveys in Geophysics* **32**: 585–602.

Church, J. A., White, N. J., Konikow, L. F., *et al.* (2011). Revisiting the Earth's sea-level and energy budgets from 1961 to 2008. *Geophys. Res. Lett.* **38**, L18601. doi: 10.1029/2011GL048794.

Clarke, A. (2010). The development of ocean climate programmes. In: *Troubled Waters: Ocean Science and Governance.* Holland, G. and Pugh, D. (eds.). Cambridge: Cambridge University Press, pp. 96–111.

Curry, R. G. and McCartney, M. S. (2001). Ocean gyre circulation changes associated with the North Atlantic oscillation. *J. Phys. Oceanogr.* **31**: 3374–3400.

Curry, R. G., Dickson, R., and Yashayaev, I. (2003). A change in the freshwater balance of the Atlantic Ocean over the past four decades. *Nature* **426**: 826–829.

Danovaro, R., Gambi, C., Dell'Anno, A., *et al.* (2008). Exponential decline of deep-sea ecosystem functioning linked to benthic biodiversity loss. *Current Biology* **18**: 1–8.

Denman, K., Christian, J. R., Steiner, N., *et al.* (2011). Potential impacts of future ocean acidification on marine ecosystems and fisheries: Current knowledge and recommendations for future research. *ICES Journal of Marine Science* **68**: 1019–1029.

Di Lorenzo, E., Schneider, N., Cobb, K. M., *et al.* (2008). North Pacific Gyre Oscillation links ocean climate and ecosystem change. *Geophysical Research Letters* **35**, L08607. doi: 10.1029/2007GL032838.

Diaz, R. J. and Rosenberg, R. (2008). Spreading dead zones and consequences for marine ecosystems. *Science* **321**: 926–929.

Durack, P. J. and Wijffels, S. E. (2010). Fifty year trends in global ocean salinities and their relationship to broad scale warming. *J. Clim.* **23**: 4342–4362.

Feely, R. A., Sabine, C. L., Lee, K., *et al.* (2004). Impact of anthropogenic CO_2 on the $CaCO_3$ system in the oceans. *Science* **305**: 362–366.

Field, J. G., Hempel, G., and Summerhayes, C. P. (eds.) (2002). *Oceans 2020: Science, Trends and the Challenge of Sustainability.* Washington DC: Island Press, 296 pp.

Fulton, E. A. (2010). Approaches to end-to-end ecosystem models. *Journal of Marine Systems* **81**: 171–183.

Gregory J. M., Church J. A., Boer G. J., *et al.* (2001). Comparison of results from several AOGCMs for global and regional sea-level change 1900–2100. *Climate Dynamics* **18**: 225–240.

Guiot J., Pons A., de Beaulieu J. L., and Reille M. (1989). A 140,000-year continental climate reconstruction from two European pollen records. *Nature* **338**: 309–313.

Hansen, B., Stefánsson, U., and Svendsen, E. (1998). Iceland, Faroe and Norwegian Coasts. In: *The Global Coastal Ocean. Regional Studies and Syntheses.* Robinson, A. R. and Brink, K. H. (eds.). *The Sea: Ideas and Observations on Progress in the Study of the Seas*, 11. New York: John Wiley & Sons, pp. 733–758.

Held, I. M. and Soden, B. J. (2006). Robust response of the hydrological cycle to global warming. *J. Climate* **19**: 5686–5699.

Hilborn, R. (1996). Risk analysis in fisheries and natural resource management. *Human Ecology and Risk Assessment* **2**: 655–659.

Hoskins, B. J. and Hodges, K. I. (2002). New perspectives on the Northern Hemisphere winter storm tracks. *Journal of the Atmospheric Sciences* **59**(6): 1041–1061.

Jansen, E., Overpeck, J., Briffa, K. R., *et al.* (2007). Palaeoclimate. In: *Climate Change 2007: The Physical Science Basis. Contribution of Working Group I to the Fourth Assessment Report of the Intergovernmental Panel on Climate Change.* Solomon, S., Qin, D., Manning, M., Chen, Z., Marquis, M., Averyt, K. B., Tignor, M., and Miller, H. L. (eds.). Cambridge: Cambridge University Press, pp. 433–497.

Johnson, G. C. and Wijffels, S. E. (2011). Ocean density change contributions to sea level rise. *Oceanography* **24**: 112–121.

Kabel, K., Moros, M., Porsche, C., *et al.* (2012). Impact of climate change on the Baltic Sea ecosystem over the past 1000 years. *Nature Climate Change* **2**(12): 871–874.

Kopp, R. E., Simons, F. J., Mitrovica, J. X., *et al.* (2009). Probabilistic assessment of sea level during the last interglacial stage. *Nature* **462**: 863–868.

Kullenberg, G. (1970). On the oxygen deficit in the Baltic deep water. *Tellus* **22**(3): 357.

Kullenberg, G. (1983). The Baltic Sea. In: *Ecosystems of the World 26: Estuaries and Enclosed Seas.* Ketchum, B. H. (ed.). Amsterdam–Oxford–New York: Elsevier, pp. 309–335.

Kullenberg, G. (2010). Ocean science, an overview. In: *Troubled Waters: Ocean Science and Governance.* Holland, G. and Pugh, D. (eds.). Cambridge: Cambridge University Press.

Latif, M. and Meincke, J. (2001). Changes in the North Atlantic. In: *Climate of the 21st Century: Changes and Risks.* Lozän, J. L., Graßl, H., and Hupfer, P. (eds.). Hamburg: Wissenschaftliche Auswertungen, pp. 196–198.

Le Quéré, C., Rodenbeck, C., Buitenhuis, E. T., *et al.* (2008). Response to comments on 'Saturation of the Southern Ocean CO_2 Sink Due to Recent Climate Change'. *Science* **319**: 570. doi: 10.1126/science.1147315.

Le Quéré, C., Andres, R. J., Boden, T., *et al.* (2012). The global carbon budget 1959–2011. *Earth Syst. Sci. Data Discuss.* **5**: 1107–1157. doi: 10.5194/essdd-5-1107-2012, 2012.

Lemke, P., Ren, J., Alley, R. B., *et al.* (2007). Observations: Changes in snow, ice and frozen ground. In: *Climate Change 2007: The Physical Science Basis. Contribution of Working Group I to the Fourth Assessment Report of the Intergovernmental Panel on Climate Change.* Solomon, S., Qin, D., Manning, M., Chen, Z., Marquis, M., Averyt, K. B., Tignor, M., and Miller, H. L. (eds.). Cambridge: Cambridge University Press, pp. 337–383.

Levitus, S., Antonov, J. I., Boyer, T. P., *et al.* (2012). World ocean heat content and thermosteric sea level change (0–2000 m), 1955–2010. *Geophysical Research Letters* **39**, L10603. doi: 10.1029/2012GL051106.

Li, W. K. W., McLaughlin, F. A., Lovejoy, C., *et al.* (2009). Smallest algae thrive as the Arctic Ocean freshens. *Science* **326**: 539.

Loeng, H., Brander, K., Carmack, E., *et al.* (2005). Marine systems. In: *Arctic Climate Impact Assessment, ACIA.* Symon, C., Arris, L., and Head, B. (eds.). Cambridge: Cambridge University Press, pp. 453–538.

Mantua, N. J., Hare, S. R., Zhang, Y., *et al.* (1997). A Pacific interdecadal climate oscillation with impacts on salmon production. *Bulletin of the American Meteorological Society* **78**: 1069–1079.

Meehl, G. A., Stocker, T. F., Collins, W. D., *et al.* (2007). Global climate projections. In: *Climate Change 2007: The Physical Science Basis. Contribution of Working Group I to the Fourth Assessment Report of the Intergovernmental Panel on Climate Change.* Solomon, S., Qin, D., Manning, M., Chen, Z., Marquis, M., Averyt, K. B., Tignor, M., and Miller, H. L. (eds.). Cambridge: Cambridge University Press, pp. 747–845.

Moloney, C. L., St John, M. A., Denman, K. L., *et al.* (2011). Weaving marine food webs from end to end under global change. *Journal of Marine Systems* **84**: 106–116.

Nicholls, R. J., Marinova, N., Lowe, J. A., *et al.* (2011). Sea-level rise and its possible impacts given a 'beyond 4 degrees C world' in the twenty-first century. *Philosophical Transactions of the Royal Society A – Mathematical, Physical and Engineering Sciences* **369**: 161–181.

North, G. R. and Duce, R. A. (2002). Climate change and the ocean. In: *Oceans 2020: Science, Trends and the Challenge of Sustainability.* Field, J. G., Hempel, G., and Summerhayes, C. P. (eds.). Washington DC: Island Press, pp. 85–108.

Occhipinti-Ambrogi, A. (2007). Global change and marine communities: Alien species and climate change. *Mar. Pollut. Bull.* **55**(7–9): 342–352. doi: 10.1016/j.marpolbul.2006.11.014.

Payne, J. L. and Clapham, M. E. (2012). End-Permian mass extinction in the oceans: An ancient analog for the twenty-first century? In: *Annual Review of Earth and Planetary Sciences Vol. 40.* Jeanloz, R. (ed.). pp. 89–111. doi: 10.1146/annurev-earth-042711-105329.

Petit, J. R., Jouzel, J., Raynaud, D., *et al.* (1999). Climate and atmospheric history of the past 420,000 years from the Vostok Ice Core, Antarctica. *Nature* **399**: 429–436.

Pörtner, H. O. and Farrell, A. P. (2008). Ecology: Physiology and climate change. *Science* **322**(5902): 690–692. doi: 10.1126/science.1163156.

Proshutinsky, A., Krishfield, R., Timmermans, M.-L., *et al.* (2009). Beaufort Gyre fresh-water reservoir: State and variability from observations. *J. Geophys. Res.* **114**. doi: 10.1029/2008jc005104.

Rahmstorf, S., Foster, G., and Cazenave, A. (2012). Comparing climate projections to observations up to 2011. *Environmental Research Letters* **7**. doi: 10.1088/1748–9326/7/4/044035.

Rayner, N. A., Brohan, P., Parker, D. E., *et al.* (2006). Improved analyses of changes and uncertainties in sea surface temperature measured *in situ* since the mid-nineteenth century: The HadSST2 dataset. *Journal of Climate* **19**: 446–469.

Robinson A. R. and Brink, K. H. (eds.) (1998). *The Global Coastal Ocean. Regional Studies and Syntheses*. The sea: Ideas and observations on progress in the study of the seas, 11. New York: John Wiley & Sons, 1062 pp.

Roemmich, D. and Gilson, J. (2009). The 2004–2008 mean and annual cycle of tempera-ture, salinity, and steric height in the global ocean from the Argo Program. *Progr. Oceanogr.* **82**: 81–100.

Rogers, A. D. and Laffoley, D. A. (2011). *International Earth System Expert Workshop on Ocean Stresses and Impacts. Summary Report*. Oxford: IPSO, 18 pp.

Royer, T. C. and Stabeno, P. J. (1998). Polar ocean boundaries. In: *The Global Coastal Ocean. Regional Studies and Syntheses*. Robinson, A. R. and Brink, K. H. (eds.). The sea: Ideas and observations on progress in the study of the seas, 11. New York: John Wiley & Sons, pp. 69–78.

Sabine, C. L., Feely, R. A., Gruber, N., *et al.* (2004). The oceanic sink for anthropogenic CO_2. *Science* **305**: 367–371.

Sarmiento, J. L., Slater, R., Barber, R., *et al.* (2004). Response of ocean ecosystems to climate warming. *Global Biogeochemical Cycles* **18**, GB 3003: 1–23.

Schanze, J. J., Schmitt, R. W., and Yu, L. L. 2010. The global oceanic freshwater cycle: A state-of-the-art quantification. *J. Marine Res.* **68**: 569–595.

Schmitt, R. (2008). Salinity and the global water cycle. *Oceanography* **21**: 12–19.

Shepherd, A., Ivins, E. R., Geruo, A., *et al.* (2012). A reconciled estimate of ice-sheet mass balance. *Science* **338**: 1183–1189.

Sherman, K., Aquarone, M. C., and Adams, S. (eds.) (2009). *Sustaining the World's Large Marine Ecosystems*. Gland: IUCN, 140 pp.

Steinacher, M., Joos, F., Frolicher, T. L., *et al.* (2009). Imminent ocean acidification in the Arctic projected with the NCAR global coupled carbon cycle-climate model. *Bio-geosciences* **6**: 515–533.

Stock, C. A., Alexander, M. A., Bond, N. A., *et al.* (2011). On the use of IPCC-class models to assess the impact of climate on living marine resources. *Progress in Oceanography* **88**: 1–27.

Stommel, H. (1958). *The Gulf Stream: A Physical and Dynamical Description*. Berkeley and Los Angeles: University of California Press and London: Cambridge University Press, 202 pp.

Stramma, L., Johnson, G. C., Sprintall, J., *et al.* (2008). Expanding oxygen-minimum zones in the tropical oceans. *Science* **320**: 655–658.

Tang, Q. (2009). Changing states of the Yellow Sea Large Marine Ecosystem: Anthropo-genic forcing and climate impacts. In: *Sustaining the World's Large Marine Ecosys-tems*. Sherman, K., Aquarone, M. C., and Adams, S. (eds.). Gland: IUCN, pp. 77–88.

US National Snow and Ice Data Center (2012). *Arctic Sea Ice News and Analysis, Monthly Archives, September 2012*, http://nsidc.org/arcticseaicenews/2012/09/.

US Ocean Carbon and Biogeochemistry (OCB) Program, the European Project on Ocean Acidification (EPOCA) and the UK Ocean Acidification Research Programme

(UKOA) 2010 (and further update). *Frequently Asked Questions About Ocean Acidification*, www.whoi.edu/website/OCB-OA/FAQs.

Vermeer, M. and Rahmstorf, S. (2009). Global sea level linked to global temperature. *Proceedings of the National Academy of Sciences of the United States of America* **106**: 21527–21532.

Voituriez, B. (2003). *The Changing Ocean: Its Effects on Climate and Living Resources.* IOC Ocean Forum Series 4. Paris: UNESCO Publishing, 170 pp.

Wassmann, P., Duarte, C. M., Agusti, S., *et al.* (2011). Footprints of climate change in the Arctic marine ecosystem. *Global Change Biol.* **17**(2): 1235–1249.

Wefer, G. and Berger, W. H. (2001). Causes of climate change in the Quaternary. In: *Climate of the 21st Century: Changes and Risks.* Lozän, J. L., Graßl, H., and Hupfer, P. (eds.). Hamburg: Wissenschaftliche Auswertungen, pp. 61–64.

Yamagata, T., Behera, S. K., Luo, J.-J., *et al.* (2004). Coupled ocean–atmosphere variability in the tropical Indian Ocean. *Geophysical Monograph Series* **147**: 189–211.

Yamamoto-Kawai, M., McLaughlin, F. A., Carmack, E. C., *et al.* (2009). Aragonite undersaturation in the Arctic Ocean: Effects of ocean acidification and sea ice melt. *Science* **326**: 1098–1100.

Yu, L. (2011). A global relationship between the ocean water cycle and near-surface salinity. *Journal of Geophysical Research* **116**, C10025. doi: 10.1029/2010JC006937.

Zebiak, S. E. (1989). Oceanic heat content variability and El Niño cycles. *J. Phys. Oceanogr.* **19**: 475–486.

Zeebe, R. E. (2012). History of seawater carbonate chemistry, atmospheric CO_2, and ocean acidification. In: *Annual Review of Earth and Planetary Sciences Vol. 40.* Jeanloz, R. (ed.). 141–165.

Notes

1 Calculated from data tables associated with Le Quéré *et al.* (2012) for the years 1995–2011 and added to the cumulative total for the period 1800–1994 estimated by Sabine *et al.* (2004).

2 Percentage ocean uptake was calculated from Sabine *et al.* (2004) and Le Quéré *et al.* (2012). Standard error estimates were calculated following standard propagation of error methodology.

4

Knowledge and implications of global change in the oceans for biology, ecology, and ecosystem services

ROBERTO DANOVARO, CRISTINA GAMBI, AND CINZIA CORINALDESI

4.1 Introduction

There is an increasing awareness that the Earth is changing, but it is still unknown whether these changes occur cyclically, stochastically, episodically, or are long-term trends. The contemporary global climate change is a reality and a result of human activities that annually release gigatonnes of carbon into the Earth's atmosphere (IPCC, 2007; Hansen *et al.*, 2007). Direct consequences of cumulative post-industrial emissions include increasing global temperature, perturbed regional weather patterns, rising sea levels, acidifying oceans, decreasing oxygen concentration, changed nutrient loads, and altered ocean circulation (Brierley and Kingsford, 2009). All of these factors cause climate change to have different effects on marine ecosystems, their function, biodiversity, and on the goods and services that they can provide. These contribute to human welfare, both directly and indirectly, and represent part of the total economic value of the planet (Costanza *et al.*, 1997); therefore, the loss of these goods and services has important social and economic implications. Also deep-sea ecosystems, which have long been thought to be extremely stable in terms of physico-chemical conditions, may experience abrupt change and climate-driven temperature shifts as a direct consequence of the prevailing surface climate conditions (Smith *et al.*, 2009; Masuda *et al.*, 2010; Fahrbach *et al.*, 2011). Contemporary climate change has the potential to perturb ocean circulation on a time-scale far shorter than that of continental drift. For example, a reduction in the North Atlantic Current could have major implications for northern Europe and beyond during this century (Cunningham *et al.*, 2007). This emphasizes the importance of regional considerations versus global generalization of the 'global warming' paradigm (Brierley and Kingsford, 2009). Climate-induced changes strongly differ across the globe, especially along a latitudinal gradient. Warming appears more pronounced at the poles than at the equator, and the responses of climate change are expected to differ for different marine habitats (Hoegh-Guldberg and Bruno, 2010; McGinty *et al.*,

2011). For example, whilst open oceans are more affected by the influence of wind on the timing and strength of stratification, coastal areas are expected to be more vulnerable to the effects of wind due to storms. Polar regions display changes in ice cover. Seas at the same latitude are expected to respond differently to climate change due to their region-specific abiotic characteristics such as basin depth and configuration, salinity regime and current pattern, and biotic attributes including biogeographic setting, biodiversity, and foodweb structure (Philippart *et al.,* 2011).

4.2 Global change in the oceans: what we know

4.2.1 Temperature

According to the fourth assessment report of the Intergovernmental Panel on Climate Change (IPCC, 2007), the global mean surface air temperature increased by 0.74 °C while the global mean sea surface temperature rose by 0.67 °C over the last century (Trenberth and Josey, 2007). Models predict that the effect of increased amounts of climate-altering gases will lead, before the end of this century, to an increase in air temperature ranging between 1.4 and 5.8 °C (IPCC, 2007). During the last 10–15 years, sea water temperatures throughout much of the globe have changed at unprecedented rates. The direct consequences of these changes have been: the rapid disappearance of sea ice covering the Arctic; the melting of glaciers; a reduction in the volume of Antarctic ice sheets; raising of sea levels and increasing storm events in the seas; high variability of precipitation and more frequent occurrences of intense rainfall events and hurricanes (IPCC, 2007; Hoegh-Guldberg and Bruno, 2010). However, temperature variations do not only affect important environmental variables such as local currents, water column stratification, nutrient cycling, and primary production, but also metabolic rates of marine organisms (McGowan *et al.,* 1998). All of these changes, induced by variability in temperature, strongly affect population and community dynamics and, over time, community structure and function. Moreover, temperature appears as the primary environmental factor in driving diversity at the large geographic scale (Tittensor *et al.,* 2010). Based on these findings, changes in the temperature of the global ocean may have strong consequences for the distribution of marine biodiversity (Worm and Lotze, 2009; Tittensor *et al.,* 2010).

4.2.2 Food supply

Climate variation is altering the production of organic carbon (OC) in the surface ocean, which ultimately comprises the primary food supply to deep-sea ecosystems (Smith *et al.,* 2008; Smith *et al.,* 2009). Changes in upper ocean temperature influence stratification, reduce vertical mixing, and can affect the availability of nutrients for phytoplankton production with the direct consequence of enhanced variability in primary production and carbon export flux to the deep-sea sediments (Smith *et al.*, 2009). The magnitude of such climate change effects is likely to be substantial because ocean warming to date appears to have caused a 6% decline in global ocean primary production (Gregg *et al.*, 2003), and climate models predict major reductions in ocean productivity over large regions within this century, especially in the tropical ocean (Bopp *et al.*, 2001). Time-series studies conducted over the past two decades in the North Pacific and the North Atlantic (at > 4000 m depth) have revealed, unexpectedly, climate-driven changes in deep-ocean ecosystems (in term of faunal abundance and composition) correlating with changes in surface ocean productivity (Smith *et al.*, 2009). Also, in the equatorial Pacific it has been documented that moderate changes in net primary production and sea surface temperature have provoked a threefold reduction in the flux of particulate organic carbon (POC) (e.g. from 1.5 to 0.5 g C m^{-2} y^{-1}; Laws, 2004). The benthic response to the reduction of food sources available in the sea bottom is a halving of microbial, nematode, and megafaunal standing stocks, a fivefold reduction in macrofaunal biomass, and two- to fourfold reductions in sediment mixed layer depth, sediment community oxygen consumption, and bioturbation intensity (Smith *et al.*, 2008). Long-term declines in POC flux, such as are expected in the equatorial abyss, are also likely to yield reductions in species diversity and body size (McClain *et al.*, 2005), as well as basic shifts in the taxonomic composition of abyssal assemblages (Smith *et al.*, 2008). All of these changes have a dramatic effect on the functioning of deep-sea ecosystems since an exponential relationship between biodiversity and ecosystem functioning has been observed in these ecosystems (Danovaro *et al.*, 2008a). Global warming may also have important implications for the fish stock as a consequence of the reduction in food sources (zooplankton for juveniles). During their larval stages, all fish consume zooplankton and some adult fish (e.g. mackerel *Scomber scombrus*) continue to be at least partly planktivorous (Hays *et al.*, 2005; Mieszkowska *et al.*, 2009). Consequently, the abundance and timing of mesozooplankton might affect fish recruitment (i.e. the amount of fish added to the exploitable stock each year), although details of this linkage are only just being revealed. Because of the plankton's sensibility to global warming, long-term changes in plankton may have an impact on commercial fish stocks (Hays *et al.*, 2005), with important social and economic consequences.

4.2.3 Ocean acidification and deoxygenation

Over the past 250 years, atmospheric carbon dioxide (CO_2) levels have increased by nearly 40%, from pre-industrial levels of approximately 280 parts per million volume (ppmv) to nearly 384 ppmv in 2007 (IPCC, 2007; Khatiwala *et al.*, 2009). This rate of increase is at least one order of magnitude faster than has occurred for millions of years (Doney and Schimel, 2007), and the current concentration is higher than experienced on Earth for at least the past 800,000 years (Lüthi *et al.*, 2008). Atmospheric CO_2 levels are predicted to continue to increase at least for the next century and probably longer, and unless emissions are substantially reduced, may well reach levels exceeding 1000 ppmv by 2100 (Royal Society, 2005). Oceans play a fundamental role in the exchange of CO_2 with the atmosphere; since pre-industrial times, the oceans have absorbed about a half of the CO_2 emissions produced from burning fossil fuels and cement manufacture. This demonstrates the integral role that oceans play within the natural processes of cycling carbon on a global scale (Royal Society, 2005). Carbon dioxide is a weak acid and the continued uptake of anthropogenic CO_2 triggers changes in ocean carbonate chemistry and pH, referred to as ocean acidification (Caldeira and Wickett, 2003). At present, the mean pH of ocean surface waters is already 0.1 units (equal to 30%) lower compared with pre-industrial times and a decrease of 0.4 units (equal to 120%) is projected by the year 2100 in response to a business-as-usual emissions pathway (Caldeira and Wickett, 2003). This change in pH drives profound changes in carbonate chemistry and is likely to affect the structure and functioning of marine ecosystems (Fabry *et al.*, 2008). Ocean acidification may interfere with shell formation in the broad suite of benthic calcifiers (e.g. molluscs, echinoderms, crustaceans, bryozoans, serpulid polychaetes, foraminifera, sponges, and corals) and weaken existing skeletons, particularly if ocean pH falls below 7.5 (Gazeau *et al.*, 2007; Kuffner *et al.*, 2008; Wood *et al.*, 2008). Ocean acidification also has the potential to alter ocean biogeochemical dynamics for OC and nutrients via several more indirect pathways. Increased carbonate dissolution in the water column could decrease the contribution of $CaCO_3$ to the ballasting of OC to the deep-sea sediments (Passow, 2004), causing more OC to remineralize in shallow waters and decreasing the ocean's CO_2 uptake efficiency. Another potentially serious consequence of global warming is a decrease in the dissolved O_2 content of the world's oceans (Keeling *et al.*, 2010). Oxygen is directly linked to carbon via photosynthesis and respiration, so the distribution of O_2 in the ocean is a diagnostic of the rate at which organic matter is produced, redistributed, and decomposed in the ocean. The loss of dissolved O_2 (deoxygenation) is predicted, not just because O_2 is less soluble in warmer water but also because global warming may increase upper ocean stratification, thereby reducing the O_2 supply to the ocean interior (Sarmiento *et al.*, 1998; Bopp *et al.*, 2001; Keeling and Garcia, 2002). Systematic deoxygenation and the extension of oxygen minimum zones are increasing throughout the world oceans and can have widespread consequences in the biogeochemical cycling of carbon, nitrogen, and many other important elements (P, Fe, Mn, etc.). Oxygen is also fundamental for all aerobic life, including organisms living in the dark ocean interior. Most organisms are not very sensitive to O_2 levels as long as concentrations are high enough or they are able to survive for a short period of their life without oxygen. However, once O_2 drops below a certain threshold, the organism suffers from a variety of stresses, leading ultimately to death if the concentrations stay too low for too long (hypoxic conditions). Thresholds for hypoxia vary greatly between marine taxa, with fish and crustaceans tending to be the most sensitive (Vaquer-Sunyer and Duarte, 2008). Anoxia is rare in the water column of the modern ocean, while deep hypersaline anoxic basins (DHABs) have been discovered in the eastern Mediterranean Sea, and in the Red Sea (Danovaro *et al.*, 2010). Suboxic and hypoxic conditions occur in all oceans and these oxygen minimum zones (OMZs) will expand in the future as documented by declines in O_2 concentrations over the past 50 years in the subpolar North Pacific and in tropical OMZs (Whitney *et al.*, 2007; . Stramma *et al.*, 2008; Levin, 2010). At very low O_2 concentrations ($< 5\ \mu\text{mol kg}^{-1}$), major changes in biogeochemical cycling occur and nitrate becomes important in respiration. Once nitrate is exhausted, the biogeochemistry then tends to be dominated by sulfate-reducing microbes, which convert sulfate to sulfide.

4.2.4 Organisms, life, and alien species

There is now ample evidence that these recent climatic changes have affected a broad range of organisms with diverse geographical distributions, from polar terrestrial to tropical marine environments (Walther *et al.*, 2001; 2002). A growing body of evidence highlights the combined effects of multiple stressors associated with global change on survivorship, metabolism, recruitment, and fecundity of benthic invertebrates (Hutchings *et al.*, 2007; Poloczanska *et al.*, 2007). These environmental changes, associated with global warming, are linked to larger ecological processes such as changes in larval dispersal and recruitment success, shifts in community structure and range extensions, and the establishment and spread of invasive species (Przeslawski *et al.*, 2008). Loss of some species might have negative effects on ecosystem function (Danovaro *et al.*, 2008a). The synergism of rapid temperature rise and other stresses, in particular habitat destruction, could easily disrupt the connectedness among species and lead to a reformulation of species communities, reflecting differential changes in species, and to numerous extirpations and possibly extinctions (Thomas *et al.*, 2004; Pounds *et al.*, 2006). Certain species of benthic invertebrates are more vulnerable to climate change than others due to their inability to shift their ranges with changing environmental conditions (Przeslawski *et al.*, 2008). Ocean warming can facilitate the establishment and spread of invasive species (e.g. ascidians, in

Stachowicz *et al.,* 2002; Agius, 2007; bryozoans, in Saunders and Metaxas, 2007) and enhance the ability of certain alien species to inhabit new regions (Occhipinti-Ambrogi and Galil, 2010). Climate change can promote the crossing of frontiers for non-native species from adjacent areas, and these non-native species become new elements of the biota (Walther *et al.,* 2002). Some invasive species also seem to be better adapted to changing and extreme environmental conditions than native species, or less recent arrivals (e.g. mussels and temperature; Fields *et al.,* 2006). Moreover, the decrease in time spent as planktonic larvae due to warmer waters may increase the larval retention time for invasive species, allowing them to establish or spread in areas where they would otherwise have drifted away (Byers and Pringle, 2006). Elevated temperatures may not only increase the spread of invasive species but also exacerbate existing invasion impacts on benthic communities by facilitating competitive exclusion, predation, and extinction (Agius, 2007). Plankton is a particularly good indicator of climate change, in particular on the shift of species distribution in relation to global warming (Hays *et al.,* 2005). Many members of the genus *Ceratium,* important primary producers in tropical and temperate waters, have expanded their range into warmer water. For example, before 1970, *C. trichoceros* was only found south of the UK, but is now found off the west coast of Scotland and in the northern North Sea. Dramatic biogeographical shifts have also been documented for warm-water assemblages of calanoid copepod, which have moved 1000 km further north in the north-east Atlantic over the past 40 years, with a concomitant retraction in the range of cold-water assemblages (Hays *et al.,* 2005). Classical examples from the benthos are the so-called Lessepsian migration of various Red Sea species through the Suez Canal to the Mediterranean, the outbreak of the green algae *Caulerpa taxifolia* and *Caulerpa racemosa* in the Mediterranean, the introduction of Ponto-Caspian species in the Baltic, and the expansion of the Japanese oyster *Crassostrea gigas* in the Wadden Sea and along the southern North Sea coastline (Barange *et al.,* 2011). The effects of an increasing number of 'new' species on existing communities are diverse, and often not well known. The role of alien species should therefore be assessed in a more integrated and dynamic context of shifting species' ranges and changing compositions and structures of communities (Walther *et al.,* 2009).

A synthesis of the various types of impacts on marine biodiversity and thus potentially on related ecosystem services is reported in Figure 4.1.

From data reported by Costello *et al.,* (2010) it appears that impacts related to global change (i.e. temperature rise, acidification, and deoxygenation) are already responsible for encouraging a large amount of marine biodiversity. In addition it has also been shown that invasive species are favoured or promoted by climate change (and particularly by temperature rise).

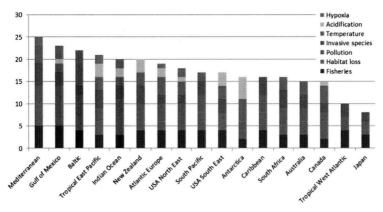

Figure 4.1 The effects of different drivers (including climate change as tempera-ture rise, acidification and hypoxia) on function of biodiversity and goods and services in different oceanic regions (based on data from Costello *et al.*, 2010). A black and white version of this figure will appear in some formats. For the colour version, please refer to the plate section.

4.3 Effects of global change on marine habitats

Global warming is an additional, more recent threat to marine habitats' integrity (Hughes *et al.*, 2003; Harley *et al.*, 2006; Lotze *et al.*, 2006; Orth *et al.*, 2006; Poloczanska *et al.*, 2007; Hawkins *et al.*, 2008). Climate change is expected to have a range of effects on marine ecosystems, their function and biodiversity. Some effects may be related to changing water temperatures, circulation and/or changing habitat, while others occur through altered pathways within biogeochem-ical cycles and foodwebs (Ramirez-Llodra *et al.*, 2010; 2011). The response of marine systems to climate change also depends on interactions with other human-induced changes in the marine environment. For example, fishing has reduced the number of large fish at higher trophic levels worldwide (Myers and Worm, 2003) whilst increasing agricultural, industrial, and household activities have resulted in nutrient enrichment of many coastal regions (Schindler, 2006). These and other global changes, such as ocean acidification (Royal Society, 2005) and the intro-duction of non-native species (Doney, 2010), are likely to result in more fragile marine ecosystems, which will challenge the effectiveness of management strat-egies that may be implemented to reduce the impacts of climate change (Hoegh-Guldberg and Bruno, 2010).

4.3.1 Coastal water ecosystems

There is a wide consensus that coastal marine ecosystems, along with the goods and services they provide, are threatened by anthropogenic global climate change

(IPCC, 2001). Coastal marine systems cover a surface $c.3,102$ ha $\times 10^8$ and the economic value of all ecosystem services provided by these areas approaches US $12.6 trillion per year globally (Costanza *et al.*, 1997). Coastal ecosystems are particularly vulnerable, because their resilience and capacity to buffer additional environmental stresses have already been undermined by human impacts such as over-harvesting, pollution, eutrophication, and habitat destruction (Hughes *et al.*, 2003). Moreover, many disturbances are acting at the terrestrial–marine interface and are predicted to increase, such as increased land run-off after floods, higher wave energies (due to increased storm frequency), and increasing turbidity. Global change severely increases the risk of flooding and subsequent loss of coastal areas as a direct consequence of an accelerating sea-level rise and also a possible increase in the frequency and intensity of storms (Airoldi *et al.*, 2005; Burcharth *et al.*, 2007). Moreover, the most obvious consequence of sea-level rise is an upward shift in species distributions and a relevant changing of ocean circulation (which drives larval transport) with important consequences for population dynamics. This shift in species composition can have negative effects on coastal biota (Harley *et al.*, 2006). Further extension, raising, and reinforcement of artificial coastal defences may protect populated areas, but are bound to result in loss of sedimentary coastal marine habitats with consequences for living marine resources (Airoldi *et al.*, 2005; Martin *et al.*, 2005; Anderson *et al.*, 2005). Such artificial coastlines may act as stepping stones for species advancing with climate change (Helmuth *et al.*, 2006; Hawkins *et al.*, 2008) and provide habitat for jellyfish polyps, contributing to the increase in jellyfish (Richardson *et al.*, 2009). Many marine organisms, including economically important fish, spend part of their life in relatively sheltered areas along the coast. Loss of these areas may affect these animals at that specific stage of their life cycle. In addition, changes in the strength and seasonality of upwelling in areas along the coast could influence the retention dispersal mechanisms of (juvenile) fish and shellfish between coastal waters and the open sea with unknown consequences for species' recruitment (Philippart *et al.*, 2011).

4.3.2 Seagrass ecosystems

Seagrasses cover about 0.1–0.2% of the global ocean, and develop highly productive ecosystems which fulfil a key role in coastal ecosystems. A conservative estimate of the total area covered by seagrass is $177,000 \, \text{km}^2$ (Spalding *et al.*, 2003), and the economic value of all ecosystem services provided by seagrass approaches US$3.8 trillion per year globally (Costanza *et al.*, 1997). Seagrass ecosystems provide key ecological services, including OC production and export, nutrient cycling, sediment stabilization, enhanced biodiversity, and trophic transfers to adjacent habitats in tropical and temperate regions (Orth *et al.*, 2006).

Changes in sea level, salinity, temperature, atmospheric CO_2, and UV radiation can alter seagrass distribution, productivity, and community composition. In turn, potential changes in distribution and structure of seagrass communities may have profound implications for local and regional biota, near-shore geomorphology, and biogeochemical cycles (Short and Neckles, 1999). Widespread seagrass loss results from direct human impacts, including mechanical damage (by dredging, fishing, and anchoring), eutrophication, aquaculture, siltation, effects of coastal constructions, and food web alterations; and indirect human impacts, including negative effects of climate change (erosion by rising sea level, increased storms, increased ultraviolet irradiance), as well as from natural causes, such as cyclones and floods (Duarte, 2002). Present losses are expected to accelerate, particularly in south-east Asia and the Caribbean, as human pressure on the coastal zone grows. Positive human effects include increased legislation to protect seagrass, increased protection of coastal ecosystems, and enhanced efforts to monitor and restore the marine ecosystem. However, these positive effects are unlikely to balance the negative impacts, which are expected to be particularly prominent in developing tropical regions, where the capacity to implement conservation policies is limited.

4.3.3 Mangrove forests

Mangrove forests currently occupy 14,650,000 ha of coastline globally (Wilkie and Fortuna, 2003), with an economic value in the order of US$200,000–900,000 ha^{-1} (UNEP–WCMC, 2006). Mangroves are important habitats and are especially often important nursery grounds and breeding sites for birds, mammals, fish, crustaceans, shellfish, and reptiles. Moreover they represent a renewable resource of wood and sites for accumulation of sediment, nutrients, and contaminants (Twilley, 1995; Kathiresan and Bingham, 2001; Manson et al., 2005). Climate change components that affect mangroves include changes in sea level, high water events, storminess, precipitation, temperature, atmospheric CO_2 concentration, ocean circulation patterns, health of functionally linked neighbouring ecosystems, as well as human responses to climate change (Gilman et al., 2008). Of all the outcomes from change in the atmospheric composition and alterations to land surfaces, relative sea-level rise may be the greatest threat (Lovelock and Ellison, 2007). Based on available evidence, among all climate change outcomes (such as rise in sea level, ocean acidification and temperature, change in frequency and intensity of precipitation/storm patterns), relative sea-level rise may be the greatest threat to mangroves (Gilman et al., 2008). The Pacific islands mangroves have been demonstrated to be at high risk of substantial reductions (Gilman et al., 2008). Recent predictions on mangrove losses reveal a 10–15% decline in global mangrove area by 2100 (Snedaker, 1995; Alongi, 2008). Reduced mangrove area and health will increase the threat to human safety and shoreline development from coastal hazards such as erosion, flooding, storm waves and surges, and tsunami, as most recently observed following the 2004 Indian Ocean tsunami (Dahdouh-Guebas *et al.*, 2005a,b; 2006). Mangrove loss also reduces coastal water quality and biodiversity, eliminates fish and crustacean nursery habitat, adversely affects adjacent coastal habitats, and eliminates a major resource for human communities that rely on mangroves for numerous products and services (Nagelkerken *et al.*, 2008). Mangrove destruction can also release large quantities of stored carbon and exacerbate global warming and other climate change trends (Kristensen *et al.*, 2008).

4.3.4 Coral reef ecosystems

A conservative estimate of the total area covered by coral reefs is 62,000 km^2 and the economic value of all ecosystem services provided by corals approaches US\$ 365 billion per year globally (Costanza *et al.*, 1997). Climate change is threatening tropical marine habitats across the world, with coral bleaching, invasive species, and community shifts linked to rising sea temperatures and acidification (Pandolfi *et al.*, 2005; Poloczanska *et al.*, 2007; Lough, 2008). Current research on the impacts of climate change on tropical fauna is mainly focused on corals (Hoegh-Guldberg, 2004; Willis *et al.*, 2006; Smith *et al.*, 2007; Lough, 2008) and associated fishes (Roessig *et al.*, 2004; Graham *et al.*, 2006; Wilson *et al.*, 2006; Munday *et al.*, 2007). Substantial impacts on community structure have been observed in coral reefs during periods of warmer than normal sea temperatures (Walther *et al.*, 2002; Przeslawski *et al.*, 2008). Poised near their upper thermal limits, coral reefs have undergone global mass bleaching events whenever sea temperatures have exceeded long-term summer averages by more than 1.0°C for several weeks (Wilkinson, 2000; Walther *et al.*, 2002). The most severe period occurred in 1998, during which an estimated 16% of the world's reef-building corals died (Wilkinson, 2000). The impact of thermal stress on reefs can be dramatic, with almost total removal of corals in some instances (Hoegh-Guldberg, 1999). In some cases branching acroporid and pocilloporid corals have bleached and/or died, leaving more massive species like *Porites spp.* intact. In other cases, all coral species have largely been removed (Mumby *et al.*, 2001; Loya *et al.*, 2001). Estimates of how ecosystem species' richness and community structure have changed after bleaching events are generally not available, but such changes are suspected to be large. When coral loss has been severe, reef fish biodiversity declines (Jones *et al.*, 2004; Graham *et al.*, 2006; Wilson *et al.*, 2006) and the loss of key species within critical functional groups, or the complete loss of entire functional groups, can have a profound effect on reef resilience (Bellwood *et al.*, 2004).

4.3.5 Kelp forests

Kelp forests occur along the majority of the world's temperate coastlines, and are among the most phyletically diverse and productive systems in the ocean (Steneck *et al.*, 2002). Climatic variables such as temperature and precipitation play an important role in controlling local and regional scale differences in population dynamics and species distributions. In many coastal areas where humans have altered chemical and biological conditions, however, kelp forests have been replaced by mats of turf-forming algae (Airoldi and Beck, 2007). While kelp canopies inhibit turfs (Russell, 2007), developing theory explains shifts from canopy to turf domination as a function of reduced water quality which enables the cover of turf to expand spatially and persist beyond its seasonal limits (Gorman *et al.*, 2009), subsequently inhibiting the recruitment of kelp and regeneration of kelp forests. Connell and Russell (2010) reveal that kelp loss may be exacerbated, on human-dominated coasts, by the positive effects of increasing CO_2 and temperature on kelp inhibitors. Altering global (i.e. CO_2) and local (i.e. eutrophication) stressors in combination can allow turfs to expand to more rapidly to occupy space made available due to the reduction of kelp forests (Russell *et al.*, 2009). As algal turfs can inhibit kelp recruitment (Connell and Russell, 2010), any phenological shift that allows turfs to persist through periods of kelp recruitment is likely to reduce the resilience of kelp forests to disturbance. When the canopy is lost, there are profound changes to local environmental conditions, with cascading effects on the ecosystem diversity and functioning. Recently, Wernberg and collaborators (2010) revealed that a successful mitigation of negative effects of global warming is strictly related to an understanding of the link between physiological and ecological responses of kelp species. Metabolic adjustment may assist Australasian kelp beds to persist and maintain abundance in warmer waters, while it is evident that there is a reduction in the physiological responsiveness of kelps to perturbation and of canopy recovery from disturbances by reducing the ecological performance of kelp recruits.

4.3.6 Submarine canyon ecosystems

Global warming is thought to influence the frequency and intensity of climate-driven episodic events in marine ecosystems. Among these, the best known event is dense shelf water cascading (DSWC) (Canals *et al.*, 2006; 2009). Submarine canyons are the main conduits for cascading shelf waters, and from this has developed the concept of 'flushing submarine canyons' (Canals *et al.*, 2006). This cascading event is seasonal, resulting from the formation of dense water by cooling and/or evaporation, and occurs on both high- and low-latitude continental margins

(Ivanov *et al.*, 2004). DSWC can transport large amounts of water and sediment, reshape submarine canyon floors, and rapidly affect the deep-sea environment. Furthermore, changes in the frequency and intensity of DSWC driven by future climate change may have a significant impact on the supply of organic matter to deep-sea ecosystems and on the amount of carbon stored on continental margins and in ocean basins (Canals *et al.*, 2006). A recent study conducted in the Catalan Sea, west of the Gulf of Lions, has shown a direct effect of DSWC on deep-sea ecosystems, and consequently, on their living resources (Company *et al.*, 2008; Pusceddu *et al.*, 2010). DSWC can cause the collapse of catches of the highly praised deep-sea shrimp *Aristeus antennatus*, but it has also shown that the population of this species was able to recover over a relatively short period (around three years after a dramatic cascading event) (Company *et al.*, 2008). The transport of particulate organic matter associated with cascading appears to enhance the recruitment of this deep-sea living resource, apparently mitigating the general trend of overexploitation. Because cascade of dense water from continental shelves is a global phenomenon, its influence on deep-sea ecosystems and fisheries worldwide should be larger than previously thought (Company *et al.*, 2008).

4.3.7 Deep-sea basins

The ocean occupies *c*.71% of the Earth's surface with a mean depth of 3700 m. Deep-sea regions of 2000 m depth cover *c*.60% of the Earth's surface and can be affected by climate change, with major implications and consequences on a global scale. Using a decadal data set (from 1989 to 1998), Danovaro *et al.* (2001; 2004) provided evidence that deep-sea nematode diversity can be strongly and rapidly affected by temperature shifts. An abrupt temperature shift (of about 0.4 °C) in the deep eastern Mediterranean Sea caused a significant change in nematode biodiversity. This temperature decrease also resulted in decreased functional diversity and species evenness, and an increase in the similarity of Mediterranean nematode assemblages to colder deep-Atlantic fauna. This evidence suggests that deep-sea fauna is highly vulnerable to environmental alteration, and that deep-sea biodiversity is also significantly affected by very small temperature changes. In the north-east Pacific, changes in megafaunal abundance appeared to be related to long-term, climate-driven variations in POC flux, and hence to productivity in the overlying surface waters (Ruhl and Smith, 2004). An increase in abundance was associated with a decrease in mean body size, suggesting the occurrence of recruitment events, while sharp declines in abundance were related to competitive faunal interactions and survivorship (Ruhl, 2007). In the north-east Atlantic, large-scale changes in the abundance of megafauna occurred in 1996 in which some species, that had been rare components of the fauna for many years, became very abundant. One of these, the holothurian *Amperima rosea*, increased in abundance (over three orders of magnitude) and became the dominant megafaunal species (Billett *et al.*, 2001). Large-scale changes in the flux of organic matter to the abyssal seafloor have been noted in the time series, particularly in 2001, and may be related to the sudden mass occurrence of *A. rosea* the following year (Billet *et al.*, 2010). In the Arctic, the Hausgarten Station near Svalbard has provided the first long-term time series of the benthos in the region. Work by the Alfred Wegener Institute has demonstrated a small but important temperature increase between 2000 and 2008 at 2500 m depth, in the Fram Strait between Svalbard and Greenland (Barange *et al.*, 2011). Within the MarBEF project DEEPSETS, a five-year (2000–2004) time-series study of nematodes at this site revealed shifts in nematode abundance and community composition, reflecting changes in food availability. For the larger organisms, a towed camera system revealed a significant decrease in megafaunal abundance at 2500 m depth (Barange *et al.*, 2011).

Impacts induced by global change are occurring in all habitat types investigated so far, although to different extents. The changes have been proven to be primarily the result of long-term trends, either in temperature change, deoxygenation, or acidification. However, some of the impacts related to climate change occur episodically. This is the case for dense shelf water formation, with the creation of a cold cascade of sediment and cold water in deeper ocean layers (Canals *et al.*, 2006), or temperature anomalies associated with the deepening of the thermocline, which induced mass mortalities in coastal shallow and mesophotic assemblages (Cerrano *et al.*, 2000).

4.4 Effects of climate change on viruses and prokaryotes

The effects of global change will likely influence all ecosystem components and a broad range of organisms including bacteria, Archaea, and Protista with diverse geographical distributions (Walther *et al.*, 2001; Genner *et al.*, 2004). Viruses are the most numerous 'lifeforms' in aquatic systems (Suttle, 2005; 2007), numbering about 15 times the total numbers of bacteria and Archaea. Recent studies revealed that viruses play a key role in deep-sea metabolism and in global biogeochemical cycles (Danovaro *et al.*, 2008b). Given that the vast majority of the biomass OC in oceans consists of microorganisms, it is expected that viruses and other prokaryotic and eukaryotic microorganisms will play important roles as agents and recipients of global climate change. As viral replication and life cycle are closely linked with host metabolism, increases in temperature will likely influence the interactions between viruses and the cells they infect (Danovaro *et al.*, 2011). Recent analysis carried out on a global scale reveals that the fraction of the total variance explained by the relationship between temperature and viruses is generally low, indicating that factors influencing virioplankton distribution are more complex than those predicted by temperature alone (Danovaro *et al.*, 2011). Viruses have the potential to interact with the climate through their contribution to the marine biogenic particles of the aerosol and by contributing to the release of DMS through lysis of their autotrophic hosts. These processes have to be quantified and included in modelling studies dealing with ocean–atmosphere interactions. The oxygen minimum zones are predicted to expand in the future ocean because of climate change, with important consequences for biogeochemical cycling of nitrogen and phosphorus and for the distribution of organisms. Because eukaryotic herbivores and bacterivores are more sensitive than prokaryotes to the reduction in oxygen levels, it can be expected that virus-induced mortality of prokaryotes will increase at the expense of protists and other bacterivores. The effects of ocean acidification on marine viruses are uncertain, but we can anticipate that the most dramatic changes will be due to the effects of pH on the host organisms that the viruses rely on, bacteria, Archaea, protists, and metazoa, which are highly pH dependent. Moreover, because some key metabolic processes of the microbial communities are highly sensitive (and inhibited) by even small decreases in the pH of the medium, ocean acidification may have a profound influence on the overall functioning of the microbial communities and on virus–host interactions (Danovaro *et al.*, 2011). A recent review of the potential effect of global warming on prokaryotes carried out in the Arctic Ocean revealed that light for phytoplankton and inorganic/organic nutrients for phytoplankton and heterotrophic bacteria are more crucial than are changes in temperature (Kirchman *et al.*, 2009). The direct consequences in polar systems, where the effects of climate changes are more severe, are that microbial processes are particularly sensitive to small changes and have potentially large impacts on carbon flows and other ecosystem functions.

4.5 Effects of climate change on marine mucilage

Worldwide, the highly productive and shallow Adriatic Sea (and particularly its northern portion) within the Mediterranean basin is the area most severely affected by the outbreak of massive marine mucilage (Danovaro *et al.*, 2009). Mucilage is made of exopolymeric compounds with highly colloidal properties that are released by marine organisms through different processes, including phytoplankton exudation of photosynthetically derived carbohydrates produced under stressful conditions (Fonda Umani *et al.*, 2005; Degobbis *et al.*, 1995). Marine mucilage floating on the surface or in the water column can display a long life span (up to 2–3 months) and once settled on the sea bottom, these large aggregates coat the sediments, extending in certain cases for kilometres and causing hypoxic and/or anoxic conditions (Precali *et al.*, 2005). The consequent suffocation of benthic organisms (including bottom associated nekton) (Danovaro *et al.,* 2005) provokes serious economic damage to tourism and fisheries (Rinaldi *et al.,* 1995). Mucilage is not closely associated with the presence of eutrophic conditions, as several mucilage outbreaks have recently been observed in oligotrophic seas, such as the Aegean Sea. Moreover, the frequency of mucilage outbreaks in the Adriatic Sea has increased in the last two decades concurrently with a significant decrease in primary production (Danovaro *et al.,* 2009). Recently an analysis based on a record of approximately 60 years of mucilage appearance in the Mediterranean Sea (1950–2008) has revealed that patterns of climate anomalies (e.g. the positive anomaly of surface temperature) explained a large proportion of variance in mucilage outbreaks, on an annual and decadal basis. If the mucilage phenomenon continues to increase in frequency and duration, and to spread around the coastal areas of the Mediterranean Sea, an increased stress for benthic ecosystems (anoxia) can occur with important consequences for the biodiversity, goods, and services of the areas influenced by these events. Moreover, marine diseases may result with potential consequences to human health (Patz *et al.,* 2005). Mucilage can be proposed as a potentially novel paradigm of ecosystem alteration caused by the synergistic effect of climate change and misuse of marine resources. Mucilage on one hand represents a symptomatic response of the marine ecosystem to direct and indirect anthropogenic impacts, and on the other, a potentially expanding carrier of viruses and bacteria, including pathogenic forms that are harmful to the health of humans and marine organisms (Danovaro *et al.,* 2009).

4.6 Global change impact on marine ecosystems' goods and services

An analysis of the cost of the loss of goods and services and the main social and economic implications for both marine and terrestrial ecosystems has been made and is being updated in the Millennium Ecosystem Assessment (IUCN, 2010). At present, although it is evident that several ecosystems are being degraded and are losing an important fraction of their production of goods and services, it is difficult to obtain a detailed analysis of the costs of this loss. In some coastal areas it is estimated that the marine habitats have lost 30 to 50% of their goods and services. Table 4.1 shows a list of the marine habitats and their value estimated as in 1994.

Taking into account the re-evaluation of these values in the last two decades or so, it can be estimated that the amount of this loss in degraded habitats can be in the order of US$10,000–20,000 ha^{-1} yr^{-1}.

The changes induced by global warming and ocean acidification are expected to have important direct/indirect consequences on marine ecosystems and human life. All marine habitats including coastal, open ocean, and deep sea provide goods and services essential for human well-being (Costanza *et al.,* 1997; Grehan *et al.,*

Table 4.1 *List of marine coastal habitats and their value estimated as in 1994 (from Costanza et al., 1997)*

Habitat type	Economic value
Estuaries	US\$ 22,832 ha^{-1} yr^{-1}
Seagrass and algal beds	US\$ 19,004 ha^{-1} yr^{-1}
Tidal marshes/mangroves	US\$ 9,990 ha^{-1} yr^{-1}
Swamps flood plains	US\$ 19,580 ha^{-1} yr^{-1}
Coral reefs	US\$ 6,075 ha^{-1} yr^{-1}
Continental shelf	US\$ 1,610 ha^{-1} yr^{-1}

2009). Marine ecosystems such as kelp beds, coral reefs, and seagrass meadows are socially and economically important, and their ecosystem services are estimated to be worth trillions of dollars to the global economy each year (Costanza *et al.*, 1997). The loss of these habitats represents a leading cause of the loss of biodiversity and ecosystem services (Lotze *et al.*, 2006; Airoldi and Beck, 2007) which may also have major socio-economic consequences. Similarly to the terrestrial biomes (Sala *et al.*, 2000), marine biodiversity is simultaneously impacted by a range of human activities such as overfishing, habitat destruction, and pollution (Dulvy *et al.*, 2003; Lotze *et al.*, 2006; Worm *et al.*, 2006). Climate change may add to and amplify these impacts on marine biodiversity (Walther *et al.*, 2002; Cheung *et al.*, 2009). The negative effects of climate change on specific habitats can have unpredictable consequences for other species/habitats and human activities (both in term of social and economic interest). Change in distributions and community structure of marine species may indeed affect fishing activities (both commercial and recreational fisheries) and have socio-economic impacts on vulnerable coastal communities (Allison *et al.*, 2009). Many fisheries are already extensively overfished and the changing environmental conditions may contribute to declining stocks of fish species. Shifts in the diversity and abundance of the bioeroding fauna may cause a weakening of reef infrastructure, exacerbating the impacts of storms and higher sea levels on reef fragmentation (Przeslawski *et al.*, 2008). This may lead to a decrease in the protective wave-reducing role of fringing reefs, leading to increased erosion of beaches and coastal structures (Hutchings and Salvat, 2000; Sheppard *et al.*, 2005). Benthic invertebrates are also a rich source of bioactive compounds with various medicinal, industrial, and commercial applications. Pharmaceuticals based on the neurotoxins from tropical cone shells are now in clinical use as the most powerful relievers of severe pain (Hogg, 2006). Notably, bioactive compounds of certain invertebrates may vary with region and reef (Fahey and Garson, 2002; Page *et al.*, 2005). All compounds with potential economic interest are likely to be influenced by food sources, changes in microbial faunas across small environmental gradients, or seasonally changing habitat conditions (Fahey and Garson, 2002; Page *et al.,* 2005). Loss of a species will mean loss of these important compounds for human life. Sponges, bryozoans, ascidians, and molluscs are the major sources of secondary metabolites in the sea and therefore have been the prime target for research (Battershill *et al.,* 2005; Page *et al.,* 2005). This evidence of the impact of global change for oceans highlights the urgency to minimize greenhouse gas emissions and human-induced global warming and to develop marine conservation strategies that account for the potential impact of climate change (Cheung *et al.,* 2009).

Since 61% of world gross national product is derived from coastal ecosystems, the importance of acting immediately to preserve this capacity which is needed for the sustainable development of humanity cannot be ignored. In addition, the evaluation made by Costanza *et al.* (1997) does not consider the economic importance of deep-sea habitats that represent approximately 95% of the ocean surface. The conservation and sustainable use of marine biological diversity is a priority for all countries, and particularly in areas beyond national jurisdiction, where the lack of appropriate laws represents a major threat to their conservation. The existing knowledge relative to coastal and deep-sea ecosystems indicates that establishing networks of marine protected areas (in both coastal areas and the open ocean) is extremely effective in maintaining marine biodiversity, without com-promising (and conversely enhancing) the values and benefits for stakeholders and for society. Therefore, addressing marine protected areas, environmental impact assessments, and access to and the benefits of sharing genetic resources is the priority for scientists and policy makers over the next decade. However, marine ecosystems have no political boundaries, and these actions have to be ensured through a wide regional and even global partnership.

References

Agius, B. P. (2007). Spatial and temporal effects of pre-seeding plates with invasive ascidians: Growth, recruitment and community composition. *Journal of Experimental Marine Biology and Ecology* **342**: 30–39.

Airoldi, L. and Beck, M. W. (2007). Loss, status and trends for coastal marine habitats of Europe. *Oceanography and Marine Biology: An Annual Review* **45**: 345–405.

Airoldi, L., Abbiati, M., Beck, M. W., *et al.* (2005). An ecological perspective on the deployment and design of low-crested and other hard coastal defense structures. *Coastal Engineering* **52**: 1073–1087.

Allison, E. H., Perry, A. L., Adger, W. N., *et al.* (2009). Vulnerability of national economies to the impacts of climate change on fisheries. *Fish and Fisheries* **10**(2): 173–196.

Alongi, D. M. (2008). Mangrove forests: Resilience, protection from tsunamis, and responses to global climate change. *Estuarine, Coastal and Shelf Science* **76**(1): 1–13.

Anderson, J. M., Åberg, P., Hawkins, S. J., *et al.* (2005) Low-crested coastal defence structures as artificial habitats for marine life: Using ecological criteria in design. *Coastal Engineering* **52**: 1053–1071.

Barange, M., Heip, C., and Meysman, F. (2011). Biological impacts. In: *Climate Change and Marine Ecosystem Research. Synthesis of European Research on the Effects of Climate Change on Marine Environments. Marine Board Special Report*, 65–79. European Union Framework Programme 7.

Battershill, C. N., Jaspars, M., and Long, P. F. (2005). Marine biodiscovery: New drugs from the ocean depths. *Biologist* **52**: 107–114.

Bellwood, D. R., Hughes, T. P., Folke, C., and Nystrom, M. (2004). Confronting the coral crisis. *Nature* **429**: 827–833.

Billett, D. S. M., Bett, B. J., Rice, A. L., *et al.* (2001). Long-term change in the megabenthos of the Porcupine Abyssal Plain (NE Atlantic). *Progress in Oceanography* **50**: 325–348.

Billett, D. S. M., Bett, B. J., Reid, W. D. K., *et al.* (2010). Long-term change in the abyssal NE Atlantic: The Amperima Event revisited. *Deep-Sea Research II* **57**: 1406–1417.

Bopp, L., Monfray, P., Aumont, O., *et al.* (2001). Potential impact of climate change on marine export production. *Global Biogeochemical Cycles* **15**: 81–99.

Brierley, A. S. and Kingsford, M. J. (2009). Impacts of climate change on marine organisms and ecosystems. *Current Biology* **19**: 602–614.

Burcharth, H. F., Hawkins, S. J., Zanuttigh, B., and Lamberti, A. (2007). *Environmental Design Guidelines for Low Crested Coastal Structures*. Oxford: Elsevier, 395 pp.

Byers, J. E. and Pringle, J. M. (2006). Going against the flow: Retention, range limits and invasions in advective environments. *Marine Ecology Progress Series* **313**: 27–41.

Caldeira, K. and Wickett, M. E. (2003). Anthropogenic carbon and ocean pH. *Nature* **425**: 365.

Canals, M., Puig, P., Durrieu de Madron, X., *et al.* (2006). Flushing submarine canyons. *Nature* **444**: 354–357.

Canals, M., Danovaro, R., Heussner, S., *et al.* (2009). Cascades in Mediterranean submarine grand canyons. *Oceanography* **22**: 26–43.

Cerrano, C., Bavestrello, G., Bianchi, C. N., *et al.* (2000). A catastrophic mass-mortality episode of gorgonians and other organisms in the Ligurian Sea (north-western Mediterranean), summer 1999. *Ecology Letters* **3**: 284–293.

Cheung, W., Vicky, W. L., Lam, W. Y., *et al.* (2009). Projecting global marine biodiversity impacts under climate change scenarios. *Fish and Fisheries* **10**(3): 235–251.

Company, J. B., Puig, P., Sardà, F., *et al.* (2008). Climate influence on deep sea populations. *PLoS ONE* **3**(1): e1431.

Connell, S. D. and Russell, B. D. (2010). The direct effects of increasing CO_2 and temperature on non-calcifying organisms: Increasing the potential for phase shifts in kelp forests. *Proceedings of Royal Society Part B: Biological Sciences* **277**: 1409–1415.

Costanza, R., d'Arge, R., deGroot, R., *et al.* (1997). The value of the world's ecosystem services and natural capital. *Nature* **387**: 253–260.

Costello, M. J., Coll, M., Danovaro, R., *et al.* (2010). A census of marine biodiversity knowledge, resources, and future challenges. *PLoS ONE* **5**(8): e12110.

Cunningham, S. A., Kanzow, T., Rayner, D., *et al.* (2007). Temporal variability of the Atlantic meridional overturning circulation at 26.5 degrees N. *Science* **317**: 935–938.

Dahdouh-Guebas, F., Jayatissa, L. P., Di Nitto, D., *et al.* (2005a). How effective were mangroves as a defense against the recent tsunami? *Current Biology* **15**: 443–447.

Dahdouh-Guebas, F., Hettiarachchi, S., Lo Seen, D., *et al.* (2005b). Transitions in ancient inland freshwater resource management in Sri Lanka affect biota and human populations in and around coastal lagoons. *Current Biology* **15**: 579–586.

Dahdouh-Guebas, F., Koedam, N., Danielsen, F., *et al.* (2006). Coastal vegetation and the Asian tsunami. *Science* **311**: 37–38.

Danovaro, R., Dell'Anno, A., Fabiano, M., *et al.* (2001). Deep-sea ecosystem response to climate changes: The eastern Mediterranean case study. *Trends in Ecology and Evolution* **16**: 505–510.

Danovaro, R., Dell'Anno, A., and Pusceddu, A. (2004). Biodiversity response to climate change in a warm deep sea. *Ecology Letters* **7**: 821–828.

Danovaro, R., Armeni, M., Luna, G. M., *et al.* (2005). Exo-enzymatic activities and dissolved organic pools in relation with mucilage development in the Northern Adriatic Sea. *Science of the Total Environment* **353**: 189–203.

Danovaro, R., Gambi, C., Dell'Anno, A., *et al.* (2008a). Exponential decline of deep-sea ecosystem functioning linked to benthic biodiversity loss. *Current Biology* **18**: 1–8.

Danovaro, R., Dell'Anno, A., Corinaldesi, C., *et al.* (2008b). Major viral impact on the functioning of benthic deep-sea ecosystems. *Nature* **454**: 1084–1087.

Danovaro, R., Fonda Umani, S., and Pusceddu, A. (2009). Climate change and the potential spreading of marine mucilage and microbial pathogens in the Mediterranean Sea. *PLoS ONE* **4**(9): e7006.

Danovaro, R., Company, J. B., Corinaldesi, C., *et al.* (2010). Deep-sea biodiversity in the Mediterranean Sea: The known, the unknown, and the unknowable. *PLoS ONE* **5**: e11832.

Danovaro, R., Corinaldesi, C., Dell'Anno, A., *et al.* (2011). Marine viruses and global climate change. *FEMS Microbiology Reviews* **35**(6): 993–1034.

Degobbis, D., Fonda Umani, S., Franco, P., *et al.* (1995). Changes in the northern Adriatic ecosystem and hypertrophic appearance of gelatinous aggregates. *Science of the Total Environment* **165**: 43–58.

Doney, S. C. 2010. The growing human footprint on coastal and open-ocean biogeochemistry. *Science* **328**: 1512.

Doney, S. C. and Schimel, D. S. (2007). Carbon and climate system coupling on timescales from the Precambrian to the Anthropocene. *Annual Review of Environmental Resources* **32**: 31–66.

Duarte, C. M. (2002). The future of seagrass meadows. *Environmental Conservation* **29**(2): 192–206.

Dulvy, N. K., Sadovy, Y., and Reynolds, J. D. (2003). Extinction vulnerability in marine populations. *Fish and Fisheries* **4**: 25–64.

Fabry, V. J., Seibel, B. A., Feely, R. A., and Orr, J. C. (2008). Impacts of ocean acidification on marine fauna and ecosystem processes. *ICES Journal of Marine Science* **65**: 414–432.

Fahey, S. J. and Garson, M. J. (2002). Geographic variation of natural products of tropical nudibranch *Asteronotus cespitosus*. *Journal of Chemical Ecology* **28**: 1773–1785.

Fahrbach, E., Hoppema, M., Rohardt, G., *et al.* (2011). Warming of deep and abyssal water masses along the Greenwich meridian on decadal timescales: The Weddell gyre as a heat buffer. *Deep-Sea Research Part II* **58**: 2509–2523.

Fields, P. A., Rudomin, E. L., and Somero, G. N. (2006). Temperature sensitivities of cytosolic malate dehydrogenases from native and invasive species of marine mussels (genus Mytilus): sequences–function linkages and correlations with biogeographic distribution. *Journal of Experimental Marine Biology and Ecology* **209**: 656–667.

Fonda Umani, S., Milani, L., Borme, D., *et al.* (2005). Inter-annual variations of planktonic food webs in the northern Adriatic Sea. *Science of the Total Environment* **353**: 218–231.

Gazeau, F., Quiblier, C., Jansen, J. M., *et al.* (2007). Impact of elevated CO_2 on shellfish calcification. *Geophysical Research Letters* **34**: L07603. doi: 10.1029/2006GL028554.

Genner, M. J., Sims, D. W., Wearmouth, V. J., *et al.* (2004). Regional climatic warming drives long-term community changes of British marine fish. *Proceedings of Royal Society Part B – Biological Sciences* **271**: 655–661.

Gilman, E. L., Ellison, J., Duke, N. C., and Field, C. (2008). Threats to mangroves from climate change and adaptation options: A review. *Aquatic Botany* **89**: 237–250.

Gorman, D., Russell, B. D., and Connell, S. D. (2009). Land-to-sea connectivity: linking human-derived terrestrial subsidies to subtidal habitat change on open rocky coasts. *Ecological Applications* **19**: 1114–1126.

Graham, N. A. J., Wilson, S. K., Jennings, S., *et al.* (2006). Dynamic fragility of oceanic coral reef ecosystems. *Proceedings of the National Academy of Sciences of the United States of America* **103**: 8425–8429.

Gregg, W. W., Conkright, M. E., Ginoux, P., *et al.* (2003). Ocean primary production and climate: Global decadal changes. *Geophysics Research Letters* **30**: 1809.

Grehan, A. J., van den Hove, S., Armstrong, C. W., *et al.* (2009). HERMES: Promoting ecosystem-based management and the sustainable use and governance of deep water resources. *Oceanography* **22**(1): 154–166.

Hansen, J., Sato, M., Ruedy, R., *et al.* (2007). Dangerous human-made interference with climate: A GISS model E study. *Atmospheric Chemistry and Physics* **7**: 2287–2312.

Harley, C. D. G, Hughes, A. R., Hultgren, K. M., *et al.* (2006). The impacts of climate change in coastal marine systems. *Ecology Letters* **9**: 228–241.

Hawkins, S. J., Moore, P. J., Burrows, M. T., *et al.* (2008). Complex interactions in a rapidly changing world: Responses of rocky shore communities to recent climate change. *Climate Research* **37**: 123–133.

Hays, G. C., Richardson, A. J., and Robinson, C. (2005). Climate change and marine plankton. *Trends in Ecology and Evolution* **20**(6): 337–344.

Helmuth, B., Mieszkowska, N., Moore, P., and Hawkins, S. J. (2006). Living on the edge of two changing worlds: Forecasting the responses of rocky intertidal ecosystems to climate change. *Annual Review Ecology, Evolution, and Systematics* **37**: 373–404.

Hoegh-Guldberg, O. (1999). Climate change, coral bleaching and the future of the world's coral reefs. *Marine and Freshwater Research* **50**: 839–866.

Hoegh-Guldberg, O. (2004). Coral reefs in a century of rapid environmental change. *Symbiosis* **37**: 1–31.

Hoegh-Guldberg, O. and Bruno, J. F. (2010). The impact of climate change on the world's marine ecosystems. *Science* **328**: 1523–1528.

Hogg, R. C. (2006). Novel approaches to pain relief using venom-derived peptides. *Current Medicinal Chemistry* **13**: 3191–3201.

Hughes, T. P., Baird, A. H., Bellwood, D. R., *et al.* (2003). Climate change, human impacts, and the resilience of coral reefs. *Science* **301**(5635): 929–933.

Hutchings, P. A. and Salvat, B. (2000). The Indian Ocean to the Pacific: French Polynesia. In: *Seas at the Millennium: An Environmental Evaluation.* Sheppard, C. (ed.). Amsterdam: Elsevier, pp. 813–826.

Hutchings, P. A., Ahyong, S., Byrne, M., *et al.* (2007). Vulnerability of benthic invertebrates of the Great Barrier Reef to climate change. In: *Climate Change and the Great Barrier Reef: A Vulnerability Assessment.* Johnson, J. E. and Marshall, P. A. (eds.).

Townsville: Great Barrier Reef Marine Park Authority and Australian Greenhouse Office, pp. 309–356.

IPCC (2001) *Climate Change 2001: The Scientific Basis. Contribution of Working Group I to the Third Assessment Report of the Intergovernmental Panel on Climate Change.* (2001) Houghton, J. T., Ding, Y., Griggs, D. J., Noguer, M., van der Linden, P. J., Dai, X., Maskell, K., and Johnson, C. A. (eds.). Cambridge and New York: Cambridge University Press, 881 pp.

IPCC (2007) *Climate Change 2007: The Physical Science Basis. Contribution of Working Group I to the Fourth Assessment Report of the Intergovernmental Panel on Climate Change.* Solomon, S., Qin, D., Manning, M., Chen, Z., Marquis, M., Averyt, K. B., Tignor, M., and Miller, H. L. (eds.). Cambridge: Cambridge University Press, 996 pp.

IUCN (2010). The Millennium Ecosystem Assessment. IUCN Policy Paper. Gland: IUCN, 3 pp.

Ivanov, V. V., Shapiro, G. I., Huthnance, J. M., *et al.* (2004). Cascades of dense water around the world ocean. *Progress in Oceanography* **60**: 47–98.

Jones, G. P., McCormick, M. I., Srinivasan, M., and Eagle, J. V. (2004). Coral decline threatens fish biodiversity in marine reserves. *Proceedings of the National Academy of Sciences of the United States of America* **101**: 8251–8253.

Kathiresan, K. and Bingham, B. L. (2001). Biology of mangroves and mangrove ecosystems. *Advances in Marine Biology* **40**: 81–251.

Keeling, R. F. and Garcia, H. (2002). The change in oceanic O_2 inventory associated with recent global warming. *Proceedings of the National Academy of Sciences of the United States of America* **99**(12): 7848–7853.

Keeling, R. F., Körtzinger, A., and Gruber, N. (2010). Ocean deoxygenation in a warming world. *Annual Review of Marine Science* **2**: 199–229.

Khatiwala, S., Primeau, F., and Hall, T. (2009). Reconstruction of the history of anthropogenic CO_2 concentrations in the ocean. *Nature* **462**: 346–349.

Kirchman, D. L., Anxelu, X., Morán, G., and Ducklow, H. (2009). Microbial growth in the polar oceans – role of temperature and potential impact of climate change. *Nature Reviews* **7**: 451–459.

Kristensen, E., Bouillon, S., Dittmar, T., and Marchand, C. (2008). Organic carbon dynamics in mangrove ecosystems: A review. *Aquatic Botany* **89**: 201–219.

Kuffner, I. B., Andersson, A. J., Jokiel, P. L., *et al.* (2008). Decreased abundance of crustose coralline algae due to ocean acidification. *Nature Geoscience* **1**: 114–117.

Laws, E. A. (2004). Export flux and stability as regulators of community composition in pelagic marine biological communities: Implications for regime shifts. *Progress in Oceanography* **60**: 343–354.

Levin, L. A. (2010). Anaerobic metazoans: No longer an oxymoron. *BMC Biology* **8**: 31.

Lotze, H. K., Lenihan, H. S., Bourque, B. J., *et al.* (2006). Depletion, degradation, and recovery potential of estuaries and coastal seas. *Science* **312**(5781): 1806–1809.

Lough, J. M. (2008). 10th anniversary review: A changing climate for coral reefs. *Journal of Environmental Monitoring* **10**: 21–29.

Lovelock, C. E. and Ellison J. C. (2007). Vulnerability of mangroves and tidal wetlands of the Great Barrier Reef to climate change. In: *Climate Change and the Great Barrier Reef: A Vulnerability Assessment*. Johnson, J. E. and Marshall, P. A. (eds.). Townsville: Great Barrier Reef Marine Park Authority and Australian Greenhouse Office, pp. 237–269.

Loya, Y., Sakai, K., Yamazato, K., *et al.* (2001). Coral bleaching: The winners and the losers. *Ecology Letters* **4**: 122–131.

Lüthi, D., Le Floch, M., Bereiter, B., *et al.* (2008). High-resolution carbon dioxide concentration record 650,000–800,000 years before present. *Nature* **453**: 379–382.

Manson, R. A., Loneragan, N. R., Skilleter, G. A., and Phinn, S. R. (2005). An evaluation of the evidence for linkages between mangroves and fisheries: A synthesis of the literature and identification of research directions. *Oceanography and Marine Biology: An Annual Review* **43**: 483–513.

Martin, D., Bertasi, F., Colangelo, M. A., *et al.* (2005). Ecological impact of coastal defence structures on sediment and mobile fauna: evaluating and forecasting consequences of unavoidable modifications of native habitats. *Ecological Engineering* **52**: 1027–1051.

Masuda, S., Awaji, T., Sugiura, N., *et al.* (2010). Simulated rapid warming of abyssal North Pacific waters. *Science* **329**: 319–322.

McClain, C. R., Rex, M. A., and Jabbour, R. (2005). Deconstructing bathymetric body size patterns in deep-sea gastropods. *Marine Ecology Progress Series* **297**: 181–187.

McGinty, N., Power, A. M., and Johnson, M. P. (2011). Variation among northeast Atlantic regions in the responses of zooplankton to climate change: Not all areas follow the same path. *Journal of Experimental Marine Biology and Ecology* **400**: 120–131.

McGowan, J. A., Cayan, D. R., and LeRoy, M. D. (1998). Climate-ocean variability and ecosystem response in the northeast Pacific. *Science* **281**: 210–217.

Mieszkowska, N., Genner, M. J., Hawkins, S. J., and Sims, D. W. (2009). Effects of climate change and commercial fishing on Atlantic Cod *Gadus morhua*. *Advances in Marine Biology* **56**: 213–273.

Mumby, P. J., Chisholm, J., Edwards, A., *et al.* (2001). Unprecedented bleaching-induced mortality in Porites spp. at Rangiroa Atoll, French Polynesia. *Marine Biology* **139**(1): 183–189.

Munday, P. L., Jones, G. P., Sheaves, M., *et al.* (2007). Vulnerability of fishes of the Great Barrier Reef to climate change. In: *Climate Change and the Great Barrier Reef: A Vulnerability Assessment*. Johnson, J. E. and Marshall, P. A. (eds.). Townsville: Great Barrier Reef Marine Park Authority and Australian Greenhouse Office, pp. 357–391.

Myers, R. A. and Worm, B. (2003). Rapid worldwide depletion of predatory fish communities. *Nature* **423**: 280–283.

Nagelkerken, I., Blaber, S. J. M., Bouillon, S., *et al.* (2008). The habitat function of mangroves for terrestrial and marine fauna: A review. *Aquatic Botany* **89**: 155–185.

Occhipinti-Ambrogi, A. and Galil, B. (2010). Marine alien species as an aspect of global change. *Advances in Oceanography and Limnology* **1**(1): 199–218.

Orth, R. J., Carruthers, T. J. B., Dennison, W. C., *et al.* (2006). Global crisis for seagrass ecosystems. *BioScience* **56**(12): 987–996.

Page, M., West, L., Northcote, P., *et al.* (2005). Spatial and temporal variability of cytotoxic metabolites in populations of the New Zealand sponge *Mycale hentscheli*. *Journal of Chemical Ecology* **31**: 1161–1174.

Pandolfi, J. M., Jackson, J. B. C., Baron, N., *et al.* (2005). Are U.S. coral reefs on the slippery slope to slime? *Science* **307**: 1725–1726.

Passow, U. (2004). Switching perspectives: Do mineral fluxes determine particulate organic carbon fluxes or vice versa? *Geochemistry Geophysics and Geosystem* **5**: Q04002.

Patz, J. A., Campbell-Lendrum, D., Holloway, T., and Foley, J. A. (2005). Impact of regional climate change on human health. *Nature* **438**: 310–317.

Philippart, C. J. M., Anadón, R., Danovaro, R., *et al.* (2011). Impacts of climate change on European marine ecosystems: Observations, expectations and indicators. *Journal of Experimental Marine Biology and Ecology* **400**: 52–69.

Poloczanska, E. S., Babcock, R. C., Butler, A., *et al.* (2007). Climate change and Australian marine life. *Oceanography and Marine Biology: An Annual Review* **45**: 407–478.

Pounds, J. A., Bustamante, M. R., Coloma, L. A., *et al.* (2006). Widespread amphibian extinctions from epidemic disease driven by global warming. *Nature* **439**: 161–167.

Precali, R., Giani, M., Marini, M., *et al.* (2005). Mucilaginous aggregates in the northern Adriatic in the period 1999–2002: Typology and distribution. *Science of the Total Environment* **353**: 10–23.

Przeslawski, R., Ahyong, S., Byrne, M., *et al.* (2008). Beyond corals and fish: The effects of climate change on noncoral benthic invertebrates of tropical reefs. *Global Change Biology* **14**: 2773–2795.

Pusceddu, A., Mea, M., Gambi, C., *et al.* (2010). Ecosystem effects of dense water formation on deep Mediterranean Sea ecosystems: An overview. *Advances in Oceanography and Limnology* **1**(1): 67–83.

Ramirez-Llodra, E., Brandt, A., Danovaro, R., De Mol, B., *et al.* (2010). Deep, diverse and definitely different: Unique attributes of the world's largest ecosystem. *Biogeosciences* **7**: 2851–2899.

Ramirez-Llodra, E., Tyler, P. A., Baker, M. C., *et al.* (2011). Man and the last great wilderness: Human impact on the deep sea. *PLoS ONE.* **6**(7): e22588.

Richardson, A. J., Bakun, A., Hays, G. C., and Gibbons, M. J. (2009). The jellyfish joyride: Causes, consequences and management responses to a more gelatinous future. *Trends in Ecology and Evolution* **24**: 312–322.

Rinaldi, A., Vollenweider, R. A., Montanari, G., *et al.* (1995). Mucilages in Italian seas: The Adriatic and Tyrrhenian Seas, 1988–1991. *Science of the Total Environment* **165**: 165–183.

Roessig, J. M., Woodley, C. M., Cech, J. J., and Hansen, L. J. (2004). Effects of global climate change on marine and estuarine fishes and fisheries. *Reviews in Fish Biology and Fisheries* **14**: 251–275.

Royal Society (2005). *Ocean Acidification Due to Increasing Atmospheric Carbon Dioxide*. Policy Document 12/05, The Royal Society, London, 60 pp.

Ruhl, H. A. (2007). Abundance and size distribution dynamics of abyssal epibenthic megafauna in the northeast Pacific. *Ecology* **88**: 1250–1262.

Ruhl, H. A. and Smith K. L., Jr. (2004). Shifts in deep-sea community structure linked to climate and food supply. *Science* **305**: 513–515.

Russell, B. D. (2007). Effects of canopy-mediated abrasion and water flow on the early colonisation of turf-forming algae. *Marine Freshwater Research* **58**: 657–665.

Russell, B. D., Thompson, J. I., Falkenberg, L. J., and Connell, S. D. (2009). Synergistic effects of climate change and local stressors: CO_2 and nutrient driven change in subtidal rocky habitats. *Global Change Biology* **15**: 2153–2162.

Sala, O. E., Chapin, F. S. III, Armesto, J. J., *et al.* (2000). Global biodiversity scenarios for the year 2100. *Science* **287**: 1770–1774.

Sarmiento, J. L., Hughes, T. M. C., Stouffer, R. J., and Manabe, S. (1998). Simulated response of the ocean carbon cycle to anthropogenic climate warming. *Nature* **393**: 245–249.

Saunders, M. and Metaxas, A. (2007). Temperature explains settlement patterns of the introduced bryozoan *Membranipora membranacea* in Nova Scotia, Canada. *Marine Ecology Progress Series* **344**: 95–106.

Schindler, D. W. (2006). Recent advances in the understanding and management of eutrophication. *Limnology and Oceanography* **51**: 356–363.

Sheppard, C., Dixon, D. J., Gourlay, M. J., *et al.* (2005). Coral mortality increases wave energy reaching shores protected by reef flats: Examples from the Seychelles. *Estuarine, Coastal and Shelf Science* **64**: 223–234.

Short, F. T. and Neckles, H. A. (1999). The effects of global climate change on seagrasses. *Aquatic Botany* **63**: 169–196.

Smith, C. R., De Leo, F. C., Bernardino, A. F., *et al.* (2008). Abyssal food limitation, ecosystem structure and climate change. *Trends in Ecology and Evolution* **23**: 518–528.

Smith, K. L., Jr., Ruhl, H. A., Bett, B. J., *et al.* (2009). Climate, carbon cycling, and deep-ocean ecosystems. *Proceedings of the National Academy of Sciences of the United States of America* **106**(46): 19211–19218.

Smith, L., Gilmour, J., and Heyward, A. (2007). Resilience of coral communities on an isolated system of reefs following catastrophic mass-bleaching. *Coral Reefs* **27**: 197–205.

Snedaker, S. C. (1995). Mangroves and climate change in the Florida and Caribbean region: Scenarios and hypotheses. *Hydrobiologia* **295**: 43–49.

Spalding, M., Taylor, M., Ravilious, C., *et al.* (2003). Global overview: The distribution and status of seagrasses. In: *World Atlas of Seagrasses*. Green, E. P. and Short, F. T. (eds.). Cambridge and Berkeley: UNEP World Conservation Monitoring Centre and University of California Press, pp. 5–26.

Stachowicz, J. J., Terwin, J. R., Whitlatch, R. B., and Osman, R. W. (2002). Linking climate change and biological invasions: Ocean warming facilitates nonindigenous species invasions. *Proceedings of the National Academy of Sciences of the United States of America* **99**: 15497–15500.

Steneck, R. S., Graham, M. H., Bourque, B. J., *et al.* (2002). Kelp forest ecosystems: Biodiversity, stability, resilience and future. *Environmental Conservation* **29**: 436–459.

Stramma, L., Johnson, G. C., Sprintall, J., and Mohrholz, V. (2008). Expanding oxygen-minimum zones in the tropical oceans. *Science* **320**: 655–658.

Suttle, C. A. (2005). Viruses in the sea. *Nature* **437**: 356–361.

Suttle, C. A. (2007). Marine viruses – major players in the global ecosystem. *Nature Review Microbiology* **5**: 801–812.

Thomas, C. D., Cameron, A., Green, R. E., *et al.* (2004). Extinction risk from climate change. *Nature* **427**: 145–148.

Tittensor, D. P., Mora, C., Jetz, W., *et al.* (2010). Global patterns and predictors of marine biodiversity across taxa. *Nature* **466**: 1098–1101.

Trenberth, K. E. and Josey, S. A. (2007). Observations: Surface and atmospheric climate change. In: *Climate Change 2007: The Physical Science Basis. Contribution of Working Group I to the Fourth Assessment Report of the Intergovernmental Panel on Climate Change*. Solomon, S., Qin, D., Manning, M., Chen, Z., Marquis, M., Averyt, K. B., Tignor, M., and Miller, H.L. (eds.). Cambridge: Cambridge University Press, pp. 235–236.

Twilley, R. R. (1995). Properties of mangrove ecosystems related to the energy signature of coastal environments. In: *Maximum Power: The Ideas and Applications of H.T. Odum*. Hall, C. A. S. (ed.). Boulder: University of Colorado Press, pp. 43–62.

UNEP–WCMC (2006). *In the Front Line: Shoreline Protection and Other Ecosystem Services from Mangroves and Coral Reefs*. Cambridge: UNEP–WCMC, 33 pp.

Vaquer-Sunyer, R. and Duarte, C. M. (2008). Thresholds of hypoxia for marine biodiversity. *Proceedings of the National Academy of Sciences of the United States of America USA* **105**: 15452–15457.

Walther, G.-R., Burga, C. A., and Edwards, P. J. (eds.) 2001. *Fingerprints of Climate Change: Adapted Behavior and Shifting Species Ranges.* New York: Kluwer Academic/Plenum, 333 pp.

Walther, G.-R., Post, E., Convey, P., *et al.* (2002). Ecological responses to recent climate change. *Nature* **419**: 389–395.

Walther, G.-R., Roques, A., Hulme, P. E., *et al.* (2009). Alien species in a warmer world: Risks and opportunities. *Trends in Ecology and Evolution* **24**(12): 686–693.

Wernberg, T., Thomsen, M. S., Tuya, F., *et al.* (2010). Decreasing resilience of kelp beds along a latitudinal temperature gradient: Potential implications for a warmer future. *Ecology Letters* **13**: 685–694.

Whitney, F. A., Freeland, H. J., and Robert, M. (2007). Persistently declining oxygen levels in the interior waters of the eastern sub-Arctic Pacific. *Progress in Oceanography* **75**: 179–199.

Wilkie, M. L. and Fortuna, S. (2003). *Status and Trends in Mangrove Area Extent Worldwide.* Forest Resources Assessment Working Paper No. 63. Rome: FAO.

Wilkinson, C. R. (ed.) (2000). *Status of Coral Reefs of the World.* Townsville: Australian Institute of Marine Science.

Willis, B. L., van Oppen, M. J. H., Miller, D. J., *et al.* (2006). The role of hybridization in the evolution of reef corals. *Annual Review of Ecology Evolution and Systematics* **37**: 489–517.

Wilson, S. K., Graham, N. A. J., Pratchett, M. S., *et al.* (2006). Multiple disturbances and the global degradation of coral reefs: Are reef fishes at risk or resilient? *Global Change Biology* **12**: 2220–2234.

Wood, H. L., Spicer, J. I., and Widdicombe, S. (2008). Ocean acidification may increase calcification rates, but at a cost. *Proceedings of the Royal Society Part B: Biological Sciences* **275**(1644): 1767–1773.

Worm, B. and Lotze, H. K. (2009). Changes in marine biodiversity as an indicator of climate and global change. In: *Climate and Global Change: Observed Impacts on Planet Earth.* Letcher, T. (ed.). Amsterdam: Elsevier, pp. 263–279.

Worm, B., Barbier, E. B., Beaumont, N., *et al.* (2006). Impacts of biodiversity loss on ocean ecosystem services. *Science* **314**: 787–790.

5

A new perspective on changing Arctic marine ecosystems: panarchy adaptive cycles in pan-Arctic spatial and temporal scales

HENRY P. HUNTINGTON, EDDY CARMACK, PAUL WASSMANN, FRANCIS WIESE, EVA LEU, AND ROLF GRADINGER

5.1 Introduction

For the first time in recent history, a new ocean is opening (Kinnard *et al.*, 2011). The retreat and thinning of the summer sea ice in the Arctic Ocean are perhaps the most visible indicators of the major physical changes underway in the Arctic Ocean (Kwok *et* al., 2009; Walsh, 2013). While rates and even causes of ice loss remain under debate (Carmack and Melling, 2011), further loss of sea ice appears inevitable, and it is likely that within only a few decades the Arctic will see a mostly ice-free summer. The biological implications of this change depend, to a large extent, upon the interplay between altered abiotic conditions (e.g. temperature, salinity, stratification, nutrient availability, wind, underwater light, climate, etc.) and the response of organisms on all trophic levels changing, for example, patterns in primary production, respiration, and diversity (e.g. timing, magnitude, and quality of algal blooms, microbial processes, etc.). Changes in these physical/biological interactions will occur across a variety of spatial and temporal scales and may be mitigated or strengthened based on widely varying rates of evolutionary adaptation. Studies, monitoring activities, and adaptive experiments that are focused on the often non-linear biophysical changes that are occurring and the linkages among them will likely offer new insights into the mechanisms, trajectories, and dynamics of ecological and evolutionary change in the Arctic and elsewhere (e.g. Overpeck *et al.*, 2005; Carmack *et al.*, 2012; McLaughlin *et al.*, 2011; Sunday *et al.*, 2011; Wassmann *et al.*, 2011; Duarte *et al.*, 2012). Here we briefly discuss some of the basic knowns about the interactions of physical change and the corresponding biological response in the Arctic Ocean, identify significant unknowns, and propose a new structural framework to guide further Arctic marine research in policy-relevant directions.

A novel way to view the interaction among various physical and biological changes and their social relevance is through the systems theory perspective of 'panarchy' proposed by Gunderson and Holling (2001). Panarchy is an interdisciplinary approach in which structures, scales, and linkages of complex-adaptive systems, including those of nature (e.g. the ocean) and of humans (e.g. economics), as well as combined social-ecological systems (e.g. institutions that govern natural resource use), are mapped across multiple space and time-scales in continual and interactive adaptive cycles of growth, accumulation, restructuring, and renewal. An essential feature of panarchy – in contrast with hierarchical systems – is to recognize the interplay between persistence and change, and between predictable and unpredictable (e.g. Walker and Salt, 2006, 2012). Importantly, conceptual models based on panarchy can incorporate, from the start, the social (human) element.

In complex-adaptive systems the dynamics of a given system at a given scale are generally dominated by a small number of key internal variables that are forced by one or more external variables (Levin, 1992; Gunderson and Holling, 2001; Walker *et al.*, 2012). The stability of such a system is characterized by its resilience, i.e. its capacity to absorb disturbance and reorganize while undergoing change, so as to retain essentially similar function, structure, identity, and feedbacks (Walker and Salt, 2012). It is in the capacity of a system to cope with pressures and adversities such as exploitation, warming, governance restrictions, competition, etc. that resilience embraces human and natural systems as complex entities continually adapting through cycles of change. The 'adaptive cycle' model of resilience (Gunderson and Holling, 2001) describes this cycle of change (Figure 5.1). Under various forms of forcing any system will have at the start a building, accumulation, or exploitation phase that leads to an increasing build-up of energy, wealth, or resources (cf. Figure 5.1). Over time, the capacity of the system to absorb an increase in forcing is reached, marking the locked-in or conservation phase (K). If forcing goes further, a tipping point is reached, followed by a breakdown in the sustainability of the system (Ω). Once the forcing ceases, the system can re-establish and reorganize itself and enter a new cycle.

However, when tipping points are also points of no return (e.g. bifurcation points), more complex dynamics can emerge (Figure 5.2). When a point of no return is reached (even if forcing is reduced or ceases) the system will enter the destruction phase Ω and then switch to a new state or regime with its own dynamics. This new regime can then develop around its own adaptive cycle, which may also reach a different point of no return (second bifurcation point) and, theoretically, switch the system back to the original regime or on to yet another new one.

To apply the panarchy frameworks to ecosystem changes in the Arctic Ocean, we consider a series of spatio-temporal scales or domains individually and through the relationships among them. Here it is important to keep in mind that the

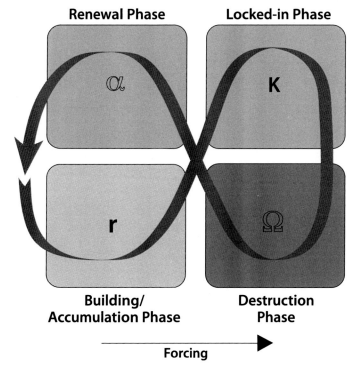

Figure 5.1 Conceptual model of the adaptive cycle, illustrating how a system, subjected to forcing, develops through two principal phases: r and K. Passing a tipping point may result in the system facing a destruction phase Ω before it renews itself, if the forcing decreases, in the α phase. Reproduced from Wassmann and Lenton (2012), which was redrawn from Gunderson and Holling (2001). A black and white version of this figure will appear in some formats. For the colour version, please refer to the plate section.

complex-adaptive system model is meant to capture the interacting state variables and relationships among them at a given (focal) scale, and not simply all possible variables occurring at that scale. The system changes as a result of both of these internal interactions and its external drivers. A resilient ecosystem, then, exhibits relatively stable functional relationships and feedbacks within and across scales, whereas ecosystem instability (regime shift) involves the alteration of such functional relationships within and across domains, often resulting in non-linear and unpredictable change (Figure 5.2).

5.2 Scales and domains in Arctic systems

In the case of pan-Arctic ecosystems (cf. overviews in Wassmann, 2006, 2011), we define four linked spatio-temporal domains that help illustrate the mechanisms of

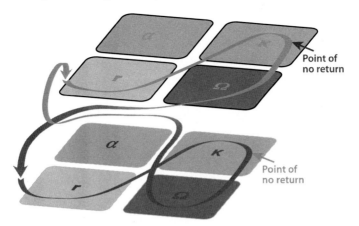

Figure 5.2 Altered conceptual model of adaptive cycles where the tipping point is now a point of no return. The original regime (lower half of the figure) can switch over to a new regime (upper half), or the system can stay within the original regime as in Figure 5.1. There can exist a multitude of regimes with potential switches between them. Here we only display two to illustrate the basic principle. Reproduced from Wassmann and Lenton (2012). A black and white version of this figure will appear in some formats. For the colour version, please refer to the plate section.

change and the difficulty of predicting the nature of the resulting new state of the Arctic marine ecosystem (see Table 5.1).

Our domain choices are, we admit, not exclusive, and other groupings may be equally valid; nonetheless, our four domains suffice to demonstrate how the concept of panarchy can be applied in the case of the rapidly changing Arctic marine environment and how panarchy is useful for assessing ecosystem structure and reaction to impacts. From this set we select one focal scale at a time for more detailed discussion. In discussing domains below, it is important to remember that Arctic marine ecosystems are strongly affected by advective processes and reflect combined pelagic and sea ice (moving) and benthic (stationary) elements and processes (cf. Carmack and Wassmann, 2006; Grebmeier *et al.*, 2006; Nelson *et al.*, 2009). We also take it as given that human activity influences all scales: we live in the Anthropocene (Crutzen and Stoermer, 2000; Zalasiewicz *et al.*, 2010; Zalasiewicz *et al.*, 2011).

The largest spatio-temporal scale is that of the Arctic Ocean itself, set within the global climate system (Domain 1 in Figure 5.3, Table 5.1). Here the extent of summer sea ice and the thickness of year-round ice have been decreasing across the Arctic Ocean for decades, a trend that, assuming today's physical forcing persists, will, within decades, result in a mostly ice-free summer season. Within this pan-Arctic and decadal scale domain, slow-variable changes in forcing (e.g. global warming, Arctic amplification, and increasing temperatures in inflowing waters

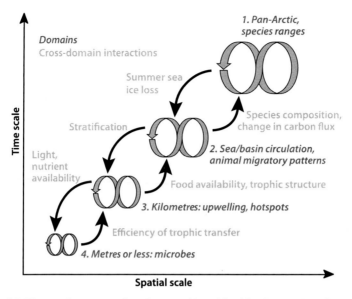

Figure 5.3 The spatio-temporal scales considered in this chapter (numbered 1–4), along with examples of cross-scale interactions. A black and white version of this figure will appear in some formats. For the colour version, please refer to the plate section.

from the sub-Arctic Atlantic and Pacific oceans (Carmack and Melling, 2011)) are reducing the extent, thickness, and internal strength of sea ice, shifting the central Arctic from a perennial ice zone (with multi-year ice) to a widening seasonal ice zone (Kinnard *et al.*, 2008). With the resulting decrease in internal ice strength and compactness, floes are able to drift more rapidly under wind forcing, with attendant effects on ocean surface currents (cf. Kwok *et al.*, 2009) and the transport and distribution of planktonic organisms. Biologically, this physical change means the loss of an entire biome associated with perennial ice (Michel *et al.*, 2012) and a northward extension of the seasonal ice season that now dominates the low Arctic or sub-Arctic. Loss of sea ice at this scale and attendant open water also have implications for atmospheric–ocean interaction, with possible implications for global weather and climatic patterns (e.g. Overland *et al.*, 2011; Francis and Vavrus, 2012), and Arctic circulation.

Biophysical implications of the loss of summer sea ice take place at the regional and annual scale (Domain 2 in Figure 5.3, Table 5.1), wherein the individual shelf regions and adjacent basins in the Arctic Ocean and associated seasonal patterns of today's primary production and food-web dynamics are found. The direction and extent of change at this scale will depend on the individual characteristics of each specific shelf and basin region (cf. Carmack and Macdonald, 2002; Carmack and Wassmann, 2006). Collectively, the projected changes include surface layer

Table 5.1 *A few examples of key characteristics of the current stable state Arctic marine ecosystem in the four domains as discussed and the changes likely to occur within and across domains in the near future. The changes across domains are the hallmark of instability in the panarchy framework.*

Today	Future (within domain)	Future (across domains)
Cold water, perennial sea ice (Domain 1)	Warmer water, ice-free summers; winters remain cold and dark	Change in circulation patterns (Domain 2), change in productivity patterns (Domain 3), change in species composition (Domain 4)
Extreme seasonal changes, but rather predictable annual cycles (Domain 2)	Extreme seasonal changes, greater interannual variability, greater stratification	Change in ice patterns (Domain 1), shifts in feeding and migratory aggregations (Domain 3), increase in flexible species at the expense of specialists (Domain 4)
Long-term adaptation of ecosystem (and key players) to short productive peak during spring (Domain 3)	Greater opportunity for flexible species, greater variation in abundance	Changes in distribution of species and use thereof by humans and other predators (Domain 2), changes in downward pressure on primary producers (Domain 4)
Highly efficient transfer of metabolic energy through a few, highly adapted specialists (e.g. *Calanus*, Arctic cod (*Boreogadus saida*)) (Domain 4)	Decrease in size of primary producers, resulting in less efficient energy transfer to upper trophic levels	Less food available for upper trophic levels (Domain 3), shifts in distribution of species and the use thereof (Domain 2)

warming and increased stratification from melting ice and river run-off. Light penetration will increase in the basin due to removal of ice and emergence of melt ponds (Nicolaus *et al.*, 2012) and decrease in shelf areas owing to greater turbidity on shallow shelves, resulting from wind mixing, river inputs of sediment, and increased permafrost melting and coastal erosion. Likewise, nutrient availability will increase in some regions due to upwelling (particularly along ice-free shelf edges and off estuaries) and decrease in other regions due to increased stratification as a result of ice melt, river inputs, and warming. Ocean acidification will increase due to increasing atmospheric CO_2 levels, increased areas of open water for air–sea exchange, high solubility of CO_2 in cold waters, reduced alkalinity associated with sea ice melt, and increased upwelling of deeper and less $CaCO_3$-saturated water (Yamamoto-Kawai *et al.*, 2009, 2011; Mathis *et al.*, 2011).

The next spatio-temporal scale (Domain 3 in Figure 5.3, Table 5.1) to consider is that of mesoscale (of order 10–20 km) oceanographic processes and forcings that exist on seasonal and sub-seasonal time-scales and which, collectively, differ between the shallow shelf and deep basin domains. Mechanisms such as wind and topographically driven upwelling, tidal mixing through passages or over topography, and frontal zone convergence can act to first supply nutrients of phytoplankton growth and then produce aggregations of zooplankton to facilitate foraging by predators (Carmack and Chapman, 2003; Rogachev *et al.*, 2008; Hannah *et al.*, 2009). For example, the ice edge where light meets nutrient-rich stratified waters in spring and early summer can be a highly productive zone, allowing for highly efficient grazing and/or enhanced vertical export, thus fuelling both pelagic and/or benthic food webs (e.g. Reigstad *et al.*, 2011).

The scales of these physical mechanisms thus impose temporal and spatial foraging scales for biota across a highly heterogeneous, though patterned, landscape. Ice edge blooms will remain a feature and may develop over the entire Arctic Ocean, but their relative importance will depend on the locally available supply of nutrients and on grazing intensity. The bloom is likely to occur earlier than it has in the past decade as ice is thinner and retreats more quickly, and incident light reaches the water column sooner (as noted in Loeng *et al.*, 2005). Melt ponds on multi-year ice floes that form sooner and freeze later will also alter the underwater light climate. Consequently, extensive under-ice pelagic blooms will become more common (Arrigo *et al.*, 2012; Mundy *et al.*, 2013). There will be, however, regional variability, depending on snow cover. The timing of ice-associated versus pelagic algal blooms relative to each other will potentially change under such conditions, decreasing the time lag between the two bloom events (Ji *et al.*, 2013). This may have far-reaching consequences for the successful recruitment, e.g. of the key pelagic grazer in Arctic shelf seas, *Calanus glacialis*, which is able to take advantage of both blooms to achieve high egg production (females grazing on sea ice algae), and optimal growth conditions for the offspring (nauplii and copepodites feeding on phytoplankton) (Søreide *et al.*, 2010; Leu *et al.*, 2011).

Changes in currents, wind mixing, and upwelling in this domain may affect the intensity and location of pelagic and benthic biological hot spots. Fish, birds, and marine mammals (as well as people who use those resources) seek out such hot spots, rather than feeding on prey dispersed over a larger spatial scale. However, marine birds that are tied to colonies during the breeding season may not be able to reach new hot spots that are farther away if they lose nearby ones. Some degree of re-organization at this scale is common within and between years, as sites of production and aggregation move, for example as the ice edge shifts its relative position or as shifting wind patterns redistribute fronts and convergence zones (e.g. Wassmann *et al.*, 2010). Although hot spots have been studied in the past, it is not clear how much the population dynamics is impacted by the presence, absence, timing, location, or intensity of hot-spots, or even how animals are able to locate these shifting areas of high productivity. As such, our understanding is far too limited to allow for reliable predictions about the potential of predators to adapt to such changes.

At the smallest spatio-temporal scale considered here (Domain 4 in Figure 5.3, Table 5.1) are the physical structures and processes that drive and regulate biological rates (e.g. scales of turbulence and convection, Langmuir cells, microlayers, melt ponds, brine channels in sea ice, etc.). To each of these abiotic temporal–spatial scales there exist corresponding biological scales suited to exploit the environment. For example, microbes and their life cycles occur in short time periods and small areas such that the adaptive and evolutionary responses will be quickest and most dramatic, as those species that can exploit new conditions rapidly outcompete those that cannot, altering the composition at the base of the food web (Li *et al.*, 2009). There is considerable variability at this scale in response to the variability of physical conditions in the Arctic, but directional change will lead to further departures from previous ranges of microbial distribution, abundance, function, and quality. The microbial food web will also continue to compete with phytoplankton for resources such as nutrients and carbon sources such as dissolved organic carbon (Thingstad *et al.*, 2008 and references therein). In a warming ocean, respiration at many trophic levels will increase much faster than primary production and that may decrease the net community production (Kritzberg *et al.*, 2010; Vaque-Sunyer *et al.*, 2010) and possibly turn the Arctic Ocean from a sink into a source of atmospheric CO_2.

5.3 Ecosystem resilience and reorganization

In general terms, an ecosystem may be considered resilient if the basic functional relationships across spatial and temporal scales remain relatively intact when exposed to perturbation. Systems with high species diversity generally have more functional redundancy, thus further increasing resilience. Large-scale physical conditions provide the basic habitat characteristics and nutrient availability to support reasonably consistent patterns of microbial and planktonic productivity at the small scale (Domain 4). This productivity, in turn, produces food that can be concentrated by physical processes in hot spots where predators feed (Domain 3), which in turn drives migration and aggregation patterns of species at the higher trophic levels, characteristic of Domain 2. As long as planktonic reorganization still results in the provision of sufficient quantity and quality of food, the rest of the food web will not be greatly affected by reorganization in Domain 4.

When the reorganization within one spatio-temporal domain begins to affect the basic structure and function of other domains, however, then the ecosystem is susceptible to an overall reorganization, perhaps resulting in a new steady state with different characteristics of production, diversity, function, distribution, and abundance of organisms (Ω phase of panarchy). In the Arctic Ocean, where we currently have a fairly simple marine ecosystem with biomass dominated by a small number of species, we appear to be at the beginning of just such a process. The large-scale loss (Domain 1) of sea ice is resulting in a new nano-plankton community (Domain 4) in many areas, dominated by species an order of magnitude smaller than some of their predecessors that were significant for primary production, e.g. diatoms and dinoflagellates (Li *et al.*, 2009; Coupel *et al.*, 2012). The smaller phytoplankton species are more efficiently grazed by other organisms in the microbial food web, resulting in less energy transported to higher trophic levels (Domain 4 to Domain 3) (cf. Parsons and Lalli, 2002), and thus potentially less prey available for predators, including human hunters and fishermen. The biomass removal in commercial fisheries is very low, however: only about 0.2% of the harvestable production in the Barents Sea is channelled through the cod fishery (Wassmann *et al.*, 2006). No future projection is able to pick up changes in such small quantities, nor to determine the spatial distribution and aggregation of prey stocks, which can be more important for predators (including fishermen) than overall biomass.

Reorganizations through each spatio-temporal scale are likely to occur on different, though overlapping, time-lines. Microbial change (Domain 4) will be far more rapid than the response of long-lived marine benthic invertebrates, mammals, and seabirds (Domains 2 and 3), and thus species at the smallest scales may go through several adaptive and evolutionary iterations before reaching a new steady state. The simplest response is likely to be adaptation of an ecosystem by species already present in the Arctic Ocean, as some are able to thrive during warmer summers. A farther-reaching response is likely to be the advection and migration by sub-Arctic species that manage to survive Arctic winters (Bluhm *et al.*, 2011). Some may just survive while some may colonize after taking advantage of favourable growing conditions in the warmer summers. Such a response is generically occurring along the inflows from sub-Arctic seas, such as the Chukchi and Barents Seas and western and northern Spitsbergen (Carmack *et al.*, 2006; Ji *et al.*, 2012). Some organisms have penetrated already along the shelves of the Arctic Ocean and later in this century they will reach more remote interior shelves and basins. This response will be mediated also by competition between Arctic and sub-Arctic species, the outcome of which may vary over time as summers become longer and warmer. As a result, initial trends in biological response may not indicate the ultimate direction in which physical change drives the system.

Although it is tempting to suppose that the reorganization of the Arctic marine ecosystem will simply be a northward shift in current species distributions and biological patterns, a panarchy perspective strongly suggests otherwise. Although some barriers to expansion – such as the Bering Sea 'cold pool' (Mueter and Litzow, 2008; Stabeno *et al.*, 2012) – may weaken or disappear under future ice deposition, cold temperatures and low light conditions during winter will remain (Stabeno *et al.*, 2012; Kaartvedt, 2008). Species that thrive on the continental shelf would be pushed northwards into deep water, but are constrained by the cost of losing their preferred habitat. Species that move northwards in summer may still be unable to survive the lethal temperatures during the Arctic winter (Irvine *et al.*, 2009; Kosobokova and Hirche, 2009). Any predictions in this context are, however, hampered by the fact that the dark period of the Arctic winter is hardly studied at all, and the few investigations to date reveal a much more active system than anticipated (e.g. Berge *et al.*, 2009).

Summer conditions are predicted to change much more drastically than winter conditions, creating a much wider range of conditions in which organisms must be able to survive; thus likely altering the relative abundance of species as well as their distributions. It is expected that the advection of warmer water zooplankton species, which enter the Arctic Ocean with the Atlantic water through the European Arctic Corridor or with Bering shelf water through the Bering Strait (Hunt *et al.*, 2013) (e.g. the calanoid copepods *Calanus finmarchicus* and *C. marshallae*), will penetrate deeper into the Arctic Ocean in decades to come.

Combining what we know about today's structure and function of the Arctic marine ecosystem at different spatial and temporal scales and using predictive physical models, we derive the following hypotheses concerning the biological response to future physical change:

- The most rapid adaptive and evolutionary responses in the Arctic marine ecosystem will be at the microbial and micro-phytoplankton level, where change will be non-uniform and non-linear over time, with disproportionally increased respiration and competition for nutrients and carbon between photo- and heterotrophs (Domain 4).
- A decrease in size and/or concentration of primary producers along with shifts in the bloom regime will yield a less efficient grazing pelagic food web and less export to the benthos, leaving less energy to support species at the highest trophic levels and thus affecting predators including humans (Domain 4 to Domain 3).
- Changes to some biological hot spots (e.g. increased upwelling along the shelf break or loss of intensive productivity along the ice edge) will result in shifts in migratory patterns and abundance of species aggregations such as bird colonies or changes in where and how commercial fisheries take place (Domain 3 to Domain 2).
- Early spring ice retreat and subsequent reduced primary production maxima at current locations such as the sea ice edge will result in reduced downward

transport of biological material, and thus an impoverishment of benthic communities in places such as the Chukchi and Barents Seas (Domain 3 to Domain 2 and vice versa) (cf. Wassmann and Reigstad, 2011).

- The combined effect of these changes over time and space will be the creation of a new ecological stable state in Arctic waters, exhibiting some characteristics of today's Arctic and some of today's sub-Arctic, perhaps fundamentally different from today's system, and accompanied by new, as yet non-existent species and the loss of some current specialist species (within and across all four domains).

A potential set of outcomes is outlined in Table 5.1. The observations made therein are largely speculation based on careful consideration of current conditions, trends, models, and expert knowledge. A coordinated, pan-Arctic programme of measurement, comparison, process and experimental studies, modelling, and synthesis can allow the testing of specific hypotheses across the four domains described therein, particularly with regard to cross-scale interactions, connected with a set of overall hypotheses such as we have proposed. The relationships across domains, essential in the concept of panarchy, are of critical importance in understanding exactly how change will be propagated through the system and remain under-studied.

5.4 Panarchy theory applied to observational strategies

It is critical to recognize and accept that 'surprise and unexpected events' are the inconvenient reality of non-linear (complex-adaptive) systems and that this constrains the ability of models to predict future outcomes with ongoing validation through observation and modelling. Put simply, one cannot manage what one does not measure. For this reason, it is suggested that a cross-scale monitoring strategy, founded in panarchy theory, can provide a means – perhaps the only means – of detecting the early signs of environmental shifts affecting marine system services and functions, and to provide an informed basis for responsive policy. Consider the following, scale-based activities:

1. Since the late 1980s the Intergovernmental Panel on Climate Change (IPCC, 2014) has taken the global lead in assessing and predicting the impacts of anthropogenic greenhouse gas emissions on global climate. The IPCC focus is *large scale* and *multidecadal*. Solid monitoring of all parts of the climate system and biota is essential to IPCC's ongoing effectiveness.

2. Large-scale oceanographic monitoring programmes follow in the spirit of the IPCC mandate by taking a *global scale* view of ocean physics, geochemistry, and biology to establish benchmarks against which future (*long time-scale*) change can be detected (cf. Carmack and McLaughlin, 2001, 2011).

3. But, many – if not most – public policy concerns are centred on climate change issues with *regional* (not global) spatial scales, *short* (not long) time-scales, and intimate (not bulk) metrics of place and change (Visbeck, 2008).
4. So, deep understanding of the local, the immediate, and the intimate lies with the people who remain connected with the land and marine environment for food, travel, and survival; that is to say, the residents of coastal communities (cf. Krupnik and Jolly, 2002; Carmack and Macdonald, 2008; Huntington, 2011; Fienup-Riordan and Carmack, 2011).

Thus, involvement of local communities could bridge the present gap between the coastal zone of the region and the rest of the planet. As stated by C. S. Holling (pers. comm.), 'We now have the planet's oceans being monitored; add to this the seas immediately offshore of the indigenous communities and the scales are bridged and the people engaged.' Thus, combining a coordinated, pan-Arctic programme of measurement, comparison, process studies, and modelling with a network of community-based observations, founded on panarchy concepts, would enable us to fingerprint ecosystem change and its northward propagation – a level of detail that current science programmes are unable to resolve.

5.5 Policy considerations

The implications of these concepts for policy and governance follow those of the panarchy framework. To understand a system like the Arctic Ocean we must rethink how and at what scales it functions, and how small changes in cross-scale interactions can, especially if unnoticed, lead to unpredicted and, from a societal point of view, undesired, large-scale changes. System services provided by the Arctic Ocean currently include reliable food, diversity, cultural services, and unique means of transportation (e.g. Eicken *et al.*, 2009). The panarchy concept provides a framework to help understanding and interpretation of sudden changes in the availability of such system services and which might help guide decision makers in dealing with the system's non-linear behaviour. As such, research and observation strategies must capture cross-scale interactions, and policy and management must accommodate these spatial and temporal interrelationships and increased human activities that are further affecting an already stressed system.

In particular, assuming that there is a desire to learn from past approaches, there is the need to bridge current barriers between disciplines, agencies, and political jurisdictions. Existing local, national, and international governance structures will not become obsolete; indeed, they will likely have larger roles to play as human activity increases in the Arctic. But any one discipline, agency, or jurisdiction cannot hope to address the full implications of change and human influence at all necessary temporal and spatial domains. Thus what is needed is more a mental shift in the way we think about systems, in the way we study them, in the ways in which we collaborate, and in the way we manage and protect them.

If indeed, as panarchy predicts, systems transitions occur as tipping points when external forces are pushed above a system's resilience/adaptive capacity and where cross-scale interactions can lead to unforeseeable and large-scale changes, then management actions and their regulatory framework need to allow for a system-wide approach where climate change, fisheries, oil and gas activity, shipping, subsistence use, and other enterprises are considered in conjunction and not in isolation, to avoid over- as well as underestimating their roles. New cooperation and governance at multiple scales, from the local to the transboundary, can lead to a new proactive era of policy, management, and science needed for the new ocean opening in front of us.

5.6 Summary

Spatial and temporal scales of natural phenomena and their associated patterns are windows through which to view the dynamics of complex-adaptive ecosystems, the non-linear behaviour of which can be bewildering. Both will differ in various geographical regions of the highly variable and rapidly changing Arctic Ocean, thus presenting huge challenges to science-informed policy making. As such, panarchy theory provides a powerful formalism for organization of data, analysis, and modelling by establishing a framework of cross-scale linkages and feedbacks that facilitates communication across intersections of ecology, society, and economics. We provide a brief and simple interpretation of panarchy as it may be applied to the urgent question of how marine ecosystems (and social-ecological systems) will respond to the rapid pace of Arctic change and how they can be managed to remain sustainable.

The human response, in terms of foraging strategy as well as management and governance of natural resources and their use, adds another dimension of multi-scale complexity as individuals, communities, and nations respond and adapt to a changing ecosystem at multiple scales over various time periods. Because humans depend on the Arctic marine ecosystem for local needs and for a substantial portion of the world's fish catch (e.g. Lindholt, 2006), the implications of change in the region are of great practical significance regionally and globally.

Panarchy views nature as a complex-adaptive system, wherein the problems of pattern and scale represent the central challenge to understanding (cf. Levin, 1992). On one hand, the progressive loss of the summer sea ice cover over the Arctic Ocean clearly poses challenges to existing ecosystems, which may involve a suite of undesirable regime shifts and even extinctions. On the other hand, it also gives humankind an unprecedented opportunity to explore an ocean changing before our eyes. The sustainable management of emerging Arctic Ocean ecosystem services depends on our ability to discover, model, and project the immediate and long-term future of life in this new ocean. The stakes are too high to take lightly. We thus advocate research that recognizes abiotic and biotic scales and cross-scale inter-actions through the panarchy framework, along with monitoring strategies designed to detect and adapt to surprise and to that which is currently unknown.

We hope that the suggestion of panarchy as a framework for further study will provide a useful tool for assessing change and for planning and carrying out an effective response by the world's top predator. This concept also highlights the potential, and even likelihood, that the Arctic Ocean might go through a series of unprecedented changes (Ω phases) leading to the evolution of new complex-adaptive Arctic ecosystems which, in turn, will require our deep understanding.

Acknowledgements

The authors wish to thank Frøydis Strand (University of Tromsø) for her support with the figures. This chapter was inspired by discussions that began at a workshop held in Fairbanks, Alaska, in February 2011, at the International Arctic Research Center, sponsored by that institution along with the Pew Charitable Trusts, and the CNSM Sidney Chapman Chair. The authors are grateful for the constructive and lively discussions, but the presentation of ideas – and any errors – herein are the responsibility of the authors alone. The contribution of RG was supported by the National Science Foundation under Grant No. 0732767.

Crawford Stanley (Buzz) Holling

We dedicate this chapter to Crawford Stanley (Buzz) Holling, whose intense lifetime of scientific inquiry has given us the fundamental bedrock of resilience theory, of multi-stable ecological states, and of complex-adaptive social-ecological systems. In this acknowledgement we recognize two sides of this most inspiring man. On the one hand there is Buzz Holling – the scholar – who has received the Mercer and the Eminent Ecologist awards from the Ecological Society of America, the Kenneth Boulding award from the International Society of Ecological Economics, the Volvo Environmental Prize in 2008, honorary doctorates from the University of Guelph, University of British Colombia, and Simon Fraser University, who served as director of the International Institute of Applied Systems Analysis, who pioneered development of adaptive management of ecological systems, is a Fellow of the Royal Society of Canada and an Officer of the Order of Canada, and who invented and founded the novel Resilience Alliance, the first truly independent internet organization of science which now flourishes in many forms around the world; he is the acknowledged Father of panarchy theory. On the other hand there is the Buzz Holling – the person – deeply valued and cherished by his friends, and whose boundless ideas and passion for science and the creative process serve as lasting inspiration for his colleagues. Born in the first third of the 20th century, the life work of Buzz Holling gives us all a much needed roadmap for the 21st century, one of hope and optimism to meet whatever surprise and uncertainty it may hold. In his own words, 'It is not a crisis, it is an opportunity!'

References

Arrigo, K. R., Perovich, D. K., Pickart, R. S., *et al.* (2012). Massive phytoplankton blooms under Arctic sea ice. *Science* **336**: 1408.

Berge, J., Cottier, F., Last, K. S., *et al.* (2009). Diel vertical migration of Arctic zooplankton during the polar night. *Biology Letters* **5**: 69–72.

Bluhm, B. A., Gebruk, A. V., Gradinger, R., *et al.* (2011). Arctic marine biodiversity: An update of species richness and examples of biodiversity change. *Oceanography* **24**: 232–248.

Carmack, E. C. and Chapman, D. C. (2003). Wind-driven shelf/basin exchange on an Arctic shelf: the joint roles of ice cover extent and shelf-break bathymetry. *Geophysical Research Letters* **30**: 1778. doi: 10 1029/2003GL017526.

Carmack, E. C. and Macdonald, R. W. (2002). Oceanography of the Canadian Shelf of the Beaufort Sea: A setting for marine life. *Arctic* **55** (Suppl. 1): 29–45.

Carmack, E. C. and Macdonald, R. W. (2008). Water and ice-related phenomena in the coastal region of the Beaufort Sea: Some parallels between native experience and Western science. *Arctic* **61**: 265–280.

Carmack, E. C. and McLaughlin, F. A. (2001). Arctic Ocean change and consequences to biodiversity: A perspective on linkage and scale. *Memoirs of National Institute of Polar Research* **54**: 365–375.

Carmack, E. C. and McLaughlin, F. A. (2011). Towards recognition of physical and geochemical change in sub-Arctic and Arctic seas. *Progress in Oceanography* **90**: 90–104. doi: 10.1016/j.pocean.2011.02.007.

Carmack, E. C. and Melling, H. (2011). Warmth from the deep. *Nature Geoscience* **4**: 7–8.

Carmack, E. C. and Wassmann, P. (2006). Food-webs and physical biological coupling on pan-Arctic shelves: perspectives, unifying concepts and future research. *Progress in Oceanography* **71**: 446–477.

Carmack, E., Barber, D., Christensen, J., *et al.* (2006). Climate variability and physical forcing of the food webs and the carbon budget on pan-Arctic shelves. *Progress in Oceanography* **71**: 145–182.

Carmack, E. C., Whiteman, G., Homer-Dixon, H., and McLaughlin, F. A. (2012). Detecting and coping with potentially disruptive shocks and flips in complex-adaptive Arctic marine systems: A resilience approach to place and people. *Ambio* **41**: 56–65. DOI 10.1007/s13280-011-0225-6.

Coupel, P., Jin, H. Y., Joo, M., *et al.* (2012). Phytoplankton distribution in unusually low sea ice cover over the Pacific Arctic. *Biogeosciences* **9**: 4835–4850.

Crutzen, P. J. and Stoermer, E. F. (2000). The 'Anthropocene'. *Global Change Newsletter* **41**: 17–18.

Duarte, C. M., Agusti, S., Wassmann, P., *et al.* (2012). Tipping elements in the Arctic marine ecosystem. *Ambio* **41**: 44–55.

Eicken, H., Lovecraft, A., and Druckenmiller, M. (2009). Sea-ice system services: A framework to help identify and meet information needs relevant for observing networks. *Arctic* **62**: 119–136.

Fienup-Riordan, A. and Carmack, E. C. (2011). The ocean is always changing: Nearshore and farshore perspectives on Arctic coastal seas. *Oceanography* **24**: 266–279.

Francis, J. A. and Vavrus, S. J. (2012). Evidence linking Arctic amplification to extreme weather in mid-latitudes. *Geophysical Research Letters* **39**: L06801. doi: 10.1029/2012GL051000.

Grebmeier, J. M., Overland, J. E., Moore, S. E., *et al.* (2006). A major ecosystem shift in the northern Bering Sea. *Science* **311** (5766): 1461–1464. doi: 311/5766/461[pii] 10.1126/science.1121365.

Gunderson, L. and Holling, C. S. (2001). *Panarchy: Understanding Transformations in Systems of Humans and Nature.* Washington DC: Island Press.

Hannah, C. G., Dupont, F., and Dunphy, M. (2009). Polynyas and tidal currents in the Canadian Arctic Archipelago. *Arctic* **62**: 83–95.

Hunt, G. L. Jr., Blanchard, A. L., Boveng, P., *et al.* (2013). The Barents and Chukchi Seas: Comparison of two Arctic shelf ecosystems. *Journal of Marine Systems* **109–110**: 43–68.

Huntington, H. P. (2011). The local perspective. *Nature* **478**: 182–183.

IPCC (2014) Intergovernmental Panel on Climate Change. Accessed on 6 March 2014. www.ipcc.ch.

Irvine, J. R., Macdonald, R. W., Brown, R. J., *et al.* (2009). Salmon in the Arctic and how they avoid lethal low temperatures. *North Pacific Anadromous Fisheries Commission Bulletin* **5**: 39–50.

Ji, R., Ashjian, C. J., Campbell, R. G. *et al.* (2012). Life history and biogeography of *Calanus* copepods in the Arctic Ocean: An individual-based modeling study. *Progress in Oceanography* **96**: 40–56.

Ji, R., Jin, M., and Varpe, Ø. (2013). Sea ice phenology and timing of primary production pulses in the Arctic Ocean. *Global Change Biology* **19**(3): 734–741.

Kaartvedt, S. (2008). Photoperiod may constrain the effect of global warming in Arctic marine systems. *Journal of Plankton Research* **30**(11): 1203–1206.

Kinnard, C., Zdanowicz, C., Koerner, R. M., *et al.* (2008). A changing Arctic seasonal ice zone: observations from 1870–2003 and possible oceanographic consequences. *Geophysical Research Letters* **35**: L02507. doi: 10.1029/2007GL032507.

Kinnard, C., Ladd, M., Zdanowicz, C., *et al.* (2011). Reconstructing sea ice extent in the Arctic over the past ~900 years using a multi-proxy approach. *Nature* **479**: 509–512.

Kosobokova, K. and Hirche, H.-J. (2009). Biomass of zooplankton in the eastern Arctic Ocean – A base line study. *Progress in Oceanography* **82**: 265–280.

Kritzberg, E. S., Duarte, C. M., and Wassmann, P. (2010). Changes in Arctic marine bacterial carbon metabolism in response to increasing temperature. *Polar Biology* **33**: 1673–1682. doi: 10.1007/s00300-010-0799-7.

Krupnik, I. and Jolly, D. (2002). *The Earth Is Faster Now: Indigenous Observations of Arctic Environmental Change.* Fairbanks, Alaska: Arctic Research Consortium of the United States.

Kwok, R., Cunningham, G. F., Wensnahan, M., *et al.* (2009). Thinning and volume loss of Arctic sea ice: 2003–2008. *Journal of Geophysical Research* **114**: C07005. doi: 10.1029/2009JC005312.

Leu, E., Søreide, J. E., Hessen, *et al.* (2011). Consequences of changing sea-ice cover for primary and secondary producers in the European Arctic shelf seas: Timing, quantity, and quality. *Progress in Oceanography* **90**: 18–32.

Levin, S. A. (1992). The problem of pattern and scale in ecology. *Ecology* **73**: 1943–1967.

Li, W. K. W., McLaughlin, F. A., Lovejoy, C., *et al.* (2009). Smallest algae thrive as the Arctic Ocean freshens. *Science* **326**: 539.

Lindholt, L. (2006). Arctic natural resources in a global perspective. In: *The Economy of the North*. Glomsrød, S., and Aslaksen, I. (eds.). Oslo: Statistics Norway, pp. 27–39.

Loeng, H., Brander, K., Carmack, E., *et al.* (2005). Marine systems. In: *Arctic Climate Impact Assessment*. Cambridge: Cambridge University Press, pp. 453–538.

Mathis, J. T., Cross, J. N., and Bates, N. R. (2011). Coupling primary production and terrestrial runoff to ocean acidification and carbonate mineral suppression in the eastern Bering Sea. *Journal of Geophysical Research* **116**: C02030. doi: http://dx.doi.org/10/1029/2010JC006453.

McLaughlin, F., Carmack, E., Krishfield, R., *et al.* (2011). The rapid response of the Canada Basin to climate forcing: From bellwether to alarm bells. *Oceanography* **24**: 146–159.

Michel, C., Bluhm, B., Gallucci, V., *et al.* (2012). Biodiversity of Arctic marine ecosystems and responses to climate change. *Biodiversity* **13**: 200–214.

Mueter, F. J. and Litzow, M. A. (2008). Sea ice retreat alters the biogeography of the Bering Sea continental shelf. *Ecological Applications* **18**: 309–320.

Mundy, C. J., Gosselin, M., Gratton, Y., *et al.* (2013). Role of environmental factors on phytoplankton bloom initiation under landfast sea ice in Resolute Passage, Canada. *Marine Ecology Progress Series* **497**: 39–49.

Nelson, R. J., Carmack, E. C., McLaughlin, F. A., *et al.* (2009). Tracking penetration of Pacific zooplankton into the western Arctic Ocean with molecular population genetics. *Marine Ecology Progress Series* **381**: 129–138.

Nicolaus, M., Katlein, C., Maslanik, J., *et al.* (2012). Changes in Arctic sea ice result in increasing light transmittance and absorption. *Geophysical Research Letters* **39**: L24501.

Overland, J. E., Wood, K. R., and Wang, M. (2011). Warm Arctic-cold continents: Climate impacts of the newly open Arctic Sea. *Polar Record* **30**: 15787. doi: 10.3402/polar.v30i0.15787.

Overpeck, J. T., Sturm, M., Francis, J. A., *et al.* (2005). Arctic system on trajectory to new, seasonally ice-free state. *EOS* **86**: 309, 312–313.

Parsons, T. R. and Lalli, C. M. (2002). Jellyfish population explosions: revisiting a hypothesis of possible causes. *La Mer* **40**: 111–121.

Reigstad, M., Carroll, J., Slagstad, D., *et al.* (2011). Intra-regional comparison of productivity, carbon flux and ecosystem composition within the northern Barents Sea. *Progress in Oceanography* **90**: 33–46.

Rogachev, K. A., Carmack, E. C., and Foreman, M. G. (2008). Bowhead whales feed on plankton concentrated by estuarine and tidal currents in Academy Bay, Sea of Okhotsk. *Continental Shelf Research* **28**: 1811–1826.

Søreide, J. E., Leu, E., Berge, J., *et al.* (2010). Timing of blooms, algal food quality and *Calanus glacialis* reproduction and growth in a changing Arctic. *Global Change Biology* **16**(11): 3154–3163.

Stabeno, P. J., Farley, E., Katchel, N., *et al.* (2012). A comparison of the physics of the northern and southern shelves of the eastern Bering Sea and some implications to the ecosystem. *Deep-Sea Research II* **65–70**: 14–30.

Sunday, J. M., Crim, R. N., Harley, C. D. G. *et al.* (2011). Quantifying rates of evolution-ary adaptation in response to ocean acidification. *PLoS ONE* **6**(8): e22881. doi: 10.1371/journal.pone.0022881.

Thingstad, T. F., Bellerby, R. G. J., Bratbak, G. *et al.* (2008). Counterintuitive carbon-to-nutrient coupling in an Arctic pelagic ecosystem. *Nature* **455**: 387–390. doi: 10.1038/nature07235.

Vaquer-Sunyer, R., Duarte, C. M., Santiago, R., *et al.* (2010). Response of respiration rates of Arctic plankton communities to warming: An experimental evaluation. *Polar Biology* **33**: 1661–1671. doi: 10.1007/s00300-010-0788-x.

Visbeck, M. (2008). From climate assessment to climate services. *Nature Geosciences* **1**: 2–3.

Walker, B. H. and Salt, D. (2006). *Resilience Thinking: Sustaining Ecosystems and People in a Changing World*. Washington, DC: Island Press.

Walker, B. H. and Salt, D. (2012). *Resilience Practice: Building Capacity to Absorb Disturbance and Maintain Function*. Washington, DC: Island Press.

Walker, B. H., Carpenter, S. R., Rockstrom, J., *et al.* (2012). Drivers, 'slow' variables, 'fast' variables, shocks, and resilience. *Ecology and Society* **17**(3): 30.

Walsh, J. E. (2013). Melting ice: What is happening to Arctic sea ice, and what does it mean for us? *Oceanography* **26**(2): 171–181.

Wassmann, P. (2006). Structure and function of contemporary food webs on Arctic shelves: An introduction. *Progress in Oceanography* **71**: 123–128.

Wassmann, P. (2011). Arctic marine ecosystems in an era of rapid climate change. *Progress in Oceanography* **90**: 1–17.

Wassmann, P. and Lenton, T. (2012). Arctic tipping points in the Earth System perspec-tive. *Ambio* **41**(1): 1–9.

Wassmann, P. and Reigstad, M. (2011). Future Arctic Ocean seasonal ice zones and implications for pelagic–benthic coupling. *Oceanography* **24**: 220–231.

Wassmann, P., Reigstad, M., Haug, T., *et al.* (2006). Food webs and carbon flux in the Barents Sea. *Progress in Oceanography* **71**: 232–287.

Wassmann, P., Slagstad, D., and Ellingsen, I. (2010). Primary production and climatic variability in the European sector of the Arctic Ocean prior to 2007: Preliminary results. *Polar Biology* **33**: 1641–1650. doi: 10.1007/s00300-010-0839-3.

Wassmann, P., Duarte, C. M., Agusti, S., *et al.* (2011). Footprints of climate change in the Arctic marine ecosystem. *Biological Global Change* **17**: 1235–1429. doi: 10.1007/s00300-010-0839-3.

Yamamoto-Kawai, M., McLaughlin, F. A., Carmack, E. C., *et al.* (2009). Aragonite undersaturation in the Arctic Ocean: Effects of ocean acidification and sea ice melt. *Science* **326**: 1098–1100.

Yamamoto-Kawai, M., McLaughlin, F. A., and Carmack, E. C. (2011). Effects of ocean acidification, warming and melting of sea ice on aragonite saturation of the Canada Basin surface water. *Geophysical Research Letters* **38**: L03601. doi: 10.1029/2010GL045501.

Zalasiewicz, J., Williams, M., Steffen, W., *et al.* (2010). The new world of the Anthropocene. *Environment Science & Technology* **44**(7): 2228–2231. doi: 10.1021/es903118j.

Zalasiewicz, J., Williams, M., Fortey, R., *et al.* (2011). Stratigraphy of the Anthropocene. *Philosophical Transactions of the Royal Society A* **369**(1938): 1036–1055. doi: 10.1098/rsta.2010.0315.

6

Ecosystem approach and ocean management

MARIO VIERROS, IAN D. CRESSWELL, PETER BRIDGEWATER,
AND ANTHONY D. M. SMITH

6.1 Introduction

The concept of an ecosystem approach has arisen largely as a management response to decline in biodiversity and natural resources, which single species management and primarily sectoral approaches had failed to stem. Because of its integrated nature, an ecosystem approach was seen as a way to better manage multiple impacts on environments holistically while maximizing long-term economic, social, and cultural benefits. The ecosystem approach also provides for the involvement of a wide range of users and other stakeholders in the management of a spatial area and resources, thus improving coordination and integration in activities.

Many different 'ecosystem approaches' exist, ranging from traditional/indigenous approaches to those more recently adopted by Western societies. The theory of how to manage using the ecosystem as the planning framework is still in its infancy and there is no one correct way to implement an ecosystem approach. However, certain principles apply to all current approaches. Perhaps the two best-known concepts are the complementary ones in use by the Food and Agriculture Organization of the United Nations (FAO) (Garcia et al., 2003) and the Convention on Biological Diversity (CBD) (UNESCO, 2000; Shepherd, 2008) with related biodiversity conventions. In practice, however, the most widely implemented approaches are in integrated coastal zone (sometimes area) management (ICZM) and integrated water resources management (IWRM), also sometimes expressed as river basin management (RBM). While not formally called 'ecosystem approaches', they espouse the use of a whole or integrated system as the base layer for all planning and management.

Approaches to management that use the ecosystem as a basis have become a central concept in the implementation of a number of international and regional agreements, such as those within the CBD (cf. CBD, 2000) and the FAO Code of Conduct for Responsible Fisheries (FAO, 2013), the Convention on the

Conservation of Antarctic Marine Living Resources (CCAMLR; cf. Fabra and Gascón, 2008), the Convention for the Protection of the Marine Environment of the North-East Atlantic (OSPAR), and the Helsinki Commission (HELCOM) – among others. OSPAR is working under the general framework of Regional Seas Programmes and Action Plans, specifically with the Abidjan Convention, and they note that 'This is an important cooperation not only for sharing of knowledge, experiences between regions but it is also required if we are to achieve *a coherent implementation of the ecosystem approach* which requires both conventions to look beyond our borders.'

The United Nations Convention on the Law of the Sea (UNCLOS) and its 1995 Agreement for the Implementation of the Provisions of the United Nations Convention on the Law of the Sea, of 10 December 1982, relating to the Conservation and Management of Straddling Fish Stocks and Highly Migratory Fish Stocks (UN Fish Stocks Agreement) also contains provisions of relevance to the ecosystem approach. The UN Division for Ocean Affairs and the Law of the Sea published a memoir on ecosystem approaches and oceans in 2007 (DOALOS, 2007). The UN Fish Stocks Agreement calls on participating states to, inter alia, adopt an ecosystems approach, whereby dependent or associated species are taken into account. In 2002, the World Summit on Sustainable Development, in Johannesburg, encouraged application of the ecosystem approach by 2010, noting the Reykjavik Declaration on Responsible Fisheries and decision V/6 of the Conference of the Parties to the CBD.

The above international commitment demonstrates that the need to employ an ecosystem approach is widely acknowledged in international policy. However, in practice, its application is still limited. This is due in large part to the considerable practical difficulties in implementation, including availability of suitable information and lack of analytical and scientific tools to support the process. It may also be, in part, due to limited understanding of what exactly constitutes an ecosystem approach, including its provisions for broad participation of all stakeholders (Vierros *et al.*, 2006). While many compatible definitions exist and are discussed in this chapter, the most important lessons about the ecosystem approach can be learned from its implementation. Given that the marine environment is generally considered 'publicly owned', and indeed the areas beyond the limits of national jurisdiction are considered the global commons, it should theoretically be easier to implement ecosystem-based management. However, practice has shown that lack of ownership can lead to a 'tragedy of the commons' and an 'out of sight, out of mind' attitude often prevails to degradation of biodiversity in ocean areas.

Whether national or regional, ecosystem approaches employ tools and techniques such as bioregional classification, coherent systems of marine protected or managed areas, ocean zoning, and fisheries management. Implementation will ideally be undertaken through successful management practices that are best suited to the particular environmental, social, cultural, and economic context of the area being managed. This chapter also discusses how many indigenous cultures hold a holistic view of the land, sea, and whose inhabitants have developed management approaches consistent with their perception of the interconnectedness of nature, and are thus also implementing the ecosystem approach. There are lessons to be learned from these traditional approaches, still practised in various parts of the world, although some have been lost due to the prevalence of contemporary sector-based management.

6.2 The ecosystem approach as adopted by the CBD

One of the earliest recognized needs in meetings of the CBD was to seek an understanding of interactions between species, and between species and their abiotic environment, at the ecosystem level. Such needs were recognized from the first meeting of the Subsidiary Body on Scientific, Technical and Technological Advice (SBSTTA), in Paris in 1995, in the context of developing the first medium-term work plan. All CBD decisions can be found at www.cbd.int/decisions/.

Later in 1995, COPII noted in decision II.8 (our emphasis):

1. Reaffirms *that the conservation and sustainable use of biological diversity and its components should be addressed in a holistic manner, taking into account the three levels of biological diversity and fully considering socio-economic and cultural factors. However,* **the ecosystem approach should be the primary framework of action to be taken under the Convention***;*

By the time SBSTTA met again in 1997, it was discussing how to operationalize the nascent marine and coastal biodiversity programme, and in recommendation III/2 SBSTTA noted, inter alia, the need to:

Promote ecosystem approaches *to the sustainable use of marine and coastal living resources, including the identification of key variables or interactions, for the purpose of assessing and monitoring:*

(a) *Components of biological diversity;*
(b) *The sustainable use of such components; and*
(c) *Ecosystem effect.*

With the above recommendation, SBSTTA placed the ecosystem approach at the heart of considerations of marine and coastal biodiversity, as well as ongoing terrestrial work.

An ad hoc expert group met to discuss the ideas around managing at ecosystem scale in Lilongwe, Malawi, early in 1998 and at the COP 4 in Bratislava, 1998, the

'Malawi Principles' were proposed, but adopted as 'The Ecosystem Approach' in decision IV.1 pro parte:

The Conference of the Parties;

Recognizing *that in several decisions adopted at the third meeting of the Conference of the Parties the ecosystem approach has been addressed as a guiding principle, although the terminology used has varied, including: 'ecosystem approach', 'ecosystem process-oriented approach', 'ecosystem management approach' and 'ecosystem-based approach',*

The CBD COP therefore recognized the large number of terms used for ecosystem approach, but attempted to consolidate these approaches as the ecosystem approach, and use of the definite article was not by chance.

Further elaboration of the ecosystem approach was undertaken by SBSTTA 5 with decision V.10 noting:

1. *The ecosystem approach is a strategy for the integrated management of land, water and living resources that promotes conservation and sustainable use in an equitable way. Thus, the application of the ecosystem approach will help to reach a balance of the three objectives of the Convention: conservation; sustainable use; and the fair and equitable sharing of the benefits arising out of the utilization of genetic resources.*
2. *An ecosystem approach is based on the application of appropriate scientific methodologies focused on levels of biological organization, which encompass the essential structure, processes, functions and interactions among organisms and their environment. It recognizes that humans, with their cultural diversity, are an integral component of many ecosystems.*
3. *This focus on structure, processes, functions and interactions is consistent with the definition of "ecosystem" provided in Article 2 of the Convention on Biological Diversity.*

This in turn led to decision V/6, in which the parties described the ecosystem approach as follows:

The ecosystem approach is a strategy for the integrated management of land, water and living resources that promotes conservation and sustainable use in an equitable way. Application of the ecosystem approach will help to reach a balance of the three objectives of the Convention. It is based on the application of appropriate scientific methodologies focused on levels of biological organization which encompass the essential processes, functions and interactions among organisms and their environment. It recognizes that humans, with their cultural diversity, are an integral component of ecosystems.

In that decision, water is not specified as either fresh or salt. Although the discussions had to that point focused on terrestrial systems, it was clear that they could be translated *mutatis mutandis* to the marine environment; COP V recommended the application of 12 principles of the ecosystem approach (see Box 6.1). These principles are complementary and linked and, where feasible and

Box 6.1
The twelve principles of the ecosystem approach

The CBD Conference of the Parties recommended application of these principles and associated guidance in decision V/6 (CBD, 2000).

Principle 1: The objectives of management of land, water, and living resources are a matter of societal choices.

Principle 2: Management should be decentralized to the lowest appropriate level.

Principle 3: Ecosystem managers should consider the effects (actual or potential) of their activities on adjacent and other ecosystems.

Principle 4: Recognizing potential gains from management, there is usually a need to understand and manage the ecosystem in an economic context. Any such ecosystem management programme should:

- Reduce those market distortions that adversely affect biological diversity
- Align incentives to promote biodiversity conservation and sustainable use
- Internalize costs and benefits in the given ecosystem to a feasible extent.

Principle 5: Conservation of ecosystem structure and functioning, in order to maintain ecosystem services, should be a priority target of the ecosystem approach.

Principle 6: Ecosystems must be managed within the limits of their functioning.

Principle 7: The ecosystem approach should be undertaken at the appropriate spatial and temporal scales.

Principle 8: Recognizing the varying temporal scales and lag-effects that characterize ecosystem processes, objectives for ecosystem management should be set for the long term.

Principle 9: Management must recognize that change is inevitable.

Principle 10: The ecosystem approach should seek the appropriate balance between, and integration of, conservation and use of biological diversity.

Principle 11: The ecosystem approach should consider all forms of relevant information, including scientific and indigenous and local knowledge, innovations, and practices.

Principle 12: The ecosystem approach should involve all relevant sectors of society and scientific disciplines.

---

appropriate, should be applied together. They should be translated flexibly to address management issues in different social, economic, and environmental contexts.

Further elaboration of these principles continued in subsequent SBSTTA and COP meetings, until in 2004 at COP 7, where 'The Ecosystem Approach' was elaborated with implementation guidelines, and formally accepted as part of the CBD rubric. At the same time, the COP also agreed that the priority at this time should be to facilitate the implementation of the ecosystem approach (decision VII/11), thus learning from practical experiences. It was also noted that 'one-size-fits-all' solutions are neither feasible nor desirable.

Soon, 20 years will have passed since the CBD first started to contemplate the ecosystem approach. Although systematic, global application of the ecosystem approach is still lacking, there are now a number of successful examples from regional, national, and local levels. There is also a lack of comprehensive documentation of implementation, despite efforts such as the CBD case study database, which currently contains only 14 case studies related to marine and coastal ecosystems. These 14 case studies refer to a variety of implementation tools and approaches, including national and regional bioregional planning processes, putting in place marine protected areas, rehabilitation of ecosystems, incorporation of sustainable use, and bottom-up activities implemented by local fishermen. This range of examples is an indication of the wide variety of actions that could be considered under the ecosystem approach.

Perhaps the most comprehensive implementation has been achieved through integrated marine and coastal area management (IMCAM), which is an approach that has been in use for over 30 years. According to the 2005 3rd National Reports of the CBD, 78.4% of all CBD parties had instituted IMCAM, including catchment management. Although more current statistics are not available, the figure is likely to be higher today, based on the fact that all coastal countries which submitted a 4th National Report indicated some IMCAM-related activities (CBD, 2010).

A relatively new trend, documented in the CBD 4th National Reports, was the development of comprehensive, large-scale (bioregional or large marine ecosystem scale) national and regional IMCAM plans that consider ecosystems, species, and habitats, as well as human uses and needs. Some examples of these include plans by Norway for integrated management of all maritime areas based on the ecosystem approach; integrated oceans management undertaken by Canada in five Large Oceans Management Areas (LOMAs); science-based bioregional plans developed by Australia which have informed the development of the National Representative System of Marine Protected Areas; and the Sulu-Sulawesi Seas Ecoregion management efforts, jointly undertaken by the Philippines, Indonesia, and Malaysia. This trend also encompasses an increasing number of marine spatial planning efforts, for example in the Baltic Sea, China, St Kitts and Nevis, Dominican Republic, Belgium, Norway, and Germany (CBD, 2010). While the examples cited above are far from comprehensive, they are a sign of a growing trend in considering larger ocean areas as the basis for management. This trend is also evident in the increasing attention afforded to improving the management of marine areas beyond national jurisdiction, including through CBD efforts to identify Ecologically and Biologically Significant Areas (EBSAs), and the complementary efforts under FAO to identify Vulnerable Marine Ecosystems (VMEs). It should be noted here that while VMEs are identified and managed by RFMOs,

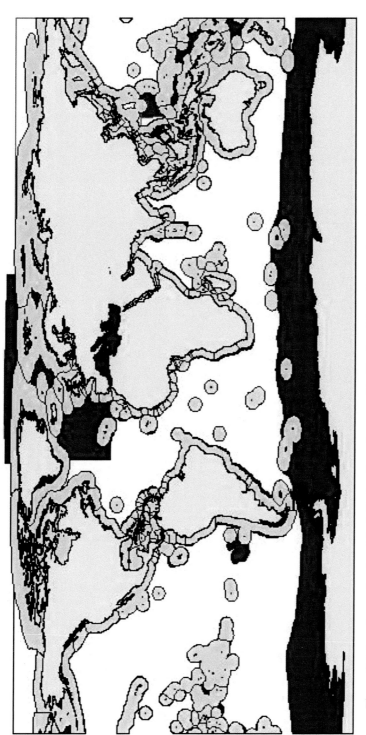

Figure 2.1 Area beyond national jurisdiction (in red) covered by Regional Seas Conventions. (*Map is for information only and expresses no opinion on boundaries.*)

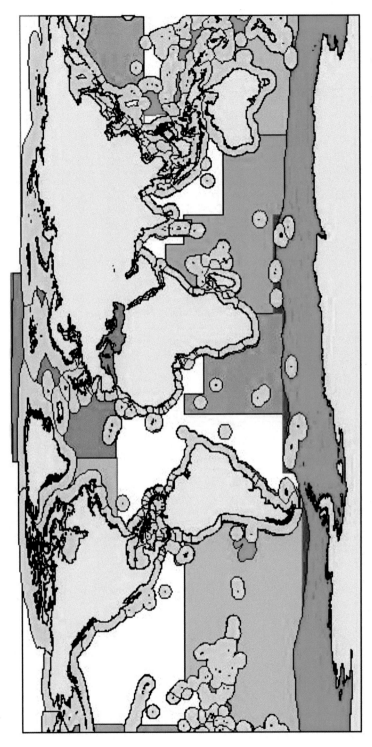

Figure 2.2 Geographical coverage of regional fisheries management organizations (RFMOs) and Regional Seas Conventions. White areas are gaps in coverage. Note that the map only includes RFMOs covering multiple species and incorporating some form of ecosystem approaches. (*Map is for information only and expresses no opinion on boundaries.*)

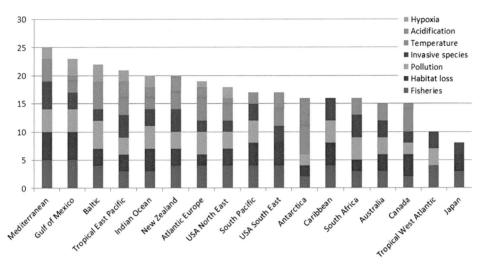

Figure 4.1 The effects of different drivers (including climate change as temperature rise, acidification and hypoxia) on function of biodiversity and goods and services in different oceanic regions (based on data from Costello *et al.*, 2010).

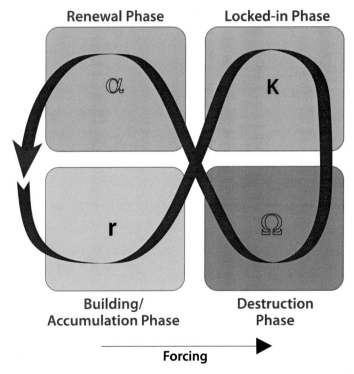

Figure 5.1 Conceptual model of the adaptive cycle, illustrating how a system, subjected to forcing, develops through two principal phases: r and K. Passing a tipping point may result in the system facing a destruction phase Ω before it renews itself, if the forcing decreases, in the α phase. Reproduced from Wassmann and Lenton (2012), which was redrawn from Gunderson and Holling (2001).

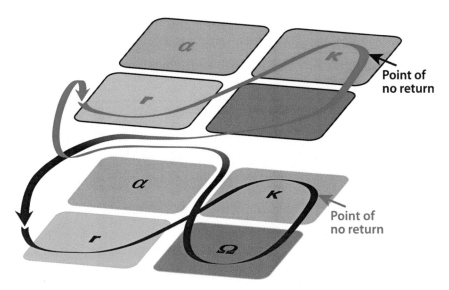

Figure 5.2 Altered conceptual model of adaptive cycles where the tipping point is now a point of no return. The original regime (lower half of the figure) can switch over to a new regime (upper half), or the system can stay within the original regime as in Figure 5.1. There can exist a multitude of regimes with potential switches between them. Here we only display two to illustrate the basic principle. Reproduced from Wassmann and Lenton (2012).

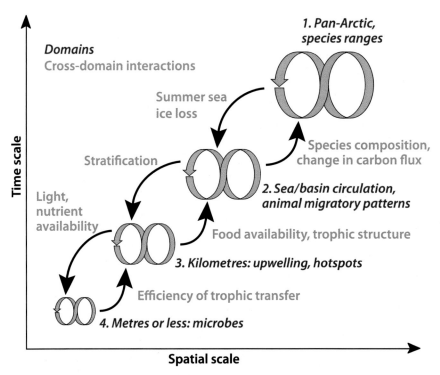

Figure 5.3 The spatio-temporal scales considered in this chapter (numbered 1–4), along with examples of cross-scale interactions.

Concept & Image of Satoumi

Coastal zone where land and coastal zone are managed in an integrated and comprehensive manner by human hands, with the result that material circulation functions are appropriately maintained and both high productivity and biodiversity are preserved.

Figure 6.1 Depiction of *satoumi*, including connections between land and sea (Yamada *et al.*, 2011).

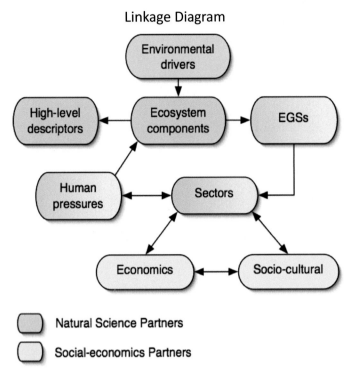

Figure 7.1 ODEMM linkage framework (modified from Koss *et al.*, 2011).

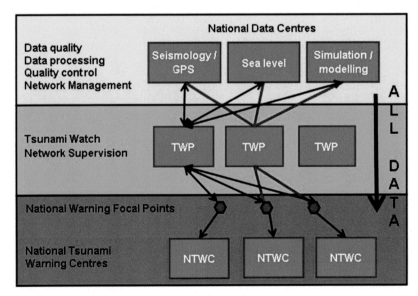

Figure 8.3 TWS architecture: example from NEAMTWS (after IOC–UNESCO, 2009).

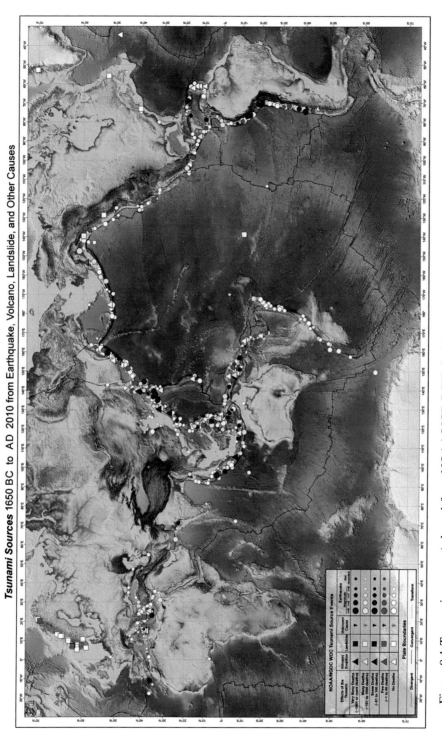

Figure 8.1 Tsunami sources (adapted from NOAA/NGDC WDC, 2010 © NOAA).

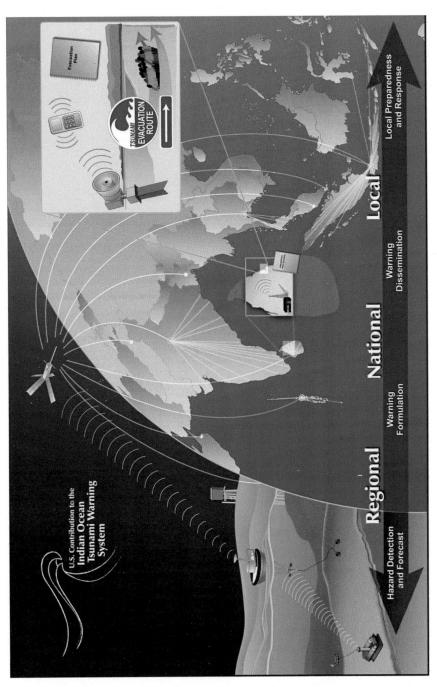

Figure 8.2 Components of an integrated tsunami warning system (© USAID).

there is, at the present time, no single designated management mechanism for EBSAs, given that the CBD mandate only extends to identifying such areas.

6.3 The FAO ecosystem approach to fisheries

Perceived failures in traditional single species focused fisheries management approaches led the FAO Committee on Fisheries (COFI) to recommend, in 1991, new approaches to fisheries management, embracing conservation and environmental, as well as social and economic, considerations. The FAO was asked to develop the concept of responsible fisheries and to elaborate a code of conduct to foster its application. This led to the development and adoption, in 1995, of the Code of Conduct for Responsible Fisheries. In 2001, the Reykjavik Declaration on Responsible Fisheries in the Marine Ecosystem was adopted, and it encouraged countries and fishing entities to achieve sustainable fisheries in the marine ecosystem. It also led to the development of technical guidelines on what became known as the ecosystem approach to fisheries, in 2003 (FAO, 2003).

The ecosystem approach to fisheries (EAF) presents a more holistic approach than traditional fisheries management methods, and strives to take into account the structure and functioning of ecosystems and their components, as well as the needs and desires of societies, in the context of sustainable use of marine resources. EAF is closely related to ecosystem based fisheries management (EBFM – Pikitch *et al.*, 2004), which aims to avoid degradation of ecosystems, minimize risk of irreversible change, obtain and maintain long-term socio-economic benefits, and improve ecosystem understanding.

The EAF evolved with and is thus complementary to the CBD ecosystem approach. The overarching principles of EAF are an extension of the conventional principles for sustainable fisheries development, to cover the ecosystem as a whole. They aim to ensure that, despite variability, uncertainty, and likely natural changes in the ecosystem, the capacity of aquatic ecosystems to produce fish food, revenues, employment, and more generally, other essential services and livelihood, is maintained indefinitely for the benefit of present and future generations. The FAO Technical Guidelines on their ecosystem approach to fisheries define EAF as follows:

An ecosystem approach to fisheries strives to balance diverse societal objectives, by taking into account the knowledge and uncertainties about biotic, abiotic and human components of ecosystems and their interactions and applying an integrated approach to fisheries within ecologically meaningful boundaries.

A primary implication is the need to cater for both human and ecosystem well-being. This implies conservation of ecosystem structures, processes, and interactions through sustainable use. Inevitably this will require consideration of a range of frequently conflicting objectives, where the needed consensus may not be readily attained without equitable distribution of benefits. In general, the tools and techniques of EAF will remain the same as those used in traditional fisheries management, but they will need to be applied in a manner that addresses the wider interactions between fisheries and the whole ecosystem. For example, catch and effort quotas, or gear design and restrictions, will be based not just on sustainable use of the target resources, but on their impacts on and implications for the whole ecosystem.

Similarly to the CBD ecosystem approach, the EAF is guided by a set of principles, and there is generally good consistency between the CBD ecosystem approach and the EAF (Vierros *et al.*, 2006). In the case of EAF, these principles are concepts that have been expressed in various instruments and conventions and, in particular, in the Code of Conduct for Responsible Fisheries. Accordingly, fisheries management should respect the following principles:

- Fisheries should be managed to limit their impact on the ecosystem to the extent possible.
- Ecological relationships between harvested, dependent, and associated species should be maintained.
- Management measures should be compatible across the entire distribution of the resource (across jurisdictions and management plans).
- The precautionary approach should be applied because the knowledge on ecosystems is incomplete.
- Governance should ensure both human and ecosystem well-being and equity.

Examples of EAF can be seen at both regional (supra-national) and national levels. Among regional fishery management organizations (RFMOs), the Commission for the Conservation of Antarctic Marine Living Resources (CCAMLR) is usually seen as the pre-eminent example (Constable *et al.*, 2000). CCAMLR includes explicit objectives at the ecosystem level, it has developed strategies for managing exploited species that take account of impacts on predators (e.g. the krill harvest strategy), and has also developed ecosystem monitoring programmes. However progress in EAF for other (mostly tuna-focused) RFMOs is much less advanced.

National progress in adopting EAF is also evident but patchy and so far mainly confined to a few OECD countries. Progress in adopting the FAO Code of Conduct (including EAF) has been reviewed by Pitcher *et al.* (2009). The five highest ranked nations were Norway, the United States, Canada, Australia, and Iceland. A recent FAO State of the World's Fisheries and Aquaculture report (FAO, 2012) highlighted ongoing efforts in the adoption of ecosystem based approaches at both the sectoral and multisectoral level, which are being pursued in various large marine ecosystems, including in the Caribbean, the Canary Current, the Benguela Current, and the Bay of Bengal. However, in most of these large marine ecosystems, efforts are concentrated on planning for an ecosystem based approach, and its full-scale implementation remains to be realized.The FAO has also noted that while many countries have made important strides towards the application of several of the principles contained in the EAF, some are partly implementing the approach without necessarily recognizing this (FAO, 2012).

6.4 Holistic marine management: indigenous systems and contemporary marine management

While the ecosystem approach is still a new concept in contemporary resource management, and particularly in the context of the marine and coastal environment, it is intrinsic to most indigenous approaches. For communities depending on and managing the resources of the sea, the ecosystem approach is not anything new, although the term 'ecosystem approach' may not have previously been used to describe their actions. Indigenous environmental knowledge and practice have been around for thousands of years, and have been effective in meeting community and ecosystem goals. Thus, communities and their knowledge can have a key role to play in application of the ecosystem approach.

Many indigenous communities have a holistic worldview that is based on the connections between all living things and their environment. For example, some indigenous peoples in Australia view the coastal landscape as an integrated cultural landscape/seascape where no distinction is made between land and sea. Barber (2005) provides the following description of Blue Mud Bay in the Northern Territory: 'In Blue Mud Bay, much of daily life and activities occur in the context of the flow of water, from freshwater rivers which flow into the increasingly salty water of the sea, and the seasonal cycles of rain and storms. These, in turn, affect the life cycles of species both on sea and land that provide food for the Aboriginal communities of Blue Mud Bay. The environments of the land and sea, their seasonality, flows and the animal and human communities that they support are all interrelated, and viewed in a holistic manner by the inhabitants of the area.' This understanding of intricate relationships in a geographical area is the basis of the ecosystem approach as it is described in international conventions, such as the Convention on Biological Diversity.

A similar understanding of water flows, the ecosystems and species they support, as well as the human communities that depend on them, is the basis for the native Hawaiian concept of *ahupua'a*. *Ahupua'a* was applied to sections of land that extended from the mountain summits down through fertile valleys to the outer edge of the reef and into the deep sea, and contained nearly everything

Hawaiians required for survival. Within the *ahupua'a*, a wise conservation system was practised by the caretakers to prevent exploitation of the land and sea, while allowing the people to use what they needed for sustenance. The ancient *ahupua'a*, the basic self-sustaining unit, extended elements of Hawaiian spirituality into the natural landscape. Amid a belief system that emphasized the interrelationship of elements and beings, the *ahupua'a* contained those interrelationships in the activities of daily and seasonal life.[1]

As noted by Ruddle and Hickey (2008), and demonstrated in the Hawaiian *ahupua'a* concept, the basic ideas contained in the ecosystem approach are inherent in most traditional systems of management that acknowledge ecological relationships. In the Pacific, these include not only *ahupua'a*, but also the Yap tabinau, the Fijian *vanua*, the Marovo (Solomon Islands) *puava*, and the Cook Islands *tapere*. Each of these are management units that run from the mountain top to the sea, and take into account connections between ecosystems, species, and their human inhabitants.

In Japan, the concept and practice of *satoumi* provide yet another expression of the ecosystem approach that has its roots in traditional ecological knowledge. Like the previous approaches, *satoumi* is centred on providing benefits both to people and to biodiversity. In Japanese, 'Sato' means the area where people live, while 'Umi' means the sea. When *satoumi* is restored in coastal waters, marine productivity and biodiversity are enhanced through the involvement of, and in harmony with, people (Yanagi, 2008). Achievement of *satoumi* relies on a long cultural heritage of fisheries knowledge and management, and an understanding of the interactions within and between ecosystems and human communities in the coastal zone. *Satoumi* is very similar in its holistic acknowledgement of connections between ecosystems and species (including human beings) to the Hawaiian *ahupua'a*. The only difference is that Japan's coastline is highly developed with a large amount of urban area (Figure 6.1).

Unlike many other management practices based on traditional cultural heritage, *satoumi* has been incorporated into Japanese national policies, including the Strategy for an Environmental Nation in the 21st Century (2007), the Third National Biodiversity Strategy of Japan (2007), and the Basic Plan on Ocean Policy (2008). The concept is being put into practice through a programme of the Japanese Ministry of Environment, which supports the efforts of local governments, residents, non-profit organizations, and universities to undertake diverse activities that include planting eelgrass to restore coastal ecosystems, re-planting of forest in watershed areas, sustainable cultivation of oysters, public education, and working with fishing communities to revive traditional fishing methods.

These examples demonstrate that the indigenous view of interconnectedness between water, land, sea, and human cultures exists in many places, and its holistic

Concept & Image of Satoumi

Coastal zone where land and coastal zone are managed in an integrated and comprehensive manner by human hands, with the result that material circulation functions are appropriately maintained and both high productivity and biodiversity are preserved.

Figure 6.1 Depiction of *satoumi*, including connections between land and sea (Yamada *et al.*, 2011). A black and white version of this figure will appear in some formats. For the colour version, please refer to the plate section.

nature is consistent with the recent emphasis on the ecosystem approach in international environmental policy. According to the CBD ecosystem approach, management should be decentralized to the lowest possible level (Principle 2). This would imply that communities are lead proponents in implementation of the ecosystem approach, as is the case within the context of traditional resource management. Governments adopting the ecosystem approach as a central component of their national biodiversity and environmental strategies can benefit to a great degree from the contribution of traditional management practices in its implementation.

Although traditional knowledge and the ecosystem approach are generally compatible, the fit is not always perfect. Not every aspect of traditional bio-cultural heritage can fit neatly into the ecosystem approach framework, nor should there be a need to conform to a particular model. With this in mind, the role of culture in the ecosystem approach may require further consideration. From a practical management viewpoint, the legitimacy of the ecosystem approach is ultimately judged by the benefits it provides to communities. Those benefits are linked not only to the health of ecosystems but also to the sensitivity of the approach to the culture of communities (Vierros *et al.*, 2010).

6.5 Implementing the ecosystem approach

One of the key challenges for management of ocean areas is to integrate various management approaches undertaken by sectors into a comprehensive and cohesive plan, with the ecosystem as its central framework. Some countries and regions are starting to develop this type of integration for their EEZs through ocean policies, and implementing them through marine spatial planning.

Marine spatial planning (MSP) is an approach or framework for providing a means to improve decision-making as it relates to the use of marine resources and space. According to the Aspen Institute, MSP is an alternative, public process that collects, analyses, and identifies where human activities occur, in order to achieve agreed upon ecological, economic, and social goals. Key features of successful MSP programmes include consideration of multiple scales; a long-term perspective; recognition that humans are an integral part of ecosystems; an adaptive management perspective; and concern for sustaining ecosystem goods and services (Ehler, 2011). The ecosystem approach and ecosystem based management are principles that underlie MSP in coastal and marine realms. MSP is forward-looking and informed by predefined goals, objectives and policies (CBD, 2012).

MSP is a framework that supports implementation of the ecosystem approach, in recognizing the connections that exist between land, freshwater, and marine ecosystems. It also addresses human uses, opportunities, and impacts in all of these systems. It focuses on three-dimensional and often dynamic ocean space, providing a planning process for the use of that space, which is aimed at delivering the goods and services society needs or desires from marine ecosystems. Ideally, this process considers both natural and political boundaries, reconciling conflicting uses of space in a fair and equitable manner, identifying and promoting synergistic uses, recognizing the intrinsic value of biodiversity, and working within the prevailing political, legal, administrative and cultural regime (CBD, 2012).

The success of MSP, similarly to the ecosystem approach, is dependent on the identification and active participation of all stakeholders. It is also dependent on the management of human uses in a way that respects the complexities and limits of ecosystems and their functioning with a view to long-term sustainability (Vierros *et al.*, 2006). By providing a means to implement the ecosystem approach, MSP has the potential to greatly improve management, reduce the loss of ecosystem services, to help to address or avoid conflict, and to create economies of scale and efficiencies for enforcement and management (CBD, 2012).

According to a study undertaken by the CBD, the equitable sharing of benefits and accountability have been key to longer lasting support for MSP by stakeholders. MSP is primarily about forward planning of space for all human uses and non-uses in the marine environment, followed by the implementation of such plans. MSP is not an end in itself, and it is not a policy. The crucial difference between MSP and other area based management is that MSP is strategic and forward looking rather than reactive and developer led (CBD, 2012).

Within the broader, multiple use framework of MSP, or even without such an overarching framework, the ecosystem approach can also be pursued at the sectoral level. For example, in Australia all fishery management agencies and jurisdictions have adopted EBFM as a policy goal, and implementation has progressed to various levels. For example, in federally managed fisheries, all have been subject to a comprehensive ecological risk assessment process (Hobday *et al.*, 2011) and this has been developed further by the regulator, the Australian Fisheries Management Authority, into environmental risk management plans (AFMA, 2012). These deal with a range of impacts of fishing beyond the target species, including impacts on by-catch, protected species, benthic habitats, and food chains. Assessment of comprehensive ecosystem management plans for fisheries is now possible through the use of ecosystem models of marine systems (Fulton *et al.*, 2011).

An example of transnational management using the ecosystem approach occurs in the Sundarban, which lies on the delta of the Ganges, Brahmaputra, and Meghna rivers on the Bay of Bengal. Sunderban is the largest contiguous mangrove area in the world, with high species richness of mangroves, and considerable structural diversity of the forest, including providing habitat for the threatened Royal Bengal tiger (*Panthera tigris tigris*). The core area (Sunderban National Park) has been designated as a World Heritage Site. All of Eastern India is dependent on fishery resources from the Sunderban. A major part of the management effort focuses on the Sunderban National Park which includes 2320 km^2 of mangroves in the Ganges delta. The park straddles the India–Bangladesh border and is called Sundarban owing to the dominance of the tree species *Heritiera fomes*, known locally as 'sundari'.

Well over three million people live in the biosphere reserve. They depend directly on forest and forest-based resources since agriculture is not an option in this area, dominated by tidal and salt water. Sales of timber, fuel wood, thatching leaves, honey, and wax are the main sources of income. Because of demographic pressures, the Sunderban is under great stress and therefore an eco-development programme exists based on a highly participative approach from the local communities, working with state and national governments. Management emphasis is given to schemes that generate additional income and economic security for the people, such as mangrove forest management, animal husbandry, popularization of energy alternatives, habitat improvement, aquaculture, honey and wax culture, development of crafts, and education – all surrounding the core (World Heritage) zone. While problems do exist from over enthusiastic, or even illegal development of inappropriate aquaculture, in general community involvement has produced a result that typifies application of the ecosystem approach.

In Australia, the management of different pressures on the Australian sea lion has evolved over the past few years from a sectoral approach to one that is now much more focused on the ecosystem in which the sea lion feeds and breeds. Because of mortalities caused by interactions with fishing gear the Australian Fisheries Management Authority has implemented permanent fishery closures around Australian sea lion colonies off South Australia, with associated adjoining zones with strict trigger levels for catch that can be closed to shark gillnetting when maximum by-catch limits are reached. Three of the seven trigger zones (73,180 km^2) have recently been closed to gillnetting for 18 months. Conversely, gillnetting is prohibited in only 14,553 km^2 (25%) of the proposed marine protected area network in this region (although most of the protected areas are at depths beyond 100 m, outside the foraging range of the sea lion). In this situation the fisheries management agency has regulated areas which are more restrictive than the marine protected area network, which only has general zoning guidelines for the same area. This illustrates the complex interplay that can occur between 'sectors' (in this case fisheries and conservation management), even where marine spatial planning has already been implemented.

In the UK there were two important developments in 2012, guided by the framework of the ecosystem approach. Firstly, in the waters around the Scottish coast, selection of Nature Conservation Marine Protected Areas has been undertaken, employing a science-led approach, while at the same time, the selection process has incorporated opportunities for stakeholders to feed in to the development of the network. Five national-level workshops were undertaken with over 80 national stakeholder representatives, providing opportunities for stakeholder interests to input scientific and other information and to discuss the development of proposed MPAs. Furthermore, regular update meetings continue to be held with key interest groups to enable more detailed discussions. These interests extend beyond Scotland, and include UK and non-UK users of Scotland's sea. Moving forward, all users of the Scottish marine environment will have important roles to play in developing the management of proposed MPAs. Full details on this activity can be found at JNCC (2012a). This approach has all of the elements necessary to work under the principles of the ecosystem approach.

The second important development occurring in 2012 was a stakeholder-led project in English, Northern Irish, and Welsh waters to attempt a balance between fisheries, renewable energy (wind farms), offshore exploration and extraction, and conservation. Key to the success of this endeavour was the identification and establishment of a series of Marine Conservation Zones (MCZs). The MCZ Project consisted of four regional MCZ projects covering the south-west (Finding Sanctuary), Irish Sea (Irish Sea Conservation Zones), North Sea (Net Gain), and south-east (Balanced Seas). Historically, designation processes have begun with nature conservation bodies such as Natural England, Countryside Council for Wales and Northern Ireland, Environment Agency, and the Joint Nature Conservation Committee (JNCC), providing advice to the government on where sites should be designated. This advice would be followed by a public consultation to allow the general public to comment on the proposals. However, the government recognized the valuable knowledge and experience sea users and interest groups have of the marine environment, and judged it to be important to consider the social and economic impacts that MCZs might have. It therefore included, in the Marine and Coastal Access Act, provision for stakeholders to be involved in making the initial recommendations for MCZs. These regional MCZ projects worked with sea users and interest groups to identify MCZs within their regions (cf. JNCC, 2012b). In September 2011 the four regional MCZ projects submitted recommendations for MCZs. These recommendations came from groups of sea users and those with an interest in the sea through a series of discussions and negotiations. Natural England and JNCC, as the government's advisers on the natural environment, reviewed these recommendations and submitted advice to government on the science behind these recommendations and the quality of the ecological data. Although this project does not explicitly state it is using the ecosystem approach, by linking socio-economic and cultural aspects with scientific analysis, the final result will be exactly that.

6.6 Conclusions

While the theory and policy settings for the ecosystem approach have developed over the past 30 years, implementation of the theory/policy has lagged considerably, with very uneven take-up around the world. Several key implementation tools exist that could be more closely and consistently brought under the banner of the ecosystem approach, such as the designation and management of biosphere reserves under UNESCO (UNESCO, 1996). A clear and simple operating framework for biosphere reserves (including a growing number of coastal and marine) has been developed, organized in a hierarchy of goals, strategies, and actions, and has three primary goals:

1. Use biosphere reserves to identify and conserve natural and cultural diversity.
2. Utilize biosphere reserves as models for land management and for approaches to sustainable development, through appropriate incentives.
3. Use biosphere reserves for research, monitoring, education, and training.

These broad goals, and their subsidiary objectives, represent a global application of the ecosystem approach. UNESCO (2000) has a fuller account of this relationship.

Many examples exist that are not named as the ecosystem approach but which are consistent with the CBD ecosystem approach principles in Box 6.1. For instance, in some sectors there has been an increase in the adoption of broader ecosystem-based management that takes into account social, economic, and environmental outcomes.

Several general points emerge in reviewing experience with the ecosystem approach in coasts and oceans to date:

- Experience has shown that integration among sectoral policies can be complementary to the reinforcement of individual sectors, hence the ecosystem approach teaches us that integration and sectoral reinforcement are not mutually exclusive (Vierros *et al.*, 2006).
- Flexibility: there is no one 'correct' way to implement the ecosystem approach. Methods may vary and will depend on the environmental, cultural, social, and economic setting of the area. However, certain principles and guidance should apply. In many instances, groups or countries may implement the ecosystem approach without calling it by that name.
- More case studies of the successful application of the ecosystem approach in coastal and deeper ocean environments should be undertaken, to add to the body of work already maintained by the CBD.
- Implementing the ecosystem approach can be demanding of information, and some of the analytical tools to assess impacts and assist planning, as well as on-going management, are still under development.
- There is a need for more comprehensive documentation of ecosystem approach implementation in ocean and coastal areas in order to assess the extent of the area under management. There is also a need to assess the effectiveness of this management for both biodiversity and human needs, leading to an adaptive process of improvement over time.
- There is a need to adopt a 'mountain to ocean trench' approach to management of the coasts and seas, which can be best achieved through application of the ecosystem approach.
- Standardization of terms will be helpful. At present there is a large range of overlapping terminology that only serves to confuse and confound the science–policy discussion.
- Implementation of the ecosystem approach can enable us to better manage multiple impacts on the marine environment. It provides the best strategy for reducing the rate of biodiversity loss in the world's oceans and for maximizing long-term economic, social, and cultural benefits (Vierros *et al.*, 2006).

Further collaboration between existing management and reporting systems that utilize ecosystem based planning and which could be viewed as adopting the ecosystem approach is required to bring together experience and learning, such that the broader philosophy of utilizing the approach can spread. The ecosystem approach is not owned by any particular sector or management or policy structure, albeit both the CBD and the FAO have set a strong and complementary policy framework for its use. It is in implementation that the issue of competing objectives must be faced. In this regard it is in fact mostly non-CBD management systems that have begun the task of implementing the ecosystem approach. Like indigenous approaches across the span of human existence, it is a process of adaptive management and practice, and the passing on of the wisdom of those elements that works. For a better future for ocean management and governance it is to be hoped that the principles and practice of the ecosystem approach will be implemented more widely.

References

Australian Fisheries Management Authority (AFMA) Ecological Risk Assessment methodology webpage: www.afma.gov.au/managing-our-fisheries/environment-and-sustainability/ecological-risk-management/ (last visited 6 November 2014).

Barber, M. (2005). Where clouds stand: Australian Aboriginal relationship to water, place and the marine environment in Blue Mud Bay, Northern Territory. A thesis submitted for the degree of Doctor of Philosophy of the Australian National University.

Constable, A. J., de la Mare, W. K., Agnew, D. J., Everson, I., and Miller, D. (2000). Managing fisheries to conserve the Antarctic marine ecosystem: Practical implementation of the Convention on the Conservation of Antarctic Marine Living Resources (CCAMLR). *ICES Journal of Marine Science* **57**: 778–791.

Convention on Biological Diversity (CBD) (2000), *Decision V/6. Ecosystem Approach.* Accessed on 1 April 2014. Available online at www.cbd.int/decision/cop/default.shtml?id=7148.

Convention on Biological Diversity (CBD) (2010), *Report on the Implementation of the Programme of Work on Marine and Coastal Biological Diversity*, UNEP/CBD/SBSTTA/14/INF/2. Accessed on 1 April 2014. Available online at www.cbd.int/doc/?meeting=sbstta-14.

Convention on Biological Diversity (2012). Decision XI/18, B of the Conference of the Parties to the Convention on Biological Diversity – Voluntary Guidelines for the Consideration of Biodiversity in Environmental Impact Assessments and Strategic Environmental Assessments in Marine and Coastal Areas. Document UNEP/CBD/COP/11/35.

DOALOS (2007), *Ecosystem Approaches and Oceans.* Accessed on 1 April 2014. www.cbd.int/doc/meetings/mar/cbwsoi-wafr-01/other/cbwsoi-wafr-01-ea-and-oceans-en.pdf.

Ehler, C. (2011). *Marine Spatial Planning in the Arctic: A First Step toward Ecosystem-Based Management. A Report of the Aspen Institute's Dialogue and Commission on Arctic Climate Change.* Washington, DC: The Aspen Institute, pp. 40–82.

Fabra, A. and Gascón, V. (2008). The Convention on the Conservation of Antarctic Marine Living Resources (CCAMLR) and the ecosystem approach. *The International Journal of Marine and Coastal Law* **23**(3): 567–598.

Food and Agriculture Organization of the United Nations (FAO) (2003), *Fisheries Management. Suppl. 2. The Ecosystem Approach to Fisheries*. FAO Technical Guidelines for Responsible Fisheries No. 4, 124 pp.

Food and Agriculture Organization of the United Nations (FAO) (2012). *The State of the World's Fisheries and Aquaculture, 2012*. Rome: FAO. Accessed on 1 April 2014, www.fao.org/docrep/016/i2727e/i2727e.pdf.

Food and Agriculture Organization of the United Nations (FAO) (2013), *The Ecosystem Approach to Fisheries Management*. Rome: FAO. Accessed on 1 April 2014, www.fao.org/fishery/topic/13261/en.

Fulton, E. A., Link, J. S., Kaplan, I. C., *et al.* (2011). Lessons in modelling and management of marine ecosystems: The Atlantis experience. *Fish and Fisheries* **12**: 171–188.

Garcia, S. M., Zerbi, A., Aliaume, C. *et al.* (2003). The ecosystem approach to fisheries. Issues, terminology, principles, institutional foundations, implementation and outlook. FAO Fisheries Technical Paper No. 443. Rome: FAO, 71 pp.

Hobday, A. J., Smith, A. D. M., Stobutzki, I. C., *et al.* (2011). Ecological risk assessment for the effects of fishing. *Fisheries Research* **108**: 372–384.

JNCC (2012a), *Advice to the Scottish Government on the Selection of Nature Conservation Marine Protected Areas (MPAs) for the Development of the Scottish MPA Network*. Scottish Natural Heritage Commissioned Report No. 547. Accessed 1 April 2014. http://jncc.defra.gov.uk/PDF/SNH%20and%20JNCC%20MPA%20network%20advice%20-%20Final%20report.pdf.

JNCC (2012b), *JNCC and Natural England's Advice to Defra on Recommended Marine Conservation Zones*. Accessed 1 April 2014. http://jncc.defra.gov.uk/PDF/120718MCZAP_JNCC_NE_MCZ%20advice_final_Executive_Summary.pdf.

Pikitch, E. K., Santora, C., Babcock, E. A., *et al.* (2004). Ecosystem-based fishery management. *Science* **305**: 346–347.

Pitcher, T., Kalikoski, D., Pramod, G. *et al.* (2009). Not honouring the code. *Nature* **457**: 658–659.

Ruddle, K. and Hickey, F. R. (2008). Accounting for the mismanagement of tropical nearshore fisheries. *Environment, Development and Sustainability* **10**(5): 565–589.

Shepherd, G. (ed.) (2008). *The Ecosystem Approach: Learning from Experience*. Gland: IUCN, 190 pp.

UNESCO (1996). *Biosphere Reserves: The Seville Strategy and the Statutory Framework of the World Network*. Paris: UNESCO.

UNESCO (2000). *Solving the Puzzle: The Ecosystem Approach and Biosphere Reserves*. Paris: UNESCO, 31 pp.

Vierros, M., Douvere, F., and Aricò, S. (2006). *Implementing the Ecosystem Approach in Open Ocean and Deep Sea Environments*. Tokyo and Yokohama: United Nations University – Institute of Advanced Studies, 39 pp.

Vierros, M., Tawake, A., Hickey, F., *et al.* (2010). *Traditional Marine Management Areas of the Pacific in the Context of National and International Law and Policy*. Darwin: United Nations University Traditional Knowledge Initiative, 87 pp.

Yamada, T., Hashimoto, H., and Muroishi, Y. (2011). *Sato-umi policy in Japan*. [Powerpoint slides]. Office for Environmental Management of Enclosed Coastal Seas, Ministry of the Environment, Japan. Available at www.emecs.or.jp/emecs9/EMECS9_Satoumi/10-2_Chino.pdf. Accessed June 2014.

Yanagi, T. (2008) 'Sato-Umi' – A new concept for sustainable fisheries. In: *Fisheries for Global Welfare and Environment*. Tsukamoto, K., Kawamura, T., Takeuchi, T., Beard, T. D. Jr., and Kaiser, M. J. (eds.). 5th World Fisheries Congress, 2008, 351–358.

Note

1 www.hawaiihistory.com.

7

Challenges in using valuation in ecosystem-based management in a marine context: the case of UK Marine Protected Area designation

S. SALMAN HUSSAIN

7.1 Introduction

Marine environments provide significant benefits to man, for example through the provision of seafood and other resources worth trillions of dollars per annum, regulation of the Earth's climate, and the modulation of global biogeochemical cycles (Holmlund and Hammer, 1999), water quality maintenance (Worm *et al.*, 2006), and also cultural and aesthetic benefits (Lewis, 2012). In spite of its clear importance, the marine environment is subject to huge pressure from anthropogenic sources. These pressures arise from inter alia the overexploitation of marine species, nutrient loading from land-based agriculture, underwater noise, the introduction of invasive species, and increasing oceanic acidification and habitat degradation. The threats are frequently interlinked. Further, in terms of biodiversity loss, Lotze *et al.* (2006) estimate that the human exploitation of marine species is responsible for 96% of species extinctions.

The ecosystem based approach (EBA) is a managed response to these anthropogenic impacts on marine ecosystems. The aim of the EBA is to facilitate the protection, recovery, and sustainable use of marine ecosystems, applying a multisector approach. Any management approach is likely to entail trade-offs, in the sense that there are some stakeholders who will realize net benefits from the management intervention and others that will incur net losses. There are winners and losers even *within* a body of stakeholders (e.g. fishermen). Further, one significant broad stakeholder group is 'wider society', and it is important to capture broader social welfare changes arising from any management intervention.

There is considerable variability in the extent to which the benefits and costs arising from applying the EBA are measurable, and also the extent to which they can be valued in monetary terms. For instance, changes in fish stocks have a relatively direct market value, whereas other benefits (such as a feeling of well-being arising from the aesthetics of the marine landscape) are much less tangible.

146

This is where the burgeoning literature on ecological economics can contribute in the policy domain. In economic terms, changes in ecosystems can realize welfare effects even if these changes do not have a direct market price. For instance, marine habitats might provide inspiration for art or provide opportunities for leisure and recreation that people do not pay for directly in the market, e.g. there is no charge to swim in the sea, and yet it is clear that people value such access – it affects their well-being. Some elements of these benefits afforded to us might be marketable (e.g. the price for a painting that has been inspired by the ocean, or the premium that we may pay for a hotel with a sea view or direct access to a beach), but the ability to use what are termed 'surrogate markets' is limited, and approaches that rely on these surrogate markets only partially capture total benefits (Edwards-Jones *et al.*, 2000).

This has led to the development of environmental valuation methodologies in economics to value such non-marketed benefits that nature provides. The focus has been on providing a monetary value for these benefits. This has in part been driven by the perceived need to apply comparisons on a like-for-like basis using a common unit of account, i.e. a monetary unit (dollars, euros etc.) The standard economic argument in favour of such an approach is that in the absence of a common metric it is not possible to assess fully the worth of one proposed policy versus another, and the only common metric that is universally accepted is money. So if we can express benefits in monetary terms then this allows us to apply cost–benefit analysis (CBA). If the rules of CBA application are adhered to then we will be making like-for-like comparisons, and therein CBA can be applied as a performance yardstick to appraise policy options.

This rationale is applied by those who advocate the application of environmental economic valuation to the EBA to marine management. There are many agents who are affected by any management measure or groups of measures, which we may term a management strategy. If we want to appraise management strategy A versus B then we ought to categorize all the benefits and costs arising from A and B respectively and make a comparison. We should allow for uncertainty in our estimates, and also consider the timing of any impacts so as to allow for what is termed society's 'positive rate of time preference', i.e. the presumption that a cost or benefit arising soon in the future affects our welfare more than that same cost or benefit arising in the distant future. Economics capture this in what is termed discounting (TEEB, 2010). Discounting is not discussed further here as it is beyond the scope of the current chapter, but the discourse is really important owing to the fact that the process systematically biases decision-making towards projects and policies that realize more immediate benefits and delay costs; this has a direct impact on inter-generational equity and thus sustainability.

This core economic modus operandi is well established and well discussed (see Hussain and Gundimeda, 2011 for instance). The aims of this chapter are as follows: (1) to discuss how valuation can be and has been applied in the EBA to marine management, (2) the criticisms and limitations of this standard approach, with a particular focus on marine ecosystems, and (3) a pragmatic example of applying the EBA in valuing the benefits of UK marine protected areas (MPAs).

7.2 Ecosystem services typologies

The concept of ecosystem services provides a mechanism for people to understand how our existence is linked with the natural environment and has been defined as 'the benefits that humans obtain from ecosystems' (MA, 2005). The need for a typology of ecosystem services arises so as to ensure that all of these benefits are made explicit, as any omission of benefit categories leads to a systemic under-representation of the benefits arising from measures aimed at conserving nature. The core principle is to make the benefits visible, to remove what is otherwise a pro-extractive, contra-conservation bias in decision-making. But there is also a corollary to this argument in that the typology must be designed so as to avoid double-counting.

The Economics of Ecosystems and Biodiversity (TEEB) study (TEEB 2010) has developed such a typology, following (but modifying) the framework used in the Millennium Ecosystem Assessment (MA, 2005). There are four main categories of services: (1) provisioning services such as sea fish for human consumption; (2) regulating services such as gas and climate regulation; (3) supporting/habitat services (e.g. seagrass beds providing a nursery habitat for juvenile fish); and (4) cultural/amenity services such as leisure and recreation.

The typologies in TEEB (2010) and MA (2005) are both constructed from a terrestrial ecosystem perspective; Beaumont *et al.* (2008) argue that marine ecosystems have been marginalized. Costanza *et al.* (1997) value global ecosystem services and natural capital at US$16–54 trillion per annum, and this remains arguably the most significant publication in the valuation literature. Notwithstanding the methodological controversies as to the actual estimates reported, the *ratio* of marine/terrestrial ecosystem service values is noteworthy, with marine ecosystems providing around two-thirds of the global aggregate. If this is coupled with the evidence to support the contention that this service provisioning is threatened (e.g. Halpern *et al.*, 2008), then the need for a service typology that is framed for marine ecosystems follows. and one that is internally consistent, i.e. a typology that avoids double-counting. Boyd and Banzhaf (2007) go so far as to argue that a lack of internal consistency, particularly as it relates to the non-provisioning services, has severely constrained the adoption and operational use of the ecosystem service typologies published to date in the literature.

In a recent review, Böhnke-Henrichs *et al.* (2013) find two extant typologies that are marine focused (Beaumont *et al.*, 2007; Atkins *et al.*, 2011). Both are valid in their own right and indeed the Beaumont *et al.* (2007) framework is applied in the case study of UK marine protected areas that follows. But the MA (2005) and TEEB (2010) typologies remain ubiquitous in the policy arena. This would not be problematic were the differences between these dominant typologies and the marine focus to be semantic, e.g. defining 'sea food' as a marine-specific ecosystem service rather than 'food', or the removal of inapplicable services such as 'pollination' and 'maintenance of soil fertility'. The more substantive issue is where to draw the boundaries around a service so as to avoid double-counting. For instance, one source of carbon sequestration is via buried organic matter; so the higher the TEEB (2010) provisioning service of 'food' (i.e. capture fisheries), the lower the available organic matter for 'climate regulation', all else being equal (Böhnke-Henrichs *et al.*, 2013). Further, extracting fish for 'food' reduces their abundance and this could affect 'opportunities for recreation and leisure', for those wishing to go snorkelling or diving. These boundary issues are not unique to marine ecosystems but the potential for double-counting is greater in some cases as compared with terrestrial ecosystems.

7.3 Marine ecosystem service valuation application: UK marine protected areas

There is a paucity of potential case studies that are specifically marine focused and show the application of the ecosystem approach using economic valuation. The case study presented in this section meets these requirements, and moreover, the outcomes of the valuation of changes in ecosystem services provisioning have led to a tangible outcome, i.e. the UK Marine and Coastal Access Act (2009).

Part of the Act is the designation of a network of marine protected areas, termed Marine Conservation Zones (MCZs) in UK legislation. Part of the benefit assessment is reported in Hussain *et al.* (2010) and McVittie and Moran (2010). What is not set out in these papers is the *process* of applying the ecosystem approach across marine ecology and economics disciplines, and linked to this the lessons learnt. Further, the methodology used has been modified and applied in a European Commission Framework Programme 7 project entitled 'Options for Delivering Ecosystem-based Marine Management' (ODEMM: see www.liv.ac.uk/odemm/), and these modifications are discussed herein.

The objective of this section is not merely to set out the specifics of the MCZ analysis and the methodological development therein, but to allow a platform for a more wide-ranging discussion of the EBA and how any appraisal of management measures for marine ecosystems might be framed to assess economic efficiency. The discussion as to whether indeed an economic efficiency assessment is appropriate and might become an *impediment* to society making appropriate choices is discussed in Section 7.4.

7.3.1 The policy context

The proposed network of MCZs is intended to complement existing European marine sites, such as Special Areas of Conservation (SACs) and Special Protected Areas (SPAs), with the aspiration to form an ecologically coherent network of marine protected areas. The policy context for this UK study appears on first inspection to be highly specific and limited in terms of global transferability: MPAs for a developed world nation with relatively good marine habitat mapping; temperate waters; and relatively few transboundary impacts. However, the methodology developed in Hussain *et al.* (2010) was designed to contend with issues which are universal in a marine ecosystem valuation context: (1) establish a business-as-usual (BAU) baseline; (2) conduct a valuation appraisal with a paucity of valuation data points; (3) ecosystem service values, where available, were expressed in aggregate form, with a need to determine marginal (incremental) impacts of the policy designation; (4) accounting for the timing of changes in ecosystem service provisioning relative to BAU; and (5) the need to apply sensitivity analysis and risk assessment.

The valuation study reported in Hussain *et al.* (2010) is an *ex ante* appraisal which was based on three proposed networks of MCZs. The basis for the selection of networks of MCZs was provided by initial research by Richardson *et al.* (2006) which developed 12 network scenarios based on varying OSPAR (Oslo and Paris Convention for the Protection of the Marine Environment of the North-East Atlantic) and other criteria, as set out in Table 7.1.

The final selection of MCZ sites is unlikely to map onto any of these networks. This is because the analysis carried out in Richardson *et al.* (2006) and used in Hussain *et al.* (2010) was for *indicative* networks. Between November 2009 and September 2011 there was a stakeholder consultation phase which was to guide final MCZ locations, moderated by scientific appraisal of the extent of ecological coherence of the proposed networks, i.e. large enough, and close enough together, to support functioning communities of marine wildlife and threatened and representative habitats.

Notwithstanding the fact that the analysis in Hussain *et al.* (2010) does not correspond with MCZ designation, in policy terms it was a necessary stepping stone so as to test the validity – in economic efficiency terms – of proceeding with

Table 7.1 *Synopsis of protection criteria for Marine Conservation Zones Network Scenarios A, G, and J (Source: Moran* et al.*, 2007)*

Scenario	% of OSPAR species and habitats included	% of UK marine landscapes included	Network size (1000 km^2)	Additional criteria
A	20	10	125.7	None
G	60	10	156	Commercial fishery species spawning and nursery areas preferred to protect areas essential to life history stages
J	60	10	147.2	Locked-out sites licensed for aggregate extraction, dredging, and dredge disposal activities.

the legislation that permits MCZ designation. But the need for stakeholder input to any policy measure is critical and I return to this discussion below.

The more general policy context was that any UK policy measure above a certain threshold is subject to an impact assessment – previously termed a regulatory impact assessment (BERR, 2008). This typically entails a strong element of (quantitative) economic cost–benefit appraisal, although there is scope to add qualitative assessment. The ecosystem approach was adopted in the sense that benefits were the additional level of ecosystem service provisioning from with-policy (MCZ designation) versus BAU.

Although it may not be mandatory to carry out such economic assessments in other legislative jurisdictions, the socio-political appetite for conservation measures is likely to be higher if such an economic evidence base supports the case for the measures.

In many marine-focused interventions, there are likely to be stakeholder groups that have a common shared interest, are well resourced in terms of the funds for lobbying, have direct access to decision-makers, and might make not just a fiscal case for continuing resource extraction and ecosystem exploitation, but also a social and cultural case. By contrast, the beneficiaries of conservation measures may be the general public at large. The ecosystem approach explicitly entails trade-off analysis wherein increases in the provisioning of a range of services are juxtaposed with decreases in the provisioning of others, weighted with respect to marginal change and the value of the service to society. As such an economic efficiency assessment of a marine focused management measure can in theory provide a level playing field and a like-for-like comparison, although there are limitations to this approach (set out in Section 7.4). Although a vocal, well-organized lobby might still prevail, irrespective of any efficiency arguments supporting a conservation measure, such economic arguments typically hold sway, especially in the light of the current global recession.

Note that it is not the contention that all conservation measures are efficient – this would clearly be erroneous. The argument is more that since decisions tend to be made – for better or for worse – using the economic unit-of-account of money then a failure to value the ecosystem service benefits generated by conservation measures in the same unit-of-account can lead to resource over-extraction and unsustainable practices.

For the MCZ case study, it is noteworthy that many such vocal and well-organized stakeholder groups are to be potentially adversely affected by the marine spatial planning designations, including the oil and gas industry, the telecommunications sector, and indeed fishermen, although there is an argument here that the off-site spillover benefits as stocks recover might exceed the short-term costs of MCZ restrictions (Dugan and Davis, 1993; Sanchirico and Wilen, 2001). The sections that follow set out the methodology applied in detail, and the results, but suffice to state here that estimates of benefits exceeded costs by a factor of at least seven and therein contributed to the evidence base supporting the Act.

7.3.2 Approach and outcomes

The methodology is split into the following elements: (1) defining a typology of ecosystem services and estimating aggregate values for these services typologies: ecosystem services and the biophysical structures and processes in the marine ecosystem that deliver these services; (2) biophysical framework; (3) biophysical impact scoring – BAU versus MCZ implementation – per habitat type; (4) assessing individual estimates of the value of the change per ecosystem service/habitat type; (5) aggregation.

Each element is treated in turn in the sections that follow. The methodology and results are discussed together for each subsection for ease of exposition, as the methodological steps are arguably more discernible and 'concrete' if outcomes are presented at each stage.

7.3.2.1 Defining typologies and estimating the value of aggregate service provisioning

Section 7.2 discussed typologies of ecosystem services. The typology adopted for the MCZ case study in Hussain *et al.* (2010) is based on Beaumont *et al.* (2006) and is summarized in Table 7.2. Note that the list of 11 ecosystem services is a subset of the 22 services in TEEB (2010). Although a theoretical case can be made for

Table 7.2 *Marine ecosystem services – categories, definitions, and values in 2007 (after Beaumont* et al., *2006; Hussain* et al., *2010)*

Ecosystem category	Ecosystem service	Monetary value	Definition
Provisioning	Food provision	£885 million	Plants and animals taken from the marine environment for human consumption
	Raw materials	£117 million	The extraction of marine organisms for all purposes, except human consumption
Supporting	Nutrient cycling	£1.3 billion	The storage, cycling, and maintenance of availability of nutrients mediated by living marine organisms
	Resilience and resistance	N/A	The extent to which ecosystems can absorb recurrent natural and human perturbations and continue to regenerate without slowly degrading or unexpectedly flipping to alternative states
Regulating	Gas and climate regulation	£8.2 billion	The balance and maintenance of the chemical composition of the atmosphere and oceans by marine living organisms
	Biologically mediated habitat	N/A	Habitat which is provided by living marine organisms
	Disturbance prevention and alleviation	£440 million	The dampening of environmental disturbances by biogenic structures
	Bioremediation of waste	N/A	Removal of pollutants through storage, dilution, transformation, and burial
Cultural	Cultural heritage and identity	N/A	The cultural value associated with the marine environment e.g. for religion, folklore, painting, and cultural and spiritual traditions
	Cognitive values	£453 million	Cognitive development, including education and research, resulting from marine organisms
	Leisure and recreation	£1.4–3.4 billion	The refreshment and stimulation of the human body and mind through the perusal and engagement with living marine organisms in their natural environment

the inclusion of other ecosystem services to supplement those in Beaumont *et al.* (2006), in pragmatic terms the typology is fit for purpose in that the designation of the service within the typology affects policy if and only if a value can be estimated, either qualitatively or quantitatively. Further, in policy terms a management option must feasibly be able to change provisioning.

Consider one potential service that is marine-specific but not included in Beaumont *et al.* (2006): marine water extracted for industrial use, an example being water for cooling power stations. This is not a form of *consumptive* use unlike, for instance, fresh water used in irrigation. Although it is likely that there would be a substantial cost to industry and society were water for cooling not to be available, it is difficult to conceive of a real-world management measure that would reduce the provisioning of this ecosystem service. As such, if we compare BAU versus with-measure, this particular service does not affect the comparison and thus in economic assessment terms (for *marginal* analysis) this service is not critical. If, however, we wish to articulate some *total* value of ecosystem service benefits provided by the marine ecosystem then it would be relevant.

One ecosystem service that is omitted from Hussain *et al.* (2010) is 'maintenance of genetic diversity' (TEEB, 2010). Historically, the total economic value (TEV) typology dominated the environmental economics discourse (Edwards-Jones *et al.*, 2000). The mapping of TEV to the TEEB (2010) classification is discussed in de Groot *et al.* (2012) and is beyond the scope of this chapter. But a significant exclusion (TEV versus MA/TEEB) is what is termed non-use or existence values, i.e. biodiversity conservation for its own sake irrespective of whether the resource is experienced in any form by humans. Although 'maintenance of genetic diversity' is unto itself anthropocentric and thus does not equate with existence value, there are associations between them. Evidence was collected in McVittie and Moran (2010) as to the value of these non-use values for the MCZ study and these are discussed below.

The third column in Table 7.2 sets out the estimates for the total value for service provisioning by UK marine and coastal waters in 2007 (the study year), for seven of the eleven services for which any estimate was available. Given the requirement in the BAU versus MCZ-adoption comparison for our need for a *marginal* value, why is service provision expressed in *aggregate* (total) terms? The answer to this question is simply because no alternative data points were available. It is also extremely likely that other economic policy appraisals applying the ecosystem approach would also only have data points expressed in aggregate, and so the approach (or a modification thereof) developed in Hussain *et al.* (2010) and discussed further here is relevant, not just to the UK MCZ case study, but more widely.

Estimates for provisioning services have the highest degree of certainty in the MCZ case study and this is likely to apply in most ecosystem services valuations.

The food provisioning estimate used UK data (Beaumont *et al.*, 2006) for ex-vessel total fish landed by UK fishing fleets, with a value-added factor of 0.45 following Pugh and Skinner (2002) for supply chain effects. The raw material estimate used market data on fish meal and seaweed following Beaumont *et al.*, (2006). Of the two supporting services, a value was only available for 'nutrient cycling,' and even this itself draws upon Costanza *et al.* (1997); a study that is both methodologically controversial and dated.

The regulating services for which estimates were possible were 'gas and climate regulation' and 'disturbance prevention and alleviation.' For the former, the estimate used in Hussain *et al.* (2010) again follows Beaumont *et al.* (2006) and uses a photosynthesis model (Smyth *et al.*, 2005) to estimate the average annual primary production (carbon sequestered by phytoplankton) in the UK to be approximately 0.07 +/− 0.004 giga ton carbon/year. This estimate of carbon sequestered was multiplied by the average value for the shadow price of carbon over the 20-year study period used by the UK government, i.e. £117.7/t C in 2007. As Beaumont *et al.* (2006) note, this is an underestimate of service provisioning as it only considers phytoplankton. For this service, the economic analysis is fairly trivial, i.e. multiplying the outputs of biophysical modelling by a value/t C. It is the biophysical modelling that is the most onerous task.

The estimate for 'disturbance prevention and alleviation' is again based on a single study that is somewhat dated (King and Lester, 1995), which estimates the value of this service just for saltmarshes. Other habitats that provide this service include littoral rock, littoral sediment, and seagrass. The approach used for this service is the 'averted cost' method. This service is concerned with reducing the intensity of disturbances such as storms, the service being provided as ecological structures and processes dissipate energy. This form of economic valuation applies the following logic: were the ecological structures and processes not to be present, what would be the value of the expected extra damage caused by these natural events? This requires a biophysical assessment of the frequency and impact of such events, coupled with the economic assessment of the financial losses occurring (damage to property etc.). An associated method for valuing this service is to estimate the cost of building and maintaining engineered flood defences; these costs might be reduced were marine ecosystems to provide more defence 'naturally', and vice versa.

In terms of cultural services, the estimate follows Beaumont *et al.* (2006), which is itself derived from Pugh and Skinner (2002), based on the value added by research and development in the marine sector in the UK in addition to education and training. For this service, it might be argued that only research pertaining to UK marine and coastal waters ought to be included, i.e. not all marine research carried out in UK research institutions, but it is difficult to partition research in this manner. Indeed, a fundamental premise of ecosystem based marine management is that an appropriate spatial scale is considered, which in many (indeed most) cases will be transboundary, e.g. 'the North-East Atlantic' for UK marine waters.

The ecosystem service of 'leisure and recreation' includes sea angling, recreational diving, and whale-watching. These activities depend on the marine environment and are carried out in the marine (as opposed to *coastal*) environment. But drawing the boundaries as to what to include and what to exclude is difficult for this service. Hussain *et al.* (2010) provide two estimates. The first is restricted to this narrower range of leisure activities, and is based on Pugh and Skinner (2002) and reported in Beaumont *et al.* (2006). The higher estimate is based on a wider definition of the service, including visits to the coast and associated expenditure. If the latter, wider definition is accepted then the estimate of £3.4 billion in 2007 is an underestimate, as the total value of the tourism activity exceeds direct spend in local coastal communities. If people travel to come to the coastline for recreation, but do not incur any costs locally they have still *signalled a preference* for this activity. One approach is to value leisure time in the context of what is termed the 'travel cost approach' (Edwards-Jones *et al.*, 2000), wherein actual expenditure on travel (fuel, vehicle depreciation, train tickets etc.) and spend on any entrance fee to the ecological site are added to some estimate of the value of leisure time per hour, typically some percentage of hourly wage rates. However, there are insufficient travel cost method studies to allow such results to be extrapolated UK-wide, as would be required for the MCZ study.

What general take-home messages can be drawn from this methodological stage of the MCZ case study? First, the UK is a reasonably data-rich environment, but in spite of this estimates were only available for seven of the eleven services, and even amongst these some use weak proxies. Valuation data gaps are likely to be more significant (relative to the UK) in other policy appraisals. Second, for many services environmental valuation entails at least as much biophysical assessment as economics per se. Valuation is not a substitute for biophysical assessment but indeed *relies* on it. Third, if we defer to the economic valuation approach, then it is important that we do not lose sight of those services which cannot be monetized (given available literature, data, and resources). I return to this issue in Section 7.4.

7.3.2.2 Biophysical framework

Marine ecosystems around the UK can be characterized and classified based on a number of geophysical attributes, including bathymetry, seabed sediments, bedforms, maximum near-bed stress, and other data. The classification system used in Hussain *et al.* (2010) was based on the UKSeaMap classification, i.e. 26 predominant habitat types and nine threatened and declining habitats (TDHs) at the time the study was being carried out. If we have an aggregate value for the provisioning of a particular ecosystem service under BAU for the UK as a whole, then this total has to be split between these landscapes and habitats.

The first question to ask is whether the extent of marine ecological scientific knowledge can support an assessment using the ecosystem approach. This subject-ive appraisal (adapted from Moran *et al.*, 2007) is set out in Table 7.3, with the list of 26 habitats in respective rows. For ease of exposition only those seven ecosystem services for which (aggregate) economic values were available are presented in Table 7.3. It is clear that there is significant variability in the extent of certainty that can be applied in biophysical estimates across different marine landscapes, with a distinct area of poor knowledge in the oceanic landscape. In any ecological eco-nomic assessment it is important to make such uncertainty transparent.

The extent of uncertainty does not necessarily correlate directly with the generic level of scientific knowledge pertaining to each habitat type. The analysis of ecosystem services is relatively new to marine ecology and thus the scientific evidence base has to be *mapped onto* final service provisioning.

This is being explored in the aforementioned ODEMM project. Whereas the Defra study on MCZs (Hussain *et al.*, 2010) attempted to map landscapes (and threatened and declining habitats) directly onto services, as per Table 7.3; ODEMM maps the link with ecosystem services via *ecological characteristics*. This categorization is based on the presumption that it is not habitat type alone that influences ecosystem service provisioning. Habitats are included as one category of characteristics, but there are two other broad categories: (1) biological features (phyto-zooplankton; bottom fauna and flora; fish; marine mammals and reptiles; seabirds; species listed under EC legislation or conventions; non-indigenous/exotic species); and (2) other features (chemicals; temperature; salinity; nutrients and oxygen; pH; pCO_2).

The linkage framework used in the ODEMM project is set out in Figure 7.1. The ecological characteristics which include (but are not limited to) habitat types are impacted upon by human pressures (which in turn are linked to industry sectors), and by environmental drivers such as climate change. The difference between the two drivers is that human pressures are 'manageable', whereas environmental drivers are not. The ecological characteristics in turn link with the high-level descriptors of Good Environmental Status (GES) in the EC Marine Strategy Framework Directive (MSFD).

Both the MCZ framework developed in Hussain *et al.* (2010) (i.e. direct link from a subset of ecological characteristics to services) and the more complex ODEMM linkage framework are valid and fit for purpose, and both allow environ-mental economic valuation to be applied in that the unit of account is changes in ecosystem service provisioning. A similar approach would likely be applicable in most ecosystem service-based evaluations of management measures.

Table 7.3 *Summary of the extent of knowledge of habitat types/ecosystem service (H=high, M=medium, L=low) (Source: Moran et al., 2007)*

	Food provision	Raw materials	Nutrient recycling	Gas and climate regulation	Disturbance prevention and alleviation	Cognitive values	Leisure and recreation
Aphotic reef	H	M	H	H	H	M	H
Oceanic cold water coarse sediment	L	L	L	L	H	M	M
Oceanic cold water mixed sediment	L	L	L	L	H	M	M
Oceanic cold water mud	L	L	L	L	H	M	M
Oceanic cold water sand	L	L	L	L	H	M	M
Oceanic warm water coarse sediment	L	L	L	L	H	M	M
Oceanic warm water mixed sediment	L	L	L	L	H	M	M
Oceanic warm water mud	L	L	L	L	H	M	M
Oceanic warm water sand	L	L	L	L	H	M	M
Photic reef	H	M	H	H	H	M	H
Shallow strong tide stress coarse sediment	H	M	M	M	H	M	H
Shallow moderate tide stress coarse sediment	H	M	M	M	H	M	H
Shallow weak tide stress coarse sediment	H	M	M	M	H	M	H
Shallow strong tide stress mixed sediment	H	M	M	M	H	M	H
Shallow moderate tide stressed mixed sediment	H	M	M	M	H	M	H
Shallow weak tide stress mixed sediment	H	M	M	M	H	M	H
Shallow mud	H	M	H	H	H	M	H
Shallow sand	H	M	H	H	H	M	H
Shelf strong tide stress coarse sediment	H	M	M	M	H	M	H
Shelf moderate tide stress coarse sediment	H	M	M	M	H	M	H
Shelf weak tide stress coarse sediment	H	M	M	M	H	M	H
Shelf strong tide stress mixed sediment	H	M	M	M	H	M	H
Shelf moderate tide stress mixed sediment	H	M	M	M	H	M	H
Shelf weak tide stress mixed sediment	H	M	M	M	H	M	H
Shelf mud	H	M	H	H	H	M	H
Shelf sand	H	M	H	H	H	M	H

Linkage Diagram

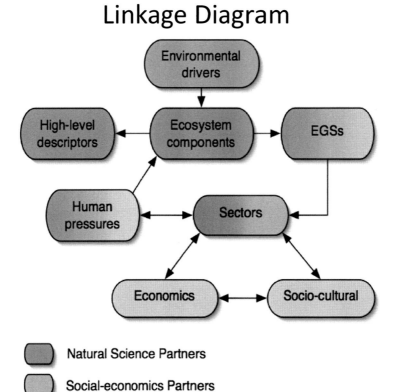

Figure 7.1 ODEMM linkage framework (modified from Koss *et al.*, 2011). A black and white version of this figure will appear in some formats. For the colour version, please refer to the plate section.

7.3.2.3 Biophysical impact scoring

As set out in Table 7.3, the assessment framework for the MCZ study was based on combinations of marine habitat/ecosystem service. Since the exact location of the MCZ study sites was not known and the networks developed in Richardson *et al.* (2006) were only indicative, the evaluation methodology used was generic, in the sense that the ecosystem provisioning of a 'typical' hectare of a habitat type (e.g. aphotic reef) was estimated under BAU and MCZ applications.

One of the critical issues in any bio-economic assessment is to set up a methodological framework that considers only *marginal* changes, and thus cross-refers the with-policy scenario with BAU. For the MCZ case study, the time horizon was 20 years from policy inception and as such, an assessment of what would happen under BAU for each habitat/service combination was required. Any extant conservation measures that would have an impact during the 20-year study period are included in BAU; these included three statutory marine nature reserves, 76 Special Areas of Conservation (for marine habitats and species), and 72 Special Protection Areas (marine habitats for birds) in the UK. Further, any predictable future trends (over 20 years) in the production levels and/or production techniques for those industrial sectors applying pressure on UK marine ecosystems also need to be embedded in BAU projections.

The marine ecological coding of cumulative impacts on each habitat/service combination was conducted independently of the economic valuation. For both BAU and MCZ designation, the coding of each cell contained three elements: (1) the percentage change in provisioning between project inception (now) and the end of the study period (20 years); (2) how quickly the ecological system would demonstrate this change in provisioning; and (3) the trajectory of this change (i.e. linear, exponential, or log).

Figure 7.2 represents three trajectories for a change in provisioning in the service 'nutrient cycling', by way of an example of how the coding worked. There are two common elements to each of the three trajectories: (1) compared with the level of nutrient cycling provided by the habitat at $t = 0$, there is a 70% increase (i.e. 0.7 on the vertical axis); (2) this change in provisioning takes 10 years, and after this point (from $t = 10$ to $t = 20$) the provisioning of nutrient cycling by this habitat plateaus at +70% of current ($t = 0$) provisioning. The three curves provide three alternative impact pathways from $t = 0$ to $t = 10$. In practice any one habitat would be coded with only one such trajectory. The highest point on the vertical axis (in this case 0.7) provides the maximum expected annual benefit of the policy versus BAU. In the coding applied in the MCZ case study (Hussain et al., 2010), 70% corresponds with a coding of 'high' impact, the mid-point of the range

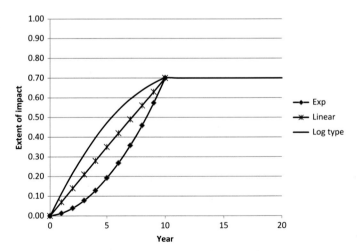

Figure 7.2 Impact trajectory for one habitat/service combination.

50%–89%. Were the coding to have been 'low' then the vertical axis would extend to 0.05, i.e. the mid-point of 1%–9%. Further categories were 'very high' (90%–100%), 'medium' (10%–49%), and 'very low' (<1%).

Figure 7.2 is drawn under the assumption that the time coding is '10/20', i.e. maximum provisioning reached at $t = 10$ and then this level persisting until $t = 20$. Other time codes included $t = 5$, $t = 8$, $t = 15$, and $t = 20$, i.e. 5/20, 8/20, 15/20, 20/20.

This coding was applied for each habitat/service combination for both BAU and MCZ designation. The BAU was used as a reference point. The with-policy MCZ coding was applied relative to this baseline to estimate additional, marginal impacts. Hussain *et al.* (2010) assessed two levels of restrictions in terms of access to MCZ designated sites. Under the higher level of restriction (general presumption against fishing of all kinds and all destructive and disturbing activities) around 47% of cells were coded as having either a 'high' or 'very high' impact from MCZ designation.

7.3.2.4 *Individual value estimates per ecosystem service/habitat type*

The impact scoring for MCZ designation allows a biophysical estimation of the additional provisioning for a typical hectare of the habitat type, measured in terms of current ($t = 0$) provisioning of that service. The total area under the respective curve in Figure 7.2 provides the benefit of MCZ application for this habitat/service combination, across the 20-year time period. But to determine what this value is in absolute (monetary) terms it is necessary to determine $t = 0$ provisioning.

For some applications of the ecosystem approach to marine management it might be possible to provide a direct estimation that is specific to the habitat and service in question. Since only aggregate provisioning estimates were available for UK marine ecosystems, it was necessary to *apportion* this total provisioning across all habitats. So for instance, what contribution does the totality of aphotic reef in the UK make to the totality of nutrient cycling provision across all UK marine ecosystems?

This apportioning was dependent on the service and there were four variants. For four of the seven services (food provisioning; raw materials; nutrient cycling; gas and climate regulation) the apportioning was based on two factors: (1) the extent of the habitat in hectares; (2) the average service provisioning for a 'typical' hectare. Each of the habitats was scored in the range 1–3 to determine the relative contribution of one hectare. For instance, 'aphotic reef' was scored '1' for all four services whereas 'shelf mud plain' was scored '3', i.e. it is assumed that one hectare of shelf mud provides around three times as much nutrient cycling as one hectare of aphotic reef.

This is then coupled with extent data. The total area of aphotic reef is 10,968km^2 or 1.8% (across all UK marine habitats), whereas the total area of shelf mud is 44,605km^2 or 7.3%. Since the respective scores were 1 and 3, it is estimated that the sum of all aphotic reef provides around 0.7% of the total nutrient cycling across all UK marine habitats, whereas shelf mud plain provides 22.0%.

For disturbance prevention and alleviation, a similar coding system was applied, but modified to allow for distance to shore. For the two remaining services (cognitive value; leisure and recreation) allocation was arbitrary, i.e. an equal allocation across all habitat types, as there was insufficient biophysical evidence to support any alternative allocation.

There is little doubt that the allocation method set out here and in Moran *et al.* (2007) can be improved upon, and indeed within the EC ODEMM project the methodology being developed and applied is more rigorous. For instance, a review of the marine ecosystem service valuation literature revealed over 50 studies on leisure and recreation, potentially allowing for meta-regression analysis to determine the link between value estimates for this service and ecological characteristics. But the point to be made again is that for the MCZ case study a viable and defensible value estimate can be and was developed, despite resource and time constraints, and this may well apply in other applications of EBA using valuation.

The outcome of the analysis described in this section thus far is a figure for overall percentage contribution of a habitat type towards a service, e.g. 22.0% for shelf mud for nutrient cycling. The next methodological step is to refer back to the aggregate value estimates (Table 7.2); 22.0% of the £1.3 billion value for nutrient cycling implies that all shallow mud in UK marine waters in total provides around £286 million in benefits/annum for nutrient cycling, and aphotic reef, around £9 million. But these figures are for *total* provisioning. For each of the three MCZ network scenarios (A, G, and J) developed in Richardson *et al.* (2006), only some proportion of each habitat type is protected; for network scenario A the percentages are 17.5% and 12.0% for aphotic reef and shelf mud respectively. In the absence of further evidence, it was assumed that the area of aphotic reef in network scenario A has the average per hectare service provisioning. Around £9 million split across 10,968km^2 implies the value of nutrient cycling per hectare of aphotic reef is around £830/hectare. The value of each service/habitat combination on a per hectare basis was then coupled with the GIS extent data for each of the three networks.

There is a further step, however, in that this is total provisioning in 2008 per hectare. The coding for impact described above uses this figure. For aphotic reef/ nutrient cycling, the coding was 'H 10 E', i.e. a high impact (50%–89%) from MCZ designation, reaching a maximum at $t = 10$, and having an exponential trajectory from $t = 0$ to $t = 10$. (This is equivalent to the bottom curve in Figure 7.2.)

The 0.7 extent of impact point in Figure 7.2 might be replaced with a value for aphotic reef/nutrient cycling of £581 (i.e. 0.7 X £830), and the area under the bottom exponential curve would then provide the per hectare benefit stream in monetary terms for MCZ designation/aphotic reef/nutrient cycling.

7.3.2.5 Aggregation

Summing together all the marginal benefits of MCZ designation provides the aggregate benefits in terms of extra ecosystem service provision. In economic appraisal, benefits are generally presented in unadjusted per annum terms and also in discounted form, i.e. allowing for the presumption that a benefit or cost accrued in the near future is worth more to society than an equivalent benefit or cost in the distant future. Following UK government guidance (BERR, 2008), a discount rate of 3.5% was applied so as to allow comparability with the cost estimates. With discounting the benefit estimate was £10.2 billion for the least extensive proposed network (scenario A) with the less restrictive management regime. The present value of the most extensive/restrictive scenario (J/more restrictive) is over double this (£23.5 billion). The undiscounted mean annual benefits range from around £0.92 billion to £1.95 billion, depending on network/restrictions.

The assessment of costs was carried out separately from Hussain *et al.* (2010). ABPMer (2007) carried out surveys of affected industry stakeholders to estimate costs of MCZ designation, and the discounted cost estimate range was £0.4 billion and £1.2 billion. As such the minimum benefit–cost ratio is around seven.

7.4 Marine ecosystem valuation: limitations

Caution should be applied when using environmental valuation, both in terms of understanding what valuation can bring to decision-making if applied appropriately (the bounds of what it can reasonably provide in terms of evidence), but equally in terms of awareness of its *inappropriate* use. Section 7.4.1 draws on a detailed analysis of the MCZ case study to highlight some of the limitations therein; Section 7.4.2 then provides a broader perspective.

7.4.1 Case study valuation: limitations

The MCZ case study has been selected here to showcase a valuation application that was successful in the sense that it contributed significantly to the evidence base supporting the Marine and Coastal Access Act by producing tangible and credible results. But in presenting the methodology and results it is apparent that there are limitations that should be borne in mind.

The MCZ studies (Hussain *et al.*, 2010; ABPMer, 2007) produced a benefit–cost ratio of at least seven, based on values for only seven of the eleven services, and the ecosystem service typology excluded non-use (existence) values. This is a strong policy-relevant outcome, but equally points to a potential limitation in commoditizing services in monetary terms, i.e. omitted benefits. If the benefit streams that can be monetized exceed costs (as was the case in the MCZ study) then this is a sufficient condition for economic efficiency. But it is not a *necessary* condition in that there are omitted benefit streams. In the aggregate service values used (Table 7.2), there was systemic undervaluation in some cases, e.g. the estimate for gas and climate regulation being based solely on the carbon sequestered by phytoplankton from photosynthesis. There is the potential for any monetized benefit estimate to dominate the policy discourse, even if it is partial and caveats as to qualitative benefits are stated.

Caution must be applied in overstating the accuracy of any valuation outcomes. For instance, the value for nutrient cycling in a typical hectare of aphotic reef has been estimated at £830/hectare. The precision of this estimate can be misleading (I provide precise figures below only so as to allow the reader to follow the methodological steps used). This is particularly the case because the level of confidence that is appropriate for different estimates varies between services, with the level of confidence depending on the economic estimation methodology and data sources, and the biophysical science, and the *interrelatedness* between the two. The extent of scientific knowledge pertaining to individual habitat/service combinations was set out in Table 7.3, revealing significant variability. In terms of the economic data, some values are dated and based on relatively poor proxies, as discussed. Further, the methodology developed for allocating the total value across habitat types is somewhat coarse. The assumption of heterogeneity in terms of equal provisioning across each hectare of a particular habitat type is also clearly a simplification, and this is an example of interrelatedness between the science and the economics, which in this case could not be dealt with.

Another example of this interrelatedness is the issue of linearity. Although the trajectory over time towards the final assumed 'plateau' of ecosystem service provisioning can be coded as non-linear (i.e. exponential or log), it is assumed that if the maximum extra provisioning is +70% compared with BAU then this is twice as valuable to society as +35%. The contribution of Barbier *et al.* (2008) is significant in this regard as it explicitly focuses on non-linearity in coastal mangroves, salt marshes, and other marine ecosystems. The authors argue that service provision does not 'scale up' uniformly with a change in habitat extent. In the case of mangroves, provisioning of the ecosystem service, termed disturbance prevention and alleviation in the MCZ study (or storm buffering), the service provisioning is lowered as successive landward zones of a mangrove forest are crossed. Such potential non-linearity is assumed not to apply in the MCZ study, although it is likely to be present.

A related issue is the occurrence of threshold effects wherein a tipping point is reached that implies an abrupt shift to a new ecosystem steady state. This discourse can be linked to the concept of safe minimum standards (TEEB, 2010). The problem in bio-economic analysis is that it is extremely difficult to define safe minimum standards, in the same way that there is considerable uncertainty in terms of the definition and evaluation of Good Environmental Status in the MSFD. With respect to the MCZ case study, the issue of threshold effects is likely to be of less concern as the valuation pertains to a management intervention that *increases* service provision, and therein reduces the proximity to ecosystem thresholds. The cost report (ABPMer, 2007) does implicitly consider reductions in ecosystem service provisioning, e.g. reduced seafood as a provisioning service owing to MCZ designation. This would, however, also potentially move the system away from a tipping point (the collapse of a fishery owing to overfishing) unless there were a displacement effect, i.e. fishermen redirecting and redoubling fishing effort in non-protected areas.

7.4.2 General limitations and caution with respect to marine ecosystem valuation

This section focuses on two elements not fully covered in the discussion of the MCZ case study and its limitations: (1) methodologies used to value ecosystem services; and (2) marine versus terrestrial valuations.

7.4.2.1 Primary valuation studies

As the case study has demonstrated, the approach used to value ecosystem services varies on a service-by-service basis. The approach used in Hussain *et al.* (2010) used extant data points with no new primary valuation study. This reliance on secondary data was in part linked to the project scope (funding and timing constraints etc.), but also to the geographical scope, as an analysis of marginal change in provisioning was required at a UK-wide level.

Notwithstanding the shift towards management at a regional (and therefore transnational) scale in marine management internationally, the vast majority of marine and coastal ecosystem valuation literature refers to study sites at much smaller spatial scale, e.g. individual strips of coastline and adjacent marine ecosystems. The most frequently applied methodologies in such primary valuation studies fall under the category of 'stated preference techniques', wherein the respondent's willingness to pay (WTP) for a defined change in the natural environment (quality, access, or both) is elicited through a structured, survey-based approach. The WTP is hypothetical in the sense that the respondent does not actually pay and therein lies a significant *potential* for bias in survey results. As Fletcher (2012) notes, one of the earliest publications in this field referred to the potential for respondents to be 'purchasing moral satisfaction' (Kahneman and Knetsch, 1992), with individuals being potentially motivated to over-represent their WTP owing to the 'warm glow' they receive from the act of (theoretically) giving without actually doing so (Andreoni, 1989). However, there is a long history of applications in applying stated preference methods, at least for terrestrial biomes.

A highly influential example of its use is in the Exxon Valdez disaster, one of the first applications to estimate non-use values (Liu *et al.*, 2010) as an input to litigation (Carson *et al.*, 2003). The National Oceanic and Atmospheric Administration (NOAA) formed a panel chaired by Nobel laureate Kenneth Arrow to comment and provide guidance on the valuation methodologies applied. Recommendations included conducting survey pre-tests, survey design so as to minimize non-responses, and a preference for face-to-face interviews (Arrow *et al.*, 1993).

A more recent methodological development in the environmental economics literature has been the use of choice experiments (Hanley *et al.*, 2001). Choice experiments (CEs) originate from market research applications (Bateman, 2002), but have been adopted in non-market valuation of ecosystems since the mid 1990s. They are based on the premise that a commodity is a composite of a series of perceived attributes (Hussain, 2000). For ecosystem-focused CEs, the 'commodity' in question might be different renewable energy production (Odam, 2011) with options as follows: (1) marine seascape visually dominated by offshore wind farms; (2) wave turbines; (3) onshore wind. These are compared with the status quo (i.e. BAU) option of 'no new renewable energy installations'. The 'attributes' within these options include differing levels of wildlife impact and different locations. Each non-BAU option has an associated cost. There is no additional cost to the BAU option, but the presumption of new fossil fuel based energy generation with higher air pollution and greenhouse gas emissions. In CE surveys, respondents are presented with different choice cards with differing attribute levels (at differing costs). Individuals must choose one option. Since a financial cost is included, a value for individual attributes can be estimated. The CE approach has recently been applied to estimate the loss in recreational benefits associated with algal bloom in the Black Sea (Taylor and Longo, 2010), fisheries management policies (Groeneveld, 2011), and changes in marine biodiversity (Ressurreição *et al.*, 2012). The contingent valuation method (CVM) used in the Exxon Valdez case (and subsequently) does not allow value estimates for individual attributes in the same manner; there is an overall pairwise comparison between two scenarios, each of which has certain attribute levels embedded within them. Both CE and CVM are valid approaches to valuation, and the assessment of the non-use values arising from the Marine and Coastal Access Act (reported in McVittie and Moran, 2010) uses both approaches.

7.4.2.2 Marine versus terrestrial ecosystem valuation

Although there is a long history of applying stated preference techniques in the terrestrial domain, there are very few applications for marine (i.e. open oceans) ecosystems per se, although there are many studies valuing coastal ecosystems. De Groot *et al.* (2012) report on the TEEB database on value estimates. Of the 665 value data points, only 14 are for 'open oceans'; there are however 94 for 'coral reefs', 28 for 'coastal systems' and 139 for 'coastal wetlands'.

This is important if valuation is to be carried out without the funding or availability of a location and a context-specific primary valuation study. This scenario is commonplace and the use of what is termed 'benefits transfer' is now routine, i.e. using value estimates from one or more study sites to transfer to one or more policy sites. There are many substantive issues associated with benefits transfer (see TEEB, 2010 for a review) and these issues are *additional* to those that pertain to methodologies for primary valuation, as the benefits transfer relies on primary studies as an input. Values derived in primary studies are context-specific in terms of the ecological characteristics of the study site and the socio-cultural and demographic features of the sample population. All else being equal, having more primary studies in valuation databases to choose from should lead to more defensible value estimates using benefits transfer. So the paucity of data points for open oceans is significant.

7.5 Discussion

There is a tendency in some quarters to think of economic valuation as either a 'silver bullet' or else a bitter pill that needs to be swallowed, typically as a means to an end in terms of promoting a conservation proposal. In some respects the content of this chapter supports this proposition, in that the principal focus has been on a particular case study (the UK Marine and Coastal Access Act) that *required* an economic assessment for impact assessment purposes (BERR, 2008) and *unequivocally supported* the conservation outcome, i.e. a network of marine protected areas.

The core questions to be addressed in this final section pertain to the replicability of the study methodology in different socio-cultural contexts around the globe and, more fundamentally, whether indeed such approaches ought to be promoted in other contexts. The latter has been touched on in Section 7.4, which considered the limitations of valuation (both specifically to the case study and also more generically), but this section aims to tie the strands together and present some conclusions.

7.5.1 Is the case study approach replicable?

In some senses the answer to the question of replicability is that, since the Hussain *et al.* (2010) study followed best practice in terms of applying economic efficiency analysis to the ecosystem approach, some of the core methodological principles can and must be followed. The first of these principles is the need to adhere to a coherent typology of ecosystem services that captures the diverse range of services provided by the habitats under examination, but avoids the pitfalls of double-counting. A second principle is the need to focus on marginal change and thus have a clear definition of BAU and with-policy, and a methodological procedure that does not conflate the two.

Notwithstanding the fact that the dominant typologies (MA, 2005; TEEB, 2010) were not designed with a marine or coastal ecosystem focus, they are to a certain extent fit for purpose as long as the valuation practitioner is aware of potential double-counting issues. This issue is discussed by Turner *et al.* (2010): the authors who cite Fisher *et al.* (2009), which presents evidence to support the contention that researchers have incorrectly summed values, and Turner *et al.* (2010) also present the suggestion by Hein *et al.* (2006) that regulating services should only be included in valuation if they impact upon external ecosystems (i.e. knock-on effects beyond the habitats being assessed), or if they provide a *direct* local benefit – not an indirect benefit from sustaining or improving the provisioning of another service.

The issue of marginality is more difficult to address fully. There is a paucity of data on the stocks and flows of ecosystem services *today*, let alone what those projected stocks and flows might be under BAU 20 years hence. It is one thing to state that researchers should be *cognizant* of the need to account for and document evidence concerning future trends in industrial output and production technologies (which together comprise 'manageable' anthropogenic drivers), as well as environmental drivers (such as climate change), and superimpose on this assessment the need to embed extant directives and other forms of regulatory intervention in defining BAU. Actually being able to *carry out* this research exercise in anything more than a rudimentary fashion is likely to be difficult in a resource and time-constrained study.

Once BAU has been defined, there is then the need to provide a more focused biophysical appraisal of with-policy impacts. Again, there may be modelling data and experimental work for some services (in all likelihood sea food provisioning, as this is the best researched area), but otherwise, the MCZ-type assessment relies on (informed) expert judgement.

These two principles are cross-cutting and apply to almost any conceivable ecosystem based assessment. What was special for the MCZ study and novel in

Hussain *et al.* (2010) was the process of disaggregating value estimates expressed in aggregate terms. Such an exercise would be required in any assessment where values are only available in aggregate terms, and as has been argued above, this scenario is likely to apply in many cases for marine assessment. Is the methodology replicable? The answer is a tentative 'yes'. The mechanics are fairly complex but certainly replicable, and are set out in detail in Moran *et al.* (2007). The coding relied on expert judgement and a response to replicability here might simply be to state that 'as expert as possible, given resource constraints and limitations of data availability' may be adequate.

But this does prompt the next question. Given the limitations of environmental economics valuation and the specific paucity of data points for open oceans (although not for coastal systems), is carrying out any economic analysis worthwhile?

7.5.2 Should we be promoting an ecosystem based approach using environmental valuation?

As alluded to previously, Costanza *et al.* (1997) find that marine and coastal ecosystems provide two-thirds of global ecosystem provisioning, and this provisioning is threatened by environmental drivers and anthropogenic drivers. The choice as to whether to pursue the use of environmental valuation is thus not a trivial one, i.e. if it leads to better decision-making, which in turn probably implies more conservation, then is it an appropriate means to an end?

Child (2009) cited in Fletcher (2012) argues that we should create a 'culture of care' rather than using economic efficiency as a yardstick to determine the extinction (through anthropogenic pressures) of a 'non economically-useful' species, trading this loss for the protection of a 'useful' habitat, with 'usefulness' being measured in utilitarian anthropocentric terms, coined around ecosystem service provisioning. It is difficult to argue against this moral position (which follows the eco-centric views of Aldo Leopold), but in pragmatic terms species and habitats are being lost, and the question remains as to whether economics (in terms of valuation) helps or hinders. The Arrow report (Arrow *et al.*, 1993) was mentioned previously: the authors state '[it is] hard to imagine that the establishment of property rights or improved pricing of natural resources could worsen the prospects of future generations'. So this is an expression of the opposing (or perhaps contrarian) position.

There is an argument, however, that economic valuation does indeed hinder rather than help the process. The commoditization of nature is very real – biodiversity offsets and payments for ecosystem services (PES) provide evidence to support this contention. Non-market valuation is a catalyst for this commoditization. If the valuation used is biased such that the value estimate used in the PES analysis is low, then it can indeed be argued that this commoditization is counter-productive. The scope for such processes of undervaluing nature is huge from omitted value categories, to transferring value estimates from one or more study sites (using benefits transfer) that understate value at the policy site. If there is an incentive for such undervaluation then a systemic problem arises; consultants may be commissioned to develop a PES scheme and as such might be incentivized to cherry-pick studies from extant literature that increase the likelihood of the PES price being acceptable to the service buyer, i.e. studies with lower value estimates. The legitimacy of such commoditization can therein be questioned.

This is particularly the case as the level of risk and uncertainty in future projections of the value of non-provisioning ecosystem services is likely to be far higher than that of provisioning services and conventional marketed goods (such as the electricity generated by a large-scale hydro project). Although the social welfare changes that might potentially arise from an inaccurate assessment of electricity generating capacity might be important, the disbenefits associated with such projections being found to be inaccurate *ex post* are of a different nature and scale to, say, irreversible habitat or species loss, or equally the destruction of cultural capital and community cohesion (TEEB, 2010).

The reason why the proposition of Arrow *et al.* (1993) in favour of valuing nature may not apply as readily to marine systems as opposed to terrestrial systems is that, for many services values are based on societal perceptions: given that we as humans are land-based creatures, this tends to limit the extent to which the attributes of marine habitats remain 'credence attributes' as opposed to 'sensory' or 'experience' attributes (Hussain, 2000). A member of the general public can see, hear, touch, and experience nature in terrestrial systems more readily than, say, the deep oceans. Might this imply that there is a tendency to undervalue marine ecosystems? It is difficult to provide evidence to either support or refute this, as there is no 'right' price per se for a non-marketed commodity.

Further, as previously discussed, the pricing of nature comes in many forms, one of which is the use of benefits transfer (BT). Given the paucity of studies on open oceans, it is quite possible that reliance on a secondary study or studies to transfer values to a policy site entails BT being applied in a vastly different socio-cultural context. A fundamental tenet of BT is that it should not be applied in such scenarios, but there is evidence that it is nonetheless (TEEB, 2010).

Returning explicitly to the question of promoting valuation versus heeding caution, as ever in such discussions there is no right or wrong answer. As a minimum, economic appraisal can inform the scientific research agenda, in the sense that research efforts might be focused in part on ecosystem services that are both valuable but also under-researched. Any economic assessment using valuation cannot substitute for biophysical modelling. Risk and uncertainty (and hence the need for sensitivity analysis) are ubiquitous in any biophysical modelling and this is potentially where the interrelatedness of the disciplines can steer a mutually beneficial course.

In terms of any valuation being fit for purpose, by definition we need to know what that purpose is. As Desvousges *et al.* (1998) aptly remark (cited in Fletcher, 2012), if the decision-making context is purely awareness-raising then accuracy is not of tantamount importance, whereas if valuation is being applied to assess potentially irreversible actions or ones that might stimulate (or indeed the converse, that is mitigate the potential for) threshold effects, including the extinction of a species, then we need to heed caution and require a significantly higher level of methodological rigour in our valuations.

References

ABPMer (2007). *Cost Impact of Marine Biodiversity Policies on Business – The Marine Bill. Final Report to Defra*. London: Defra.

Andreoni, J. (1989). Giving with impure altruism – Applications to charity and Ricardian equivalence. *Journal of Political Economy* **97**: 1447–1458.

Arrow, K., Solow, R., Portney, P. R. *et al.* (1993). *Report of the NOAA Panel on Contingent Valuation: National Oceanic and Atmospheric Administration*. Washington DC: NOAA.

Atkins, J. P., Burdon, D., Elliott, M. *et al.* (2011). Management of the marine environment: Integrating ecosystem services and societal benefits with the DPSIR framework in a systems approach. *Marine Pollution Bulletin* **62**: 215–226.

Barbier, E. B., Koch, E. W., Silliman, B. R., *et al.* (2008). Coastal ecosystem-based management with nonlinear ecological functions and values. *Science* **319**(5861): 321–323.

Bateman, I. (2002). *Economic Valuation with Stated Preference Techniques: A Manual*. Cheltenham: Edward Elgar.

Beaumont, N., Townsend, M., Mangi, S., *et al.* (2006). *Marine Biodiversity: An Economic Valuation. Final Report to Defra*. London: Defra.

Beaumont, N. J., Austen, M. C., Atkins, J. P., *et al.* (2007). Identification, definition and quantification of goods and services provided by marine biodiversity: Implications for the ecosystem approach. *Marine Pollution Bulletin* **54**: 253–265.

Beaumont, N. J., Austen, M. C., Mangi, S. C., *et al.* (2008). Economic valuation for the conservation of marine biodiversity. *Marine Pollution Bulletin* **56**: 386–396.

BERR (2008). *Code of Practice on Guidance on Regulation*. London: Department for Business, Enterprise and Regulatory Reform.

Böhnke-Henrichs, A., Baulcomb, C., Koss. R., *et al.* (2013). Typology and indicators of ecosystem services for marine spatial planning and management. *Journal of Environmental Management* **130**: 135–145.

Boyd, J. and Banzhaf, S. (2007). What are ecosystem services? The need for standardized environmental accounting units. *Ecological Economics* **63**: 616–626.

Carson, R. T., Mitchell, R. C., Hanemann, M., *et al.* (2003). Contingent valuation and lost passive use: Damages from the Exxon Valdez oil spill. *Environmental & Resource Economics* **25**: 257–286.

Child, M. F. (2009). The Thoreau ideal as a unifying thread in the conservation movement. *Conservation Biology* **23**: 241–243.

Costanza, R., d'Arge, R., de Groot, R., *et al.* (1997). The value of the world's ecosystem services and natural capital. *Nature* **387**: 253–260.

Desvouges, W. H. F., Johnson, R., and Banzhaf, H. S. (1998). *Environmental Policy Analysis With Limited Information, Principles and Applications of the Transfer Method*. Edward Elgar Publishing Limited: Northampton, MA.

Dugan, J. E. and Davis, G. E. (1993). Applications of marine refugia to coastal fisheries management. *Canadian Journal of Fisheries and Aquatic Sciences* **50**: 2019–2042.

Edwards-Jones, G., Davies, B., and Hussain, S. S. (2000). *Ecological Economics: An Introduction*. Oxford: Blackwell Science.

Fisher, B., Turner, R. K., and Morling, P. (2009). Defining and classifying ecosystem services for decision making. *Ecological Economics* **68**: 643–653.

Fletcher, R. (2012). Qualitative identification and quantitative valuation for marine cultural ecosystem services in Turkey. Unpublished MSc dissertation in Ecological Economics. Edinburgh: University of Edinburgh.

Groeneveld, R. A. (2011). Quantifying fishers' and citizens' support for Dutch flatfish management policy. *ICES Journal of Marine Science* **68**: 919–928.

de Groot, R., Brander, L., der Ploeg, S., *et al.* (2012). Global estimates of the value of ecosystems and their services in monetary units. *Ecosystem Services* **1**(1): 50–61.

Halpern, B. S., Walbridge, S., Selkoe, K. A., *et al.* (2008). A global map of human impact on marine ecosystems. *Science* **319**: 948–952.

Hanley, N., Mourato, S., and Wright, R. E. (2001). Choice modelling approaches: A superior alternative for environmental valuation? *Journal of Economic Surveys* **15**: 435–462.

Hein, L., van Koppen, K., de Groot, R. S., *et al.* (2006). Spatial scales, stakeholders and the valuation of ecosystem services. *Ecological Economics* **57**: 209–228.

Holmlund, C. M. and Hammer, M. (1999). Ecosystem services generated by fish populations. *Ecological Economics* **29**: 253–268.

Hussain, S. S. (2000). Green consumerism and eco-labelling: A strategic behavioural model. *Journal of Agricultural Economics* **51**(1): 77–89.

Hussain, S. S. and Gundimeda, H. (2011). Tools for valuation and appraisal of ecosystem services in policy making. In: *The Economics of Ecosystems and Biodiversity in Local and Regional Policy and Management*. Wittmer, H., and Gundimeda, H. (eds.). London and Washington: Earthscan, pp. 57–94.

Hussain, S. S., Winrow-Giffen, A., Moran, D., *et al.* (2010). An *ex ante* ecological economic assessment of the benefits arising from marine protected area designation in the UK. *Ecological Economics* **68**(4): 828–838.

Kahneman, D. and Knetsch, J. L. (1992). Valuing public goods: The purchase of moral satisfaction. *Journal of Environmental Economics and Management* **22**: 57–70.

King, S. E. and Lester, J. N. (1995). The value of saltmarsh as a sea defence. *Marine Pollution Bulletin* **30**: 180–189.

Koss, R. S., Knights, A. M., Eriksson, A. *et al.* (2011). ODEMM Linkage Framework Userguide. ODEMM Guidance Document Series No.1. EC FP7 project (244273) Options for Delivering Ecosystem-based Marine Management, University of Liverpool.

Lewis, A. (2012). Deriving cultural values from choice experiments: A case study in the Baltic Sea. Unpublished MSc dissertation in Ecological Economics. Edinburgh: University of Edinburgh.

Liu, S., Costanza, R., Farber, S., *et al.* (2010). Valuing ecosystem services: Theory, practice, and the need for a transdisciplinary synthesis. *Ecological Economics Reviews*: 1185.

Lotze, H. K., Lenihan, H. S., Bourque, B. J., *et al.* (2006). Depletion, degradation, and recovery potential of estuaries and coastal seas. *Science* 23 June 2006: 1806–1809.

MA – the Millennium Ecosystem Assessment (2005). *Ecosystems and Human Well-Being: Synthesis*. Washington DC: Island Press.

McVittie, A. and Moran, D. (2010). Valuing the non-use benefits of marine conservation zones: An application to the UK Marine Bill. *Ecological Economics* **70**(2), 413–424.

Moran, D., Hussain, S. S., Fofana, A., *et al.* (2007). *Marine Bill – Marine Nature Conservation Proposals – Valuing the Benefits. Final Report to Defra*. London: Defra.

Odam, N. J. (2011). Developing infant technologies in mature industries: A case study on renewable energy. Unpublished PhD thesis. Stirling: University of Stirling.

Pugh, D. T. and Skinner, L. (2002). A new analysis of marine-related activities in the UK economy with supporting science and technology. ICMST Information Document No. 10. www.oceannet.org/library/publications/documents/marine_related_activities. pdf.

Ressurreição, A., Zarzycki, T., Kaiser, M., *et al.* (2012). Towards an ecosystem approach for understanding public values concerning marine biodiversity loss. *Marine Ecology Progress Series* **467**: 15–28.

Richardson, E. A., Kaiser, M. J., Haddink, J. G., *et al.* (2006). *Developing Scenarios for a Network of Marine Protected Areas: Building the Evidence Base for the Marine Bill. Final Report to Defra*. London: Defra.

Sanchirico, J. N. and Wilen, J. E. (2001). A bioeconomic model of marine reserve creation. *Journal of Environmental Economics and Management* **42**: 257–276.

Smyth, T. J., Tilstone, G. H., and Groom, S. B. (2005). Integration of radiative transfer into satellite models of ocean primary production. *Journal of Geophysical Research Oceans* **110**, C10014.

Taylor, T. and Longo, A. (2010). Valuing algal bloom in the Black Sea coast of Bulgaria: A choice experiments approach. *Journal of Environmental Management* **91**: 1963–1971.

TEEB (2010). *The Economics of Ecosystems and Biodiversity. Ecological and Economics Foundations*. London and Washington DC: Earthscan.

Turner, R. K., Morse-Jones, S., and Fisher, B. (2010). Ecosystem valuation. *Annals of the New York Academy of Sciences* **1185**: 79–101.

Worm, B., Barbier, E. B., Beaumont, N., *et al.* (2006). Impacts of biodiversity loss on ocean ecosystem services. *Science* **314**: 787–790.

8

The contribution of international scientific cooperation and related institutions to effective governance for the oceans: the cases of regional tsunami early warning systems and the Argo Project

STEFANO BELFIORE AND AURORA MATEOS

8.1 Introduction

Good ocean governance through marine science and technology means promoting the use of the oceans and seas as part of efforts to 'eradicate poverty, to ensure food security and to sustain economic prosperity and the well-being of present and future generations' (United Nations, 2000, para. 39). Therefore, it is necessary to ensure access for decision-makers to advice and information on marine science and technology, the appropriate transfer of technology, and support for the production and diffusion of factual information and knowledge for end-users through appropriate policies.

Disaster risk reduction has since long been recognized as a fundamental contribution to development (see, recently, UNISDR and WMO, 2012) and ocean observation plays a key role in fostering the knowledge needed to manage marine resources sustainably (see, recently, United Nations, 2012c). In this light, the present contribution focuses on two interrelated issues for which an institutional framework for international cooperation and coordination has emerged, but is still incomplete and may require further guidance within existing mechanisms or the development of additional instruments. The organization of regional tsunami warning and mitigation systems (TWS), and a global framework to address ocean hazards related to sea level, entail a complex set of multilateral and bilateral arrangements for the exchange of real-time seismic and sea-level data and the issuing of advisory and warning messages across the countries of shared ocean basins (for a comprehensive review of the subject of tsunamis see Bernard and Robinson, 2009). The international Argo Project is an array of over 3500 free-drifting profiling floats that allow continuous monitoring of the temperature, salinity, and velocity of the upper 2000 m of the ocean. Both the TWS and Argo are relevant cases that prove how effective ocean governance can be achieved through the design and implementation of appropriate international scientific research and observational programmes.

8.2 Ocean observation and marine scientific research

The United Nations Convention on the Law of the Sea (UNCLOS) devotes its Part XIII (Articles 238–265) to Marine Scientific Research and its Part XIV to Development and Transfer of Marine Technology. Part XIII provides a comprehensive set of provisions for the conduct of marine scientific research (MSR), in particular in relation to the consent needed of a coastal state for other states or international organizations to carry out MSR in its exclusive economic zone (EEZ) or continental shelf (Art. 246, UNCLOS). During the preparatory works for UNCLOS, several definitions of MSR were proposed without, however, agreement being reached (UN/DOALOS, 2010). As a result, Part XIII of UNCLOS does not define MSR and provides only a few references to the nature of MSR in relation to:

- the phenomena and processes occurring in the marine environment (Art. 243, UNCLOS);
- the advance of scientific knowledge of the marine environment for peaceful purposes, in particular in relation to (a) exploration and exploitation of natural resources, living or non-living, (b) drilling into the continental shelf, use of explosives, or introduction of harmful substances into the marine environment, and (c) construction, operation, or use of artificial islands, installations, and structures (Art. 246, UNCLOS).

A number of provisions relate to the role of competent international organizations in the promotion of international cooperation, in particular for the creation of favourable conditions and publication and dissemination of information and knowledge (Arts. 242–244), as well as in relation to projects undertaken by them or under their auspices.

Article 251 calls on competent international organizations to establish general criteria and guidelines to assist states in ascertaining the nature and implications of marine scientific research. A comprehensive guide to the implementation of the provisions of UNCLOS related to MSR was published by the United Nations in 2010 (UN/DOALOS, 2010). Concerning the transfer of marine technology covered by Part XIV of UNCLOS, Article 271 reads as follows:

States, directly or through competent international organizations, shall promote the establishment of generally accepted guidelines, criteria and standards for the transfer of marine technology on a bilateral basis or within the framework of international organizations and other fora, taking into account, in particular, the interests and needs of developing States.

In implementation of this Article, criteria and guidelines for the transfer of marine technology were also published (IOC, 2005a). These guidelines provide a fairly comprehensive definition of marine technology, which 'refers to instruments, equipment, vessels, processes and methodologies required to produce and use knowledge to improve the study and understanding of the nature and resources of the ocean and coastal areas'.

It has been pointed out (Pugh, 2001; Ryder, 2003; cf. also Schiller and Brassington, 2011) that, since the final text of UNCLOS was adopted there have been important developments in MSR, most notably the use of technologies and infrastructures such as satellites, aircraft, ships of opportunity, autonomous vessels, buoys, and floats, while UNCLOS assumes research vessels are the primary platform for MSR (Art. 248, UNCLOS). Therefore, the nature and continuous development of MSR may not be covered by the provisions of UNCLOS in an adequate way, leaving space for additional guidance to be provided under Article 251.

It should also be recalled (Ryder, 2003) that Part XIII does not apply to the collection of meteorological information in the marine environment. In 1979, during the negotiations for the final text of UNCLOS, the members of the World Meteorological Organization (WMO), through Resolution 16 (Cg-VIII) (WMO, 1995; WMO, 1999), expressed the hope that the provisions of UNCLOS related to MSR would not 'result in restrictions to operational meteorological and related oceanographic observations carried out in accordance with international programmes such as World Weather and the Integrated Global Ocean Station System'. As defined by the Resolution, relevant activities included operational activities such as (a) the collection of meteorological information from voluntary observing ships, buoys, other ocean platforms, aircraft, and meteorological satellites, and (b) research activities, both meteorological and oceanographic, such as those carried out during the Global Weather Experiment. The Resolution considered that 'an adequate marine meteorological data coverage from ocean areas, in particular from those areas in the so-called "exclusive economic zone", is indispensable for the issue of timely and accurate storm warnings for the safety of life at sea and the protection of life and property in coastal and off-shore areas' and that 'the International Convention for the Safety of Life at Sea of 1960 specifies that the contracting Governments undertake, inter alia, to issue warnings of gales, storms and tropical storms and to arrange for selected ships to take meteorological observations', as well as that 'members of the World Meteorological Organization have undertaken the responsibility of issuing warnings for the high seas and coastal waters according to internationally agreed procedures'.

Resolution 16 (Cg-VIII) was reported to the 46th meeting of the Third Committee of the Third United Nations Conference on the Law of the Sea (20 August 1980) by its Chairman, Ambassador A. Yankov from Bulgaria, who noted that the activities referred to in the Resolution 'had already been recognized as routine observations and data collecting which were not covered by Part XIII of the negotiating text. Furthermore, they were in the common interest of all countries and had undoubted universal significance' (United Nations, 1979a, paras. 4 and 5). This opinion was reported also in the 134th Plenary meeting of the Third Conference (United Nations, 1979b, para. 43), without dissent. It should be added that the Convention of the World Meteorological Organization pre-dates UNCLOS, as it was adopted by the Washington Conference of Directors of the International Meteorological Organization (IMO) on 11 October 1947, following the experience of IMO, founded in 1873. The Convention entered into force on 23 March 1951, thus establishing WMO the following year. WMO policy and practice in relation to meteorological and related data and products is currently provided for by Resolution 40 (Cg-XII) (WMO, 1995; cf. WMO, 1999), whose Annex I specifies data and products to be exchanged without charge and with no conditions on use, including:

- *in situ* observations from the marine environment (reports of surface observation from sea stations; reports of buoy observations; reports of bathythermal observations; temperature, salinity and current report from sea stations; etc.);
- data from upper air sounding networks (including upper-level pressure, temperature, humidity and wind reports from sea stations);
- reports from the network of stations recommended by the regional associations as necessary to provide a good representation of climate (including report of monthly means and totals from ocean weather stations; reports of monthly aerological means from ocean weather stations);
- severe weather warnings and advisories for the protection of life and property targeted upon end-users;
- those data and products from operational meteorological satellites that are agreed between WMO and satellite operators (including data and products necessary for operations regarding severe weather warnings and tropical cyclone warnings).

It should be added that the collection of meteorological data for weather forecasts and warnings, preferably in conformity with WMO technical regulations and recommendations, is foreseen by the International Convention for the Safety of Life at Sea (SOLAS) 1974, as amended in 2004 (Regulation 5, Meteorological services and warnings). The Convention specifies that '[f]orecasts, warnings, synoptic and other meteorological data intended for ships shall be issued and disseminated by the national meteorological service in the best position to serve various coastal and high seas areas, in accordance with mutual arrangements made by Contracting Governments, in particular as defined by the World Meteorological Organization's system for the preparation and dissemination of meteorological forecasts and warnings for the high seas under the global maritime distress and safety system (GMDSS)'. The GMDSS is an integrated communications system using satellite and terrestrial radiocommunications to ensure that no matter where a ship is in distress, aid can be dispatched. This system also ensures provision of maritime safety information (MSI), both meteorological and navigational, on a global basis at sea.

The United Nations Framework Convention on Climate Change (UNFCCC) 1992 (Article 5, Research and systematic observation) calls on the parties to '[s]upport and further develop, as appropriate, international and intergovernmental programmes and networks or organizations aimed at defining, conducting, assessing and financing research, data collection and systematic observation, taking into account the need to minimize duplication of effort'. In the following provision, the Convention calls on the parties to 'promote access to, and the exchange of, data and analyses thereof obtained from areas beyond national jurisdiction'. While the Convention does not specify from where and how climate data should be obtained, it appears to place some restrictions on the exchange of data and related analysis in relation to the jurisdiction of the parties.

As a field that is in continuous evolution thanks to technological developments and scientific advancements, MSR requires adaptive regulatory mechanisms to accompany its development. As stated by one delegation at the second meeting of the United Nations Open-ended Informal Consultative Process on Oceans and the Law of the Sea (UNICPOLOS) (New York, 7–11 May 2001), which addressed, inter alia, marine science and the development and transfer of marine technology: 'There is a risk that [the] marine science regime, as defined in Part XIII of UNCLOS, will remain an empty shell unless concrete policies and results-oriented initiatives are formulated and implemented' (United Nations, 2001a, echoed in United Nations, 2001b, Annex I). Subsequent meetings of UNICPOLOS have addressed issues related to MSR as applied to different domains and, to give an example, it is noteworthy that at the 12th meeting (New York, 20–24 June 2011) one expert emphasized 'the need to consider how the Convention could evolve to address new issues such as a regime for marine genetic resources and the establishment of a network of marine protected areas, as well as environmental impact assessments, capacity-building and the transfer of marine technology' (United Nations, 2011, para. 55), calling for attention to the need to adapt the existing regulatory framework provided by UNCLOS to new needs and developments. Regarding MSR, the adoption of principles, codes of conduct, and regulations based on a common understanding would provide clarity and predictability for the scientific community, facilitate the introduction of standards procedures in accordance with international practice, and ensure better flow of information through authorized organizations and channels to facilitate dissemination of data and information and create acceptability of results. Therefore the design of common policies for conducting MSR is crucial for ocean governance.

The case studies presented in this chapter concern international initiatives in MSR and operational oceanography (Schiller and Brassington, 2011) that illustrate how the evolution of practices and the expansion of need for information to manage the oceans sustainably, also require developments in international arrangements for the conduct of operations and the exchange of data. Here, the issue of exchange of data is particularly relevant, as operational activities, especially when aimed at forecasts and early warnings, normally require exchange of data in real or near-real time, while this is not necessarily the case with MSR.

8.3 Tsunami early warning and mitigation systems: the emergence of new regional regimes?

On 26 December 2004 an earthquake of magnitude 9.2 on the Richter scale, with its epicentre off the west coast of Sumatra, Indonesia, triggered a tsunami that affected several countries of the Indian Ocean, including Indonesia, Sri Lanka, India, Thailand, the Maldives, Somalia, Myanmar, Malaysia, the Seychelles, and others. The earthquake was the third most powerful earthquake recorded since 1900 and the death toll from the tsunami reached about 200,000 people, including tourists from Australia and Europe (Japan, UNESCO and UNU, 2012).

Since 1965, the Intergovernmental Oceanographic Commission of the United Nations Educational, Scientific and Cultural Organization (UNESCO/IOC) has been in charge of coordinating the International Pacific Tsunami Warning System (PTWS) (IOC, 1965), established after the tsunami that on 22 May 1960 devastated the countries of the Pacific Rim following an earthquake with magnitude 9.5 on the Richter scale, off southern Chile, which caused the deaths of more than 1600 people around the Pacific (NOAA/NGDC WDC, 2012). In 2005, the 23rd session of the IOC Assembly adopted three resolutions concerning the establishment of early warning and mitigation systems through intergovernmental processes in the Indian Ocean (IOC, 2005b), the Caribbean and adjacent seas (IOC, 2005c), and the north-eastern Atlantic, the Mediterranean, and connected seas (IOC, 2005d). In this context, the IOC Assembly also launched a process for defining the core elements of a global framework for early warning systems for tsunamis and other ocean-related hazards (IOC, 2005e).

Of over 2100 recorded tsunami events between 1410 BC and 2011 (Figure 8.1), it should be noted that 20% of all tsunamis recorded are considered 'very doubtful' and 24%, 'doubtful'. The Pacific Ocean accounts for 61%, the Mediterranean and Black Sea, 22%, the Atlantic Ocean and the Caribbean Sea, 11%, and the Indian Ocean, 6%. The distribution of over 15,700 runups (see IOC, 2008a) is as follows: 83% Pacific Ocean, 10% Indian Ocean, 4% Mediterranean Sea, 3% Atlantic Ocean, and <1% Red Sea and Black Sea. The distribution of fatalities is 74% in

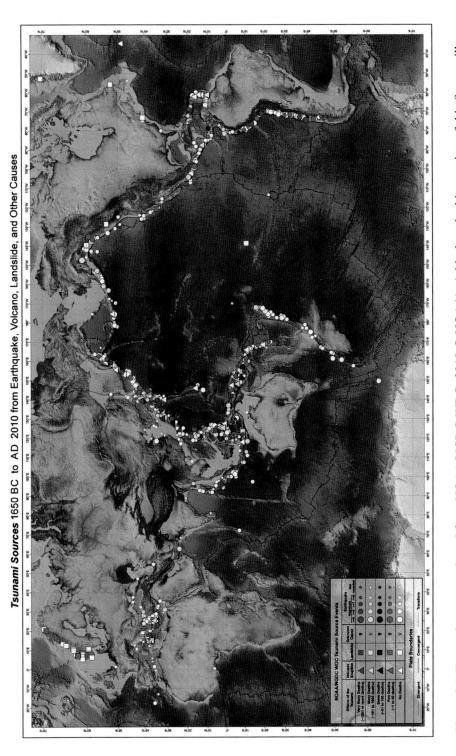

Figure 8.1 Tsunami sources (adapted from NOAA/NGDC WDC, 2010 © NOAA). A black and white version of this figure will appear in some formats. For the colour version, please refer to the plate section.

the Pacific Ocean, 10% in the Indian Ocean and the Red Sea, 9% in the Mediterranean Sea, and 8% in the Atlantic Ocean and the Caribbean Sea (NOAA/NGDC WDC, 2012).

Hazard and risk assessment, monitoring and detection networks, forecasting and warnings, awareness, education, and preparedness form the backbone of these systems which benefit from existing infrastructure for ocean-related observations, data management, forecasting, dissemination of forecasts and warnings, and capacity-building. Through networks of seismic stations, submarine earthquakes or slides that may generate a tsunami are detected and located (Figures 8.2 and 8.3).

The institutional arrangements put in place for regional tsunami warning systems include a governance mechanism, operational standards and requirements, and separate agreements for the provision of tsunami watch services. The four intergovernmental coordination groups (ICGs) for tsunami early warning and mitigation systems established by IOC for the Pacific Ocean, the Indian Ocean, the Caribbean, and the north-eastern Atlantic and Mediterranean form the governance bodies: Intergovernmental Coordination Group for the Pacific Tsunami Warning and Mitigation System (ICG/PTWS), Intergovernmental Coordination Group for the Indian Ocean Tsunami Warning and Mitigation System (ICG/IOTWS), Intergovernmental Coordination Group for the Tsunami and other Coastal Hazards Warning System for the Caribbean and Adjacent Regions (ICG/CARIBE EWS), and Intergovernmental Coordination Group for the Tsunami

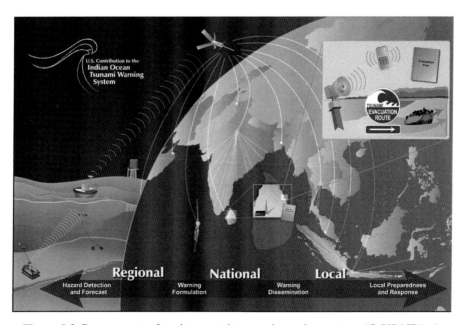

Figure 8.2 Components of an integrated tsunami warning system (© USAID). A black and white version of this figure will appear in some formats. For the colour version, please refer to the plate section.

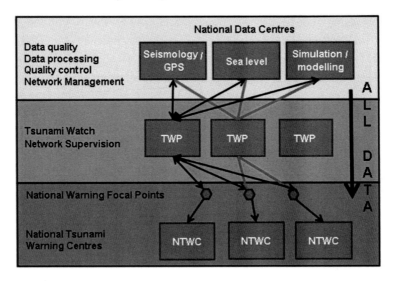

Figure 8.3 TWS architecture: example from NEAMTWS (after IOC–UNESCO, 2009). A black and white version of this figure will appear in some formats. For the colour version, please refer to the plate section.

Early Warning and Mitigation System in the North-Eastern Atlantic, the Mediterranean and connected seas (ICG/NEAMTWS). The coordination of the PTWS, originally ensured by the International Coordination Group for the Tsunami Warning System in the Pacific (ITSU), is entrusted to ICG/PTWS, established in 2006 (IOC, 2006). To ensure coherent development of the regional tsunami warning systems, the IOC Assembly, in 2005, also established a working group to define a global framework for tsunami and other ocean-related hazards early-warning systems (IOC, 2007a). This working group was subsequently reorganized as the Working Group on Tsunamis and Other Hazards Related to Sea-Level Warning and Mitigation Systems (TOWS-WG).

8.3.1 Formation, scope, and membership

The ICGs were formed around a single, regionally delimited, issue area: the protection of lives and properties from tsunamis through the establishment of TWS in four ocean basins, addressing 'a tsunami capable of destruction in a particular geographic region, generally within 1,000 km or 1–3 hours' tsunami travel time from its source' (IOC, 2008a). The expansion of the TWS to address different types of hazards from the sea is being considered in the context of a multi-hazard approach and in cooperation with other competent intergovernmental organizations. Such hazards would include storm surges or other destructive water levels that are detected or confirmed by the same monitoring networks and may involve comparable warning procedures (IOC, 2009; cf. WMO's Flood Forecasting Initiative).

No precise definition of the geographical application area of the tsunami early warning and mitigation systems has been given by the IOC, but the limits of oceans and seas defined by the International Hydrographic Organization (IHO) (IHO, 1953) provide a useful reference. In the case of the north-eastern Atlantic, on consideration of historical records of tsunami runups, the southern limits were set by the IOC Secretariat at 15° North, that is, covering up to the Mauritanian coast (for consideration of an eruption of the Cumbre Vieja Volcano on the Island of La Palma which may cause a catastrophic failure of its west flank and generate a tsunami that would affect large portions of the north-western and north-eastern Atlantic see, e.g. Ward and Day, 2001).

The regional tsunami warning systems do not rely on compliance with binding agreements for their development and implementation, but rather on commitment to a series of principles embedded in IOC policies, among which is the Oceanographic Data Exchange Policy (IOC, 2003), which stipulates that IOC 'Member States shall provide timely, free and unrestricted access to all data, associated metadata and products generated under the auspices of IOC programmes'. The Policy also encourages IOC member states to apply the same approach to 'relevant data and associated metadata from non-IOC programmes that are essential for application to the preservation of life, beneficial public use and protection of the ocean environment, the forecasting of weather, the operational forecasting of the marine environment, the monitoring and modelling of climate and sustainable development in the marine environment'.

Full membership of the four ICGs is limited to the IOC member states bordering the concerned ocean or sea basin, while other IOC member states can participate as observers; other organizations, including non-governmental, programmes, and projects may also be invited as observers. The members to the ICGs commit to establish national institutions, the Tsunami National Contacts (TNCs), tasked with national coordination, and Tsunami Warning Focal Points (TWFPs), that is, the entities tasked to receive tsunami event information by the regional system and forward it to the established National Tsunami Warning Centres (NTWCs), or disaster management organizations such as civil protection agencies.

For the purposes of the present analysis, membership of the ICGs and tsunami warning systems can be divided into three categories: (a) members that have established TNCs and/or TWFPs; (b) members that have attended at least one session of the ICGs, their working groups and task teams, and/or communication tests, exercises and drills; and (c) members that do not have established TNCs

Table 8.1 *Membership of the ICGs*

ICG	Members	TWFP	TNC	Participating	Non-participating
ICG/PTWS (1965/2006)	46	19*	19*	44 (96%)	2 (4%)
ICG/IOTWS (2005)	28	24 (86%)	24 (86%)	23 (82%)	5 (18%)
ICG/CARIBE-EWS (2005)	24†	23‡	23‡	21 (91%)	2 (9%)
ICG/NEAMTWS (2005)	39	27 (69%)	29 (74%)	31 (79%)	5 (13%)

* Focal points of PTWS members were established prior to the creation of the TWFP/TNC definition in 2005, hence the apparently low number.
† 24 IOC member states including France, the Netherlands and the United Kingdom, which have dependent territories, and excluding St Vincent and the Grenadines, which is not a member of the IOC.
‡ Including dependent territories.
Source: IOC Secretariat (as of August 2012)

and/or TWFPs and do not participate in the meetings of the ICGs for their working groups and task teams (Table 8.1). The latter category essentially represents potential members that have not yet engaged in the ICG. In a few cases, member states have established TWFPs and/or TNCs without attending sessions of the ICGs.

The active membership and engagement is highest in the most mature system (PTWS) and in the one that has obtained most attention and resources in past years (IOTWS).

8.3.2 Institutional arrangements and decision-making

As primary subsidiary bodies of the IOC (IOC, 2005f), the ICGs provide scientific and technical advice and guidance to the IOC governing bodies, the Assembly and the Executive Council, and determine technical policies for the implementation of its programme of activities. The ICG has officers (a chair and one or two vice-chairs) and a steering committee that coordinates and integrates the work of the ICG in the intersessional periods.

For the technical development and implementation of the programme of activities, the ICGs establish standing working groups and task teams of a limited duration. All ICGs have established working groups related to (a) hazard assessment and modelling; (b) seismic and sea-level measurements; and (c) advisories, public awareness, and preparedness.

The ICGs are assisted by the IOC Secretariat, through its Paris headquarters and through decentralized offices and/or outposted staff in Apia, Samoa (PTWS), Jakarta, Indonesia and Perth, Australia (IOTWS), and Kingston, Jamaica and Port au Prince, Haiti (CARIBE-EWS). The PTWS relies also on the essential technical expertise of the US International Tsunami Information Center (ITIC), established by the US National Oceanic and Atmospheric Administration (NOAA) in Honolulu, Hawaii. On the model of ITIC, tsunami information centres (TICs) were established in Jakarta, Indonesia for the Indian Ocean, Paris, France for the northeastern Atlantic and Mediterranean, and Bridgetown, Barbados for the Caribbean: Jakarta Tsunami Information Centre (JTIC) (see Table 8.2 for expansion of scope to the Indian Ocean); Tsunami Information Centre for the North-Eastern Atlantic and the Mediterranean (NEAMTIC); and Tsunami Information Centre for the Caribbean (CTIC).

The role of the TICs is to collect information on the international warning activities for tsunamis and other sea-level related hazards, to provide to regional and national stakeholders information on tsunamis and other sea-level related hazards, and on tsunami warning and mitigation systems, to foster identification and exchange of best practices and assist national stakeholders in the establishment of regional and national components of the TWS, and to act as information resources for the development and distribution of awareness, educational and preparedness materials, event data collection, and the fostering of research and its application, with the aim of mitigating loss of life and property from tsunamis and other sea-level related hazards.

In accordance with the general rules of procedure of IOC (IOC, 2001), the ICGs make decisions by consensus. They pass resolutions on internal matters and can submit recommendations to the IOC governing bodies. At the level of IOC, resolutions are the formal expression of a mandate or opinion or a declaration of a decision to act made by a governing body; resolutions are normally adopted to reflect major policy, programmatic, external relations, financial, and institutional decisions. As in the case of its parent body, UNESCO, IOC resolutions are not binding and no mechanisms for enforcement are available.

8.3.3 Programmatic activities and operation

The ICGs and their subsidiary bodies promote and coordinate activities aimed at three main areas of work: (1) architecture of the monitoring networks and the tsunami warning system and related standard operating procedures; (2) assessment and development of capacities in preparedness for tsunamis, including thorough carrying out of communication tests, exercises, and drills; and (3) public awareness and education, and promotion of research.

Table 8.2 *Changes in the life of the ICGs*

ICG / Major changes	Changes in key concepts and principles	Changes in the group of leading actors	Expansions in the functional scope
ICG/PTWS	2014: Change from the issuance of warnings to advisory tsunami wave forecast products by the USA PTWC	1965: Establishment of the Tsunami Warning System in the Pacific (TWSP), with warnings issued by USA Pacific Tsunami Warning Center (PTWC), and tsunami monitoring, capacity-building, historical data collection, and awareness services provided by the UNESCO/IOC–USA NOAA International Tsunami Information Center (ITIC) 1982: West coast of North America coverage provided by the West Coast and Alaska Tsunami Warning Center (WC/ATWC, established in 1967). Expanded to East Coast of North America in 2005 2005: Establishment of the Japan Northwest Pacific Tsunami Advisory Center, as a regional advisory centre in cooperation with PTWC	2014: Change from the issuance of warnings to advisory tsunami wave forecast products by the USA PTWC
ICG/IOTWS	2005: ICG decides to develop a network of regional tsunami service providers rather than a single Indian Ocean Warning Centre 2008: Implementation plan envisages phased implementation of the RTSPs and transition from the Interim Advisory Service (IAS) provided by PTWC and JMA	2005: Through the provision of interim tsunami watch services, Japan and USA lead actors 2011: Operational phase reached and tsunami watch service provided by the countries of the region: Australia, India, and Indonesia	

Working Group			
	2009: ICG decides that RTSPs should provide detailed tsunami threat information to NTWCs only. Information available to the public would not contain details of threat, as this is a national responsibility to determine 2011: ICG supports the expansion of the role of the Jakarta Tsunami Information Centre (JTIC) into an Indian Ocean Tsunami Information Centre (IOTIC)		
ICG/CARIBE-EWS		2012: Establishment of the Caribbean Tsunami Information Center hosted by Barbados	2012: Recommended extension of the area of responsibility to the whole western Atlantic. The recommendation was not accepted by the 45th session of the IOC Executive Council (IOC, 2012)
ICG/NEAMTWS	2011: Abandonment of the concept of 'regional tsunami watch centre with defined area of responsibility' in favour of the concept of 'tsunami service provider' on the example of the IOTWS	2007: Leading actors emerging are France, Greece, Italy, Portugal, and Turkey, with the offer to establish 'regional tsunami watch centres' 2011 on: Increased collaboration with the European Commission, which supports a project for the establishment of the Tsunami Information Centre for the North-Eastern Atlantic and the Mediterranean (NEAMTIC) 2012: France, Greece, and Turkey start the operational phase as national tsunami warning centres, distributing data to other countries upon request	Between 2009 and 2011: Attempt to enlarge the scope of the ICG to multi-hazard approach to coastal inundation. Limited follow-up

Exchange of real-time seismic data occurs through existing international and regional networks (besides IMS, e.g. Global Seismographic Network (GSN), Mediterranean Network (MedNet), GEOFON Global Seismic Network). In addition, an agreement signed in 2009 between UNESCO and the Preparatory Commission for the Comprehensive Nuclear-Test-Ban Treaty Organization (CTBTO) enables tsunami warning centres that are recognized by UNESCO to obtain from CTBTO data from the International Monitoring System (IMS), composed of 337 seismic station facilities, for the purpose of producing tsunami warnings (UNESCO, 2009).

Exchange of real-time sea-level data is facilitated by IOC's Sea Level Station Monitoring Facility, established in cooperation between two IOC core programmes, the Global Sea Level Observing System (GLOSS) and the International Oceanographic Data Exchange Programme (IODE), with support from the Flanders Marine Institute (VLIZ) in Ostend. Transmission of data from seismic and sea-level stations and of warning occurs via WMO Global Telecommunications System (GTS) or the Internet. It should be added that technology in ocean observations had improved vastly since the creation of the PTWS in 1965 and the 2004 tsunami event encouraged countries to take advantage of this opportunity, and this also impacted in some sense the governance arrangements. For example, demonstration gauges have become more affordable, data transmission opportunities have improved, and general communication has improved via email and the Internet. One particular development concerns 'tsunameters', ocean observing platforms with sea floor pressure sensors to detect tsunami waves that transmit data to a surface moored buoy which in turn transmits the collected observations in real time via satellite. Networks of tsunameters are particularly useful for large-scale or regional tsunamis and are normally deployed in international waters or at the external limit of EEZs. The possibility of using submarine telecommunications cables for ocean and climate monitoring and early detection of tsunamis was explored in 2011–2012 by a joint task force of the International Telecommunications Union (ITU), WMO and UNESCO/IOC (ITU, 2012a, 2012b, 2012c).

8.3.4 Effectiveness: operational arrangements, behavioural changes, and goal attainment

The goal of the ICGs is to coordinate the development of tsunami warning systems that can deliver tsunami watches and warnings. In the context of UNESCO's programme and budget for 2012–2013 (UNESCO, 2011), the expected result has been characterized as follows: 'Effective end-to-end early warning systems for tsunamis and other sea level-related hazards established by Member States at national and regional levels, including disaster preparedness and mitigation measures'. Performance indicators and benchmarks at the end of 2013 focused on

(a) improved lapse time for regional watch messages after tsunami-generating earthquakes (benchmark: less than 20 minutes in all regional warning systems); and (b) number of countries at risk from tsunamis developing gender-sensitive tsunami community preparedness programmes (benchmark: ten).

For an analysis of the of the evolution of the four ICGs, it will be useful to explore major changes in their lives according to three categories: (a) changes in key concepts and principles, (b) changes in the group of leading actors, and (c) expansions in the functional scope (Breitmeier, H. *et al.*, 2006). Table 8.2 summarizes the changes, which are briefly reviewed below.

The TWS are built around the concept that a limited number of operational tsunami watch providers (TWPs) would distribute tsunami watches to the national TWFPs of the region (see example from NEAMTWS in IOC, 2007b). These, in turn, would issue tsunami warnings to their respective competent agencies, most typically, disaster management organizations. The TWPs do not have defined sub-regional responsibilities, as data redundancy and overlaps in coverage actually reinforce data availability and accuracy.

Within each region covered by a TWS, the role played by countries hosting TWPs and TICs is particularly important. In terms of changes in the group of leading actors, it is noteworthy that in the Indian Ocean, the interim phase with the provision of tsunami watch services by US and Japanese agencies was positively concluded in 2011 with the launch of TWPs in Australia, India, and Indonesia. In the case of the Mediterranean and north-eastern Atlantic, since the mid 2000s, the European Commission (EC) is playing an increasing role both through funding in support of research projects (EC, 2006a, 2006b, 2006c, 2007a, and 2007b) and the NEAMTIC (EC, 2011), but also through the technical work of the Joint Research Centre (JRC), for example in wave propagation models. No particular expansions of the functional scope of the TWS can be identified, also because only PTWS and IOTWS are currently at a fully operational stage. As explored by ICG/NEAMTWS, the incorporation of TWS in multi-hazards warning systems, also addressing other types of coastal flooding such as storm surges, may be achieved in the future (see (WMO, 2009, paragraph 4.6.13). An evaluation of the performance and effectiveness of TWS is outside the scope of the present contribution (for studies on the development and performance of individual TWS, see, e.g. Srinivasa Kumar *et al.*, 2010; Srinivasa Kumar *et al.*, 2012; Hanka *et al.*, 2010; Lauterjung *et al.*, 2010; Ozaki, 2011; and Committee on the Review of the Tsunami Warning and Forecast System and Overview of the Nation's Tsunami Preparedness, National Research Council, 2011). Applying, *mutatis mutandis*, Henkin's hypothesis (1979) that the majority of member states comply with the majority of international environmental rules most of the time, it can be seen that, with few exceptions, the ICGs are enjoying great participation by their members (Table 8.1), who are contributing to the development and implementation of tsunami early warning and mitigation systems through seismological networks and the application of the IOC Oceanographic Data Exchange Policy to sea-level data.

8.4 Ocean observation through the Argo Project

The Argo Project is an array of over 3500 active free-floating ocean monitoring devices. Argo builds on existing observation networks of observing ships and ships of opportunity, tide gauges, surface drifters, subsurface drifters, moored buoys, and profiling floats. All data are relayed and made publicly available within hours after collection. As the Argo floats may drift into EEZs, a notification system has been put in place. Argo is a component of the Integrated Global Observing Strategy (IGOS), supported by intergovernmental programmes such as the Global Climate Observing System (GCOS), the Global Ocean Observing System (GOOS), and the Global Terrestrial Observing System (GTOS), sponsored by multiple international and UN agencies and programmes. Its worldwide scope calls for some form of international legal regulation, because it employs thousands of voluntary observing ships and ships of opportunity, as well as tide gauges, surface drifters, subsurface drifters, moored buoys, and profiling floats that may drift into national EEZs. Argo floats are deployed to collect a large database of ocean signals related to climate change and to provide *in situ* satellite observations of the Earth System as a whole, while protecting life and property, predicting climate variations and severe weather, collecting, storing, and distributing data and information freely to all interested users in near-real time. Argo provides free, unrestricted, and operational data distribution. Data are delivered in real time to WMO, GTS, and Argo Global Data Centres in USA and France. A long-term archival centre has been established at the US National Oceanographic Data Center. Regional data centres have also been created to promote data use on a regional basis (Argo Information Centre, 2006).

8.4.1 Subjects involved

The Argo array is composed of member states and intergovernmental organizations like UNESCO and WMO, and is otherwise undefined, using indistinct terms like 'programme' or 'project' (see IOC, 1999 and 2008b). Data sharing (on the two distribution channels) is *sine qua non* to be part of Argo. Following IOC Resolution IOC-XLI-4, the subjects involved in Argo activities are the Argo Information Centre (AIC), located in Toulouse, acting as coordinator and providing technical assistance, and the UNESCO/IOC–WMO member states, divided into implementers (i.e. deployers) and coastal states. The implementers play an active role: they own the floats and provide the information to the AIC which then disseminates the information; the coastal states play a passive role by letting the Implementer collect date from waters under their jurisdiction. The floats belong to the Implementer and may operate on the high seas (UNCLOS, Art. 86), or the territorial sea (UNCLOS, Art. 2), or the EEZ (UNCLOS, Art. 55) of the implementer or any other coastal state.

Even though there is no mention of ownership on the label affixed to the floats, it states: 'This is scientific equipment; it is not military material and has no commercial value. This float is used for ocean predictions and provides valuable information to many communities including fishermen. For your safety do not open this instrument. If found please contact the Argo Information Centre ...'. The AIC is in charge of locating the owner and initiating the operation to recover the float. However, the United Nations Secretariat through WMO and UNESCO/IOC is progressively increasing its role in Argo, to the point that it may become an Implementer in the near future.

8.4.2 Why Argo matters

The broad-scale global array of temperature/salinity profiling floats, known as Argo, has already grown to be a major component of the ocean observing system. Argo provides a quantitative description of the changing state of the upper ocean, and an ocean–atmosphere forecast. It also documents seasonal to decadal climate variability, to aid our understanding of its predictability. Therefore, Argo is a major instrument for observing and analysing climate change as well as for marine and ocean forecasting.

8.4.3 Argo legal framework

In 1989, the IOC adopted the main Resolution XX-6 (IOC, 1999) regulating the Argo project, providing the following: 'the Argo project shall be fully consistent with UNCLOS'. This line subscribes all Argo activities to UNCLOS, despite the fact of 'the absence of a specific international legal instrument regulating profiling floats, drifting buoys, and other similar objects deployed in the oceans' (IOC, 1999). Therefore, it is uncertain whether UNCLOS provisions for MSR are applicable to Argo since many states refuse to accept that the Argo activities fall under Part XIII of UNCLOS.

IOC Resolution IOC-XLI-4 confirmed once again that 'the Argo project shall be fully consistent with UNCLOS' and it added 'without prejudice to the sovereign rights of the coastal states over their continental shelf and exclusive economic zone'. However, the application of UNCLOS has been questioned by states.

The interpretation of Argo activities with regard to UNCLOS is very controversial. While some IOC member states view the Argo project as an activity better characterized as operational oceanography, not governed by the marine scientific research provisions under UNCLOS 1982 Part XIII, other member states understand that Argo activities are governed by Part XIII of UNCLOS (marine scientific research).

To date, the IOC Assembly has been unable to reconcile the positions on marine scientific research regulation versus freedom from marine scientific research regulation, on other instruments deployed at sea (gliders, buoys, etc.), and it may suffer the same fate as the IOC's failed attempt to adopt an Ocean Data Acquisition Systems (ODAS) Convention in the 1970s and 1990s (IOC, 1993).

Given the lack of specific provision in UNCLOS, provisions to be applied, *mutatis mutandis*, are Articles 259–262, in addition to Article 263, which provides for compensation for damage resulting from measures in contravention of UNCLOS (Churchill and Lowe, 1988).

Argo works in a responsible way, fully aware of its obligations under UNCLOS. As stated by the Argo Steering Team (AST): 'Argo does its best to minimize the probability of incident or hazard. Instruments that come ashore are safely recovered. The AST agreed that the cost of, and responsibility for, the retrieval should be borne by the float owner. While this may be expensive, it is a necessary overhead of the Argo project' (Argo Information Centre, 2006).

In June 2008, the Executive Council of IOC, after extensive debate by members of the IOC Advisory Body of Experts on the Law of the Sea (IOC/ABE-LOS), adopted the latest instrument on Argo, the Guidelines for the legal regulation of Argo Profiling Float Deployments on the High Seas (IOC, 2008b). The document provides the procedure regulating the communication mechanism between states and data policy. However, it reiterates the line 'fully consistent with UNCLOS' but does not shed any further light on the interpretation of UNCLOS with regard to Argo.

However, the 2008 guidelines provide a general communication mechanism between the Argo float Implementer, and the Argo Focal Point (AFP) in the coastal state for the case where an Argo Programme float deployed in the high seas might drift into the EEZ of an IOC member state that has requested notification. This mechanism weakens the relevant international organizations while benefitting bilateral relations between states.

8.5 Concluding remarks

The Future We Want, the outcome document adopted at the United Nations Conference on Sustainable Development, Rio+20 (Rio de Janeiro, 20–22 June 2012), calls on states to 'support marine scientific research, monitoring and observation of ocean acidification and particularly vulnerable ecosystems, including through enhanced international cooperation in this regard' (United Nations, 2012b). In spite of this, the Rio+20 conference did not emphasize the importance of MSR and operational oceanography for the sustainable development of the oceans. A programme like Argo plays a major role in this task, but is also a key instrument in ocean predictability and climate change. Argo is not specifically regulated in UNCLOS or related legal instruments; however, it is an example on how to collaborate internationally in MSR and how this has evolved since the adoption of UNCLOS. Indeed, activities falling under operational oceanography, aimed at systematic measurements of the seas, the oceans, and the atmosphere, rapid interpretation and dissemination of products in the form of real-time descriptions of the state of the sea and continuous forecasts of the conditions of the sea, are among the most important developments in oceanography in the last 40 years. Such activities rely on the use of multiple platforms, often automated – drifting surface buoys, moored buoys, measurements from volunteer ships, tide gauges, profiling floats, aircraft, and satellites – whose scope is significantly wider than that of traditional research vessels. Some of these platforms are also vital in the collection of data essential for the production of marine weather forecasts and tsunami advisories and warnings.

The outcome document of Rio+20 also underlined the need to foster collaboration among academic, scientific, and technological communities, in particular in developing countries, to 'close the technological gap between developing and developed countries, strengthen the science–policy interface' (United Nations, 2012a). However, the Rio de Janeiro conference has not focused sufficiently the attention of states and other stakeholders on the need to increase international cooperation in scientific programmes, failing in what promised to be 'a unique opportunity to speed up implementation and to take action on ocean-related emerging issues' (United Nations, 2012b). The competent international organizations in MSR are the only acceptable means of moving forward MSR programmes through international cooperation, as they provide a guarantee for developing countries to participate on an equitable basis. To this end, further guidance should be developed and provided under Article 251 of UNCLOS for different applications of MSR, together with financial support and technical assistance to developing countries. The 'Oceans Compact' launched by the UN Secretary-General in 2012 (United Nations 2012c), which sets out a strategic vision for the UN system to deliver on its ocean-related mandates, calls for, among several objectives, reducing vulnerability to ocean hazards, including tsunamis, and strengthening ocean knowledge and the management of oceans, including through addressing the need for robust ocean observation and relevant infrastructure, including capacity development in ocean and coastal areas. It is expected that such an initiative, building on the goals of Rio+20, may foster interagency collaboration to meet countries' needs in these critical subjects.

References

Argo Information Centre (2006). *Argo Project & IOC Resolution XX-6 Implementation*, IOC/ABE-LOS-VI/Inf.1. Paris: IOC.

Bernard, E. N. and Robinson, A. (2009). *Tsunamis,* The Sea 15. Princeton, MA: Harvard University Press.

Breitmeier, H., Young, O. R., and Zürn, M. (2006). *Analyzing International Environmental Regimes*. Cambridge, MA: The MIT Press.

Churchill, R. and Lowe, V. (1988). *Law of the Sea*. Manchester: Manchester University Press.

Committee on the Review of the Tsunami Warning and Forecast System and Overview of the Nation's Tsunami Preparedness, National Research Council (2011). *Tsunami Warning and Preparedness: An Assessment of the US Tsunami Program and the Nation's Preparedness Efforts*. Washington DC: National Academy of Science.

EC (2006a). Seismic and Tsunami Risk Assessment and Mitigation Scenarios in the Western Hellenic Arc (SELLARC) (online): www.seahellarc.gr (4 February 2014).

EC (2006b). Tsunami Risk and Strategies for the European Region (TRANSFER) (online): www.transferproject.eu (4 February 2014).

EC (2006c). Integrated Observations from Near Shore Sources of Tsunamis (NEAREST) (online): http://nearest.bo.ismar.cnr.it (4 February 2014).

EC (2007a). Distant Early Warning System (DEWS) (online): www.dews-online.org) (4 February 2014).

EC (2007b). Scenarios for Hazard-induced Emergencies Management (SCHEMA) (online): www.schemaproject.org (4 February 2014).

EC (2011). Tsunami Information Centre for the North-Eastern Atlantic and the Mediterranean (NEAMTIC) (online): http://neamtic.ioc-unesco.org (4 February 2014).

Hanka, W., Saul, J., Weber, W. *et al.*, GITEWS Seismology Group (2010). Real-time earthquake monitoring for tsunami warning in the Indian Ocean and beyond. *Nat. Hazards and Earth Syst. Sci.* **10**: 2611–2622.

Henkin, L. (1979). *How Nations Behave: Law and Foreign Policy*. New York: Columbia University Press.

IHO (1953). *Limits of Oceans and Seas*, Special Publication 23. Montecarlo: IHO.

IOC (1965). Resolution IV-6, International Coordination Group for the Tsunami Warning System in the Pacific, adopted by the IOC Assembly at its fourth session (Paris, 3–12 November 1965). Paris: IOC.

IOC (1993). Draft Convention on the Legal Status of Ocean Data Acquisition Systems, Aids and Devices (ODAS). Second Revision, IOC-XVII/Inf. 1. Paris: IOC.

IOC (1999). Resolution XX-6, The Argo Project, adopted by the IOC Assembly at its twentieth session (Paris, 29 June–9 July 1999). Paris: IOC.

IOC (2001). Rules of Procedure, *IOC/INF-1166*. Paris: UNESCO.

IOC (2003). Resolution XXII-6, IOC Oceanographic Data Exchange Policy, adopted by the IOC Assembly at its twenty-second session (Paris, 9–13 July 2003). Paris: IOC.

IOC (2005a). *IOC Criteria and Guidelines on the Transfer of Marine Technology*, IOC/INF-1203. Paris: UNESCO.

IOC (2005b). Resolution XXIII-12, Intergovernmental Coordination Group for the Indian Ocean Tsunami Warning and Mitigation System, adopted by the IOC Assembly at its twenty-third session (Paris, 21–30 June 2005). Paris: IOC.

IOC (2005c). Resolution XXIII-13, Intergovernmental Coordination Group for the Tsunami and Other Coastal Hazards Warning System for the Caribbean and Adjacent Regions, adopted by the IOC Assembly at its twenty-third session (Paris, 21–30 June 2005). Paris: IOC

IOC (2005d). Resolution XXIII-14, Intergovernmental Coordination Group for the Tsunami Early Warning and Mitigation System in the North-Eastern Atlantic, the Mediterranean and Connected Seas, adopted by the IOC Assembly at its twenty-third session (Paris, 21–30 June 2005). Paris: IOC.

IOC (2005e). Resolution XXIII-15, Establishment of a framework for a global tsunami and other ocean-related hazards early-warning system, Global Ocean-related Hazards Warning and Mitigation System Framework (GOHWMS), adopted by the IOC Assembly at its twenty-third session (Paris, 21–30 June 2005). Paris: IOC.

IOC (2005f). *Guidelines for the Structure and Responsibilities of the Subsidiary Bodies of the Commission, and for the Establishment of Decentralized Offices*, IOC/INF-1193. Paris: IOC.

IOC (2006). Resolution EC-XXXIX.8, Intergovernmental Coordination Group for the Pacific Tsunami Warning and Mitigation System, adopted by the IOC Executive Council at its thirty-ninth session (Paris, 21–28 June 2006). Paris: IOC.

IOC (2007a). *Framework Document for the Global Ocean-related Hazards Warning and Mitigation System (GOHWMS)*, IOC-XXIV/2 Annex 10. Paris: IOC.

IOC (2007b). Intergovernmental Coordination Group for the Tsunami Early Warning and Mitigation System in the North Eastern Atlantic, the Mediterranean and Connected Seas (ICG/NEAMTWS), Third Session, Bonn, Germany, 7–9 February 2007, ICG/NEAMTWS-III/3. Paris: IOC.

IOC (2008a). *Tsunami Glossary*, IOC Technical Series 85. Paris: IOC.

IOC (2008b). Resolution EC-XLI.4 on the Guidelines for the implementation of Resolution XX-6 of the IOC Assembly regarding the deployment of profiling floats in the High Seas within the framework of the Argo Programme, adopted by the IOC Executive Council at its forty-first session (Paris, 24 June–1 July 2008). Paris: IOC.

IOC (2009). *Hazard Awareness and Risk Mitigation in Integrated Coastal Area Management*, IOC Manuals and Guides 50. Paris: IOC.

IOC (2012). *Forty-fifth Session of the IOC Executive Council (Paris, 26–28 June 2012): Summary Report*, IOC/EC-XLV/3. Paris: IOC.

ITU (2012a). *Using Submarine Cables for Climate Monitoring and Disaster Warning: Strategy and Roadmap*. Geneva: ITU.

ITU (2012b). *Using Submarine Cables for Climate Monitoring and Disaster Warning: Opportunities and Legal Challenges*. Geneva: ITU.

ITU (2012c). *Using Submarine Cables for Climate Monitoring and Disaster Warning: Engineering Feasibility Study*. Geneva: ITU.

Japan, UNESCO and UNU (2012). Japan–UNESCO/UNU Symposium on the Great East Japan Tsunami on 11 March 2011 and Tsunami Warning Systems: Policy Perspectives (Tokyo, 16–17 February 2011) (online): www.ioc-tsunami.org/tohokusymposium (accessed 4 February 2014).

Lauterjung, J., Koltermann, P., Wolf, U., and Sopaheluwakan, J. (2010). The UNESCO–IOC framework – Establishing an international early warning infrastructure in the Indian Ocean region. *Nat. Hazards and Earth Syst. Sci.* **10**: 2623–2629.

NOAA/NDBC (2012). Locations of NDBC DART® stations. National Data Buoy Center (online): www.ndbc.noaa.gov (4 February 2014).

NOAA/NGDC WDC (2010). *Tsunami Sources in the World*. Washington DC: NOAA.

NOAA/NGDC WDC (2012). Global historical tsunami database. *National Geophysical Data Center* (online): www.ngdc.noaa.gov/nndc/struts/form? t=101650&s=70&d=7 (4 February 2014).

Ozaki, T. (2011). Outline of the 2011 off the Pacific coast of Tohoku earthquake (Mw 9.0) – Tsunami warnings/advisories and observations. *Earth Planets Space* **63**: 827–830.

Pugh, D. (2001). *Towards an Implementation of Article 251 of UNCLOS.* IOC/ABE-LOS I/7. Paris: IOC.

Ryder, P. (2003). A possible migration from marine scientific research to operational oceanography in the context of the United Nations Convention on the Law of the Sea (UNCLOS). In *Building the European Capacity in Operational Oceanography, Proceedings of the Third International Conference on EuroGOOS (Athens, Greece, 3–6 December 2002)*. Dahlin, H., Flemming, N. C., Nittis, K., and Petersson, S. E. (eds.). Oceanography Series 69. Amsterdam: Elsevier, pp. 25–35

Schiller, A. and Brassington, G. B. (2011). *Operational Oceanography in the 21st Century*. Dordrecht/Heidelberg/London/New York: Springer.

Srinivasa Kumar, T., Kumar, C. P., and Nayak, S. (2010). Performance of the Indian Tsunami Early Warning System. *Int. Arch. Photogram. Rem. Sens. Spatial Inform. Sci.* **38**(8): 271–274.

Srinivasa Kumar, T., Nayak, S., Kumar, C. P., *et al.* (2012). Performance of the tsunami forecast system for the Indian Ocean. *Current Science* **102**(1): 110–114.

UN/DOALOS (2010). *Marine Scientific Research: A Revised Guide to the Implementation of the Relevant Provisions of the United Nations Convention on the Law of the Sea.* New York: United Nations.

UNESCO (2009). *Relations with the Preparatory Commission for the Comprehensive Nuclear-Test-Ban Treaty Organization (CTBTO) and Draft Memorandum of Understanding between UNESCO and that Organization*, 182 EX/64. Paris: UNESCO.

UNESCO (2011). *Programme and Budget 2012–2013*, 36 C/5 Add. Paris: UNESCO.

UNISDR and WMO (2012). *Disaster Risk and Resilience*. Geneva: UNISDR–WMO.

United Nations (1979a), *Official Records of the Third United Nations Conference on the Law of the Sea, Volume XII (Summary Records, Plenary, General Committee, First and Third Committees, as well as Documents of the Conference, Resumed Eighth Session)*, A/CONF.62/80, Resolution 16 (Cg-VIII) adopted by the World Meteorological Organization at its eighth congress at Geneva in April/May 1979. New York: United Nations.

United Nations (1979b), *Official Records of the Third United Nations Conference on the Law of the Sea, Volume XIV (Summary Records, Plenary, General Committee, First and Third Committees, as well as Documents of the Conference, Resumed Ninth Session)*, A/CONF.62/SR.134, 134th Plenary meeting. New York: United Nations.

United Nations (2000). *Report on the work of the United Nations Open-ended Informal Consultative Process on Oceans and the Law of the Sea at its first meeting*, A/55/274. New York: United Nations.

United Nations (2001a). *United Nations Open-ended Informal Consultative Process on Oceans and the Law of the Sea, Second meeting (7–11 May 2001)*, A/AC.259/4, Marine science and the development and transfer of marine technology, including capacity-building (submitted by the delegation of Norway). New York: United Nations.

United Nations (2001b). *Report on the work of the United Nations Open-ended Informal Consultative Process established by the General Assembly in its Resolution 54/33 in order to facilitate the annual review by the Assembly of developments in ocean affairs at its second meeting*, A/56/121, letter dated 22 June 2001 from the Co-Chairpersons of the Consultative Process addressed to the President of the General Assembly. New York: United Nations.

United Nations (2011), *Report on the work of the United Nations Open-ended Informal Consultative Process on Oceans and the Law of the Sea at its twelfth meeting*, A/66/186. New York: United Nations.

United Nations (2012a). *The Future We Want*, A/66/L.56. New York: United Nations.

United Nations (2012b). *Rio+20 UNCSD: Oceans*, Rio 2012 Issues Brief 4. New York: United Nations.

United Nations (2012c). *The Oceans Compact: Healthy Oceans for Prosperity*. New York: United Nations.

Ward, S. N. and Day, S. J. (2001). Cumbre Vieja Volcano; Potential collapse and tsunami at La Palma, Canary Islands. *Geophys. Res. Lett.* **28**(17): 3397–3400.

WMO (1995). Resolution 40, WMO policy and practice for the exchange of meteorological and related data and products including guidelines on relationships in commercial meteorological activities, adopted by the twelfth session of the World Meteorological Congress (Geneva, 30 May–21 June 1995). Geneva: WMO.

WMO (1999). Resolution 25, Exchange of hydrological data and products, adopted by the thirteenth session of the World Meteorological Congress (Geneva, 4–26 May 1999). Geneva: WMO.

WMO (2009). *Regional Association VI (Europe), Fifteenth Session (Brussels, 18–24 September 2009): Abridged Final Report with Resolutions*, WMO No. 1046. Geneva: WMO.

9

Emerging and unresolved issues: the example of marine genetic resources of areas beyond national jurisdiction

MARJO VIERROS,* CHARLOTTE SALPIN,* CLAUDIO CHIAROLLA,
AND SALVATORE ARICÒ*

9.1 Introduction

While much of the ocean still remains to be explored, it is now known that it is extremely rich in biological diversity, including organisms which are host to unique genetic resources.

While the exact number of marine species is unknown, scientists estimate that there may be 0.7 to 1.0 million marine species, of which approximately 226,000 eukaryotic species have been described (Appeltans *et al.*, 2012). More species were described in the past decade (*c.*20,000) than in any previous decade. It has been reported that there are approximately 170,000 synonyms; that 58,000–72,000 species are collected but not yet described; and that 482,000–741,000 more species have yet to be sampled. Molecular methods may add tens of thousands of cryptic species. Thus, there may be 0.7–1.0 million marine species. Past rates of description of new species indicate that there may be 0.5 ± 0.2 million marine species. On average, 37% (median 31%) of species in over one hundred recent field studies around the world might be new to science.

Thus, a very large proportion of marine species are yet to be discovered. It is thought that these species live in remote and hard-to-reach environments, such as the deep sea and the seabed, or are microscopic. In fact, the Census of Marine Life estimated that more than a billion types of microbes might live in the oceans (Census of Marine Life, 2010). One drop of seawater may contain as many as 350,000 bacteria and other microorganisms (Knowlton, 2010).

Article 2 of the Convention on Biological Diversity (CBD) defines genetic resources as genetic material of actual or potential value. Genetic material is defined as any material of plant, animal, microbial, or other origin containing functional units of heredity. It follows that marine genetic resources are material from marine plants, animals, and microbial or other organisms, and parts thereof containing functional units of heredity of actual or potential value. It is noteworthy that the 2010 Nagoya Protocol on Access to Genetic Resources and the Fair and Equitable Sharing of Benefits Arising from their Utilization to the Convention on Biological Diversity, which applies to genetic resources within the scope of Article 15 of the CBD, also includes within its scope derivatives, which are defined as naturally occurring biochemical compounds resulting from the genetic expression or metabolism of biological or genetic resources, even if they do not contain functional units of heredity (Article 2). This chapter follows the definition of the CBD, although it must be noted that the scientific understanding of genetic resources may not be uniform (see Aricò and Salpin, 2005).

From an applied research perspective, including with a commercial intent, the deep sea and the seabed contain genetic resources of actual or potential value and interest. The ratio of potentially useful natural compounds is higher in marine than in terrestrial organisms, which implies a higher probability of commercial success with material sourced from the marine environment. Organisms from the deepest areas of the oceans, such as hydrothermal vents and cold seeps, and from other environments such as polar areas, are subject to extremes of pressure and temperature, and sometimes toxicity. As a result, these organisms, which are often referred to as 'extremophiles', have evolved unique molecular and metabolic characteristics that allow their survival in such conditions. The combination of extreme conditions and potential for new discoveries make extreme environments potentially one of the largest reservoirs of genetic resources of major interest for industrial and other applications (Yooseph *et al.*, 2007).

Marine genetic resources hold significant promise for helping meet some of the major current development challenges such as combating disease through the development of new drugs; helping to combat pollution of the marine environment through bioremediation; achieving food security through sustainable mariculture practices; and meeting growing energy demands through the development of alternative energy sources. Interest in these resources may also stimulate the creation of public–private partnerships in the area of research and development, for the benefit of advancing marine science and knowledge.

Yet, in spite of the increasing attention paid to issues related to marine genetic resources by scientists, industry, and policy makers, available information on research and development related to marine genetic resources, as well as on the modalities of the partnerships established between scientific institutions – largely public – and companies – largely private – is still poor. Discussions are also ongoing regarding the legal regime applicable to marine genetic resources of areas beyond national jurisdiction. The maritime zones beyond national jurisdiction are: the 'Area', which is the seabed and ocean floor and subsoil thereof beyond the limits of national jurisdiction (UNCLOS, Article 1(1)(1)); and the high seas, which are all areas of the water column that are not included in the EEZ (exclusive economic zone), the territorial sea or the internal waters of a state, or in the archipelagic waters of an archipelagic State (UNCLOS, Article 86).

Some states believe that the United Nations Convention on the Law of the Sea adequately covers activities related to those resources, as well as related equity issues, while others are of the view that a new regime to address the conservation and sustainable use of, including access to and sharing of benefits resulting from, these resources is needed (see, in particular, discussions at the General Assembly Ad Hoc Open-ended Informal Working Group to study issues relating to the conservation and sustainable use of marine biological diversity beyond areas of national jurisdiction. Cf. United Nations, 2008, 2010, 2011, 2012a, and 2013b; Germani and Salpin, 2011). The limited information available, coupled with uncertainty regarding legal aspects and policy, may hamper the utilization of marine genetic resources in a manner that can be socially and economically beneficial and equitable, as well as environmentally sound. The need for legal clarity for investors was generally highlighted at the eighth meeting of the United Nations General Assembly Informal Consultative Process on Oceans and the Law of the Sea, which focused its discussions on marine genetic resources in 2007 (United Nations, 2007a).

This chapter presents recent knowledge and information related to marine genetic resources, with a particular focus on those from areas beyond national jurisdiction. It also provides a brief overview of developments in research and applications related to these resources, as well as recalling some of the legal aspects and policy discussions. Finally, the chapter attempts to place questions related to marine genetic resources in the broader context of the conservation and sustainable use of marine biodiversity.

9.2 New scientific research on, and commercial uses of, marine genetic resources

The exploration of the deep seabed is still a recent endeavour, which started at the end of the 19th century with the British research vessel *Challenger* (1872–1876). However, it was not until 1977 that hydrothermal vents were discovered, with the help of the submersible Alvin, during a survey of the Galapagos Rift in the eastern Pacific Ocean at depths of more than 1000 metres. Today, a host of research activities are undertaken to study the ecology, biology, and physiology of the deep-sea and seabed ecosystems and species, including large scientific collaborations such as the Census of Marine Life, which culminated in the publication of its significant findings in 2010, as well as the Tara expeditions (Tara Expeditions, 2014).

In spite of these programmes, very little of the deep sea and seabed has been explored thus far, owing primarily to the costs associated with deep-sea research and technological constraints. It is estimated that less than 0.1% of abyssal plains and less than 200 of an estimated 30,000 seamounts have been studied (Kitching-man and Lai, 2004).

While it appears that the majority of the research activities related to the deep sea and seabed are not directly of a commercially oriented nature, they help to generate the scientific information necessary for applied research on marine genetic resources. The involvement of commercial interests in sampling from the deep sea and open ocean, where it exists, is often limited to contributing to the funding necessary for research dives carried out by national scientific research organiza-tions or academic institutions and/or research collaboration in laboratories once samples have been collected (Leary, 2007). Commercial developments may also be based on organisms that have been deposited in culture collections by research institutions, and sourced from there by other entities for further study. Investi-gations of the unusual characteristics of organisms from the deep sea and seabed have already resulted in a number of patented inventions covering a diverse range of applications, from the development of enzymes for industrial processes to pharmaceuticals and skincare products (Leary *et al.*, 2009). These patents are held by private companies as well as by governmental and academic institutions, and most are based on organisms collected from within national jurisdiction. The number of patents filed based on marine organisms in general has grown steadily by 12% per year, with 95% of patent claims having been filed after 2000 (Arnaud-Haond *et al.*, 2011). However, not all patents result in commercialized products and subsequent profits. For example, in the field of pharmaceuticals, thus far, only 1–2% of pre-clinical candidates have become commercial products (Aricò and Salpin, 2005). Where companies are successful, though, there is a potential for high financial rewards. For example, sales in the USA of the pain medication Prialt, which is based on a synthetic derivative from marine cone shell (*Conus magus*) venom from the shallow waters of the Indo-Pacific, were approximately $20 million and $18 million in 2009 and 2008, respectively (The Pharma Letter, 2010; PR Newswire, 2010).

Patented inventions related to marine organisms are based on a range of source material, from invertebrates to microbes. While not all inventions are patented and information is therefore scarce on the organisms of particular interest, available information indicates that most patented inventions involve molluscs (sea slugs, sea hares, cone snails), corals (particularly soft corals), sponges, tunicates, worms, bryozoans, algae, and vertebrates such as fish, including sharks. Marine sponges have been a common source of pharmaceuticals, with compounds extracted from sponges having shown promise for the treatment of cancer, asthma, Alzheimer's, arthritis, inflammation, and other ailments. Several are in clinical trials, and at least one potential cancer drug from a deep-sea sponge from the Gulf of Mexico is under investigation. One example of a drug from a marine source is Yondelis®, which is used for treating soft tissue sarcoma and ovarian cancer. It was originally sourced from the sea squirt *Ecteinascidia turbinata* by the Spanish company PharmaMar, and is currently licensed for use in the European Union (PharmaMar, 2014; Leary *et al.*, 2009). Some fish and other organisms from polar areas have yielded anti-freeze proteins used for the control of cold-induced damage in medical, food, and cosmetic products. Marine algae are a common source of cosmetic compounds, including for anti-ageing products (Aricò and Salpin, 2005; Leary *et al.*, 2009). Marine microbes have also been an important source of patented inventions. This is particularly true for bacteria and other microbes from hydrothermal vent environments. For example, the figures below provide a number of examples of patented inventions related to organisms collected from hydrothermal vent environments. Box 9.1 also provides examples of some important products from deep-sea bacteria.

While the majority of source organisms originate from areas within national jurisdiction (Oldham *et al.*, 2014), there are some documented cases of samples collected from marine areas beyond the limits of national jurisdiction which have resulted in patented inventions. These patents have been filed by entities from both developed and more advanced developing countries such as India and China. For example, a patent related to an enzyme from a marine fungus (*Aspergillus* sp.) for use in laundry detergents (patent EP1692296, Council of Scientific and Industrial Research, India) was originally collected from a depth of 5000 m in the Central Indian Basin, and likely comes from beyond national jurisdiction. Similarly, mapping of coordinates extracted either from a patent application or the referenced scientific literature indicates that patents filed for a polysaccharide from bacterium

---

Box 9.1
**Examples of enzymes developed from deep-sea genetic resources
(Leary *et al.*, 2009)**

Fuelzyme™ enzyme was developed on the basis of samples collected from a hydrothermal vent, likely from the Mid-Atlantic Ridge. This enzyme, which is currently marketed by Verenium (USA), is used in ethanol production from corn.

Vent polymerase is a thermostable enzyme sourced from a hydrothermal vent archaebacteria (*Pyrococcus* sp.) in Italy. It is marketed by New England Biolabs (USA) for use in DNA cloning, sequencing, and amplification.

Venuceane™ is an enzyme showing antioxidant properties, based on the bacteria *Thermus thermophilus* collected in the Guaymas Basin, Gulf of California, at 2000 m depth. The enzyme is marketed by the French company Sederma for use in cosmetics.

Vibrio diabolicus for use in bone repair and other medical purposes (US patent 7,015,206, Institut Français de Recherche pour l'Exploitation de la Mer, France), a polymerase from hyperthermophilic bacterium *Pyrolobus fumarii* for use in bio-technology (US patent 7,781,198, Verenium Corporation – formerly Diversa), and an enzyme from the thermophilic bacterium *Thermodesulfatator indicus* (US patent 20110008848 (application) by GeneSys Ltd, UK) originated from beyond national jurisdiction. The source organisms were collected from the Pacific, Atlantic, and Indian Oceans respectively, indicating a wide geographic range of sampling.

These examples demonstrate that there is ongoing activity, including through the assertion of patents, to use marine genetic resources that originate in areas beyond national jurisdiction for commercial purposes. While such patents are still few, they are an indication of the commercial potential of these organisms. It should be noted, though, that it is unclear how many patents result in actual commercialization or widespread application. Tracking the path from a patent to a product or process is difficult for a number of reasons: a product or process may be marketed under a different name; there may have been a series of additional transformations and applications of the original material; or because companies often do not disclose information relating to product development.

In general, it is also often difficult to tell whether a given sample or source organism was collected from within or beyond national jurisdiction. Patent appli-cations seldom include the geographical coordinates of sampling locations, and may instead contain general descriptions of a location, such as 'the Mid-Ocean Ridge' or 'the East Pacific Rise', which may or may not be beyond national jurisdiction. In addition, many patents are based on bacterial strains or specimens sourced from culture collections, sometimes complicating further the link between field research (through which the sample was collected), and later commercial use. Discerning the exact collection location of an organism that is the object of a patent and later commercial development usually requires a careful and in-depth study of the scientific literature related to the discovery of that organism to identify the geographic coordinates. While time-consuming, it is most often the only way to map collection locations of organisms that have resulted in patented inventions and commercialized products or processes.

9.2.1 Recent developments in scientific research related to marine genetic resources

Reaching deep-sea environments and maintaining the sampled organisms alive and bioactive, as well as analysing and culturing them, requires sophisticated and expensive technologies. Typically, this technology includes oceanographic vessels equipped with manned or unmanned submersible vehicles and *in situ* sampling tools, as well as technology and techniques related to molecular biology and DNA sequencing (Aricò and Salpin, 2005). With the exception of basic molecular biology techniques, most of the technology necessary for accessing the deep sea and studying and isolating its organisms is therefore owned and operated by research institutions in a very few countries (Arnaud-Haond *et al.*, 2011).

Efforts are ongoing to develop improved methods of purification, isolation, screening, and identification of novel bioactive compounds from marine organisms, and to better understand the role and functions of marine microorganisms in ocean ecosystems. In particular, work is seeking to: improve metagenomic screening and libraries; better understand the patterns of distribution of genetic diversity; reduce genetic erosion in the context of relevant conservation programmes encompassing the genetic level of biodiversity; improve the capacity to culture microorganisms, most of which are currently unculturable in laboratories; shorten lead times in drug discovery; and scale-up natural product purification from analytical to pilot level. New identification methods for marine genetic resources include those given in Box 9.2. Work also continues on developing ways to track and monitor genetic resources through the use of persistent global unique identifiers. A tracking system would allow each genetic resource and its derivatives, such as sequence data, to be tracked from the point of origin through one or more users (United Nations, 2009). However, this raises issues of cost and effectiveness. In that regard, including disclosure requirements in databases may provide a more cost-effective option.

Breakthroughs in high-throughput DNA sequencing and bioinformatics have contributed to rendering genetic resources into informational forms such as gene or

---

Box 9.2
Most current identification methods for marine genetic resources

Real Time PCR is a method used in the identification of single genes known to science.

High-Throughput Genomic Sequences methods are used to sequence entire genomes or any DNA to be found within entire ecological systems (oceans, soils, and sediments, even intestines). Examples of related technologies are: 454, Illumina, Solid.

Proteomics can be considered as an alternative or complementary method to conventional methods used to identify genes that have codified for proteins (gene expression). It is a long and difficult method, which is not self-sufficient but merely complements DNA methods in that it allows us to identify proteins corresponding with specific genes. Proteomics can be applied to different samples.

Barcoding is a technique for species identification based on the sequence of 16S RNA (mitochondrial RNA in eukaryotes).

protein sequence data held in databases. Practices in open science, technological advances, and the digitization of genetic and other information have rendered genetic resources and derived information more accessible. Research and development on genetic resources is therefore not only a field biologist's domain, but also one increasingly dominated by computational biologists and bioinformaticists (Glowka, 2010).

DNA identification methods are becoming increasingly accessible. For example, sequencing an entire bacterial genome currently costs around 300 euros, but the analysis remains complex. The scientific capacity to do so is not evenly distributed among those countries with an interest in marine genetic resources, including in terms of the capacity to handle significant amounts of data. It is also noteworthy that advances in genomics have resulted in a steep increase in available data and the capacity to handle such data is lagging behind.

Bioinformatics is often utilized as a support to analyse the data and, increasingly, to reduce the period of a clinical trial. Although bioinformatics is increasingly present as an emerging discipline and field of expertise in developing countries, capacity in bioinformatics also needs to be built in order to provide equal opportunities for the study of marine genetic resources (Thambisetty, 2002).

9.2.2 Environmental impact of marine scientific research

There has been some discussion concerning the potential environmental impacts of *in situ* research aimed at sampling marine organisms. While generally thought to be minimal, such impacts may, at times, be difficult to assess because the scale and location of all research activities related to marine genetic resources are not always known, and information about the population and life history parameters of the source organisms is not always available.

In general, initial sampling of biological material for research purposes consists of relatively small quantities (e.g. an individual sponge or other organism). However, some organisms, such as sponges and microbial fauna, are largely unculturable in laboratory conditions (Molinski *et al.*, 2009). Repeated collection of promising organisms may therefore be necessary and, cumulatively, have environmental impacts. Examples of cases where marketing of a specific product requires continuous harvesting lie in krill oil and other products from krill extracts, as well as proposals to use oceanic Sargassum weed from the Sargasso Sea in the production of biofuels (Patent US7479167 and patent application US20090119978). Furthermore, many marine species live in essential symbiotic relationships, and all partner organisms need to be sequenced to understand the symbiosis and the factors that govern them (Allen and Jaspars, 2009). Also, in some cases, the quantities required to produce active agents may be fairly substantial. For example, producing 1 gram of an anti-cancer agent required close to 1 metric ton (wet weight) of *Ecteinascidia turbinata* (Leary, 2011).

Sensitive and pristine habitats and rare species, as well as species with limited distribution, may be particularly at risk in such cases. Adopting a precautionary approach would therefore be important. Harvesting of the octocoral *Pseudoptero-gorgia elisabethae* in the Bahamas, where tailored collection strategies and export regulations are in place (Goffredo and Lasker, 2008), provides an example of the management of the risks associated with bioprospecting.

The further development and improvement of laboratory techniques to produce synthetics and derivatives may lessen the eventual environmental impacts and the costs of research (field work is generally the most expensive part of any scientific endeavour). In the future, our capacity to rely on the contribution of microbial cultures to the production of active compounds could increase.

Scientific research conducted in sensitive environments, including in the deep sea, and on vulnerable ecosystems such as seamounts, may call for voluntary codes of conduct for responsible marine scientific research. Examples are provided by the InterRidge and Diversa (now Verenium) Corporation codes of conduct.

The InterRidge Statement of Commitment to Responsible Research Practices at Deep-sea Hydrothermal Vents is a voluntary code of conduct for responsible research practices by scientists involved in research on deep-sea hydrothermal vent fauna and environments. The statement includes guidelines on: avoiding research activities with deleterious impacts on the sustainability of populations of hydrothermal vent organisms; avoiding research activities that lead to long-lasting and significant alteration and/or visual degradation of vent sites; avoiding collections that are not essential to the conduct of scientific research; avoiding transplanting biota or geological material between sites; being familiar with the status of current and planned research (through the InterRidge and other public domain databases) in an area and avoiding activities that will compromise experiments or observations of other researchers; facilitating the fullest possible use of all biological, chemical, and geological samples collected through collaborations and cooperation amongst the global community of scientists; and reaffirming commitment to open international sharing of data, ideas, and samples in order to avoid unnecessary re-sampling and impact on hydrothermal vents, and to further our global understanding of these habitats (InterRidge, 2014).

In 2005, the San Diego-based private company Diversa Corporation developed a framework model for 'ethical bioprospecting'. The model entailed participating countries, institutions, and stakeholders sharing equitably the benefits of the bioprospecting operations, on the basis of securing legal access to both terrestrial and marine biodiversity, including through the application of Prior Informed Consent (PIC) (see Christoffersen and Mathur, 2005). Subsequently, Diversa

Corporation became Verenium Corporation. The latter continues to apply the ethical model for accessing genetic resources, including in relation to deep-sea organisms (both Diversa Corporation and Verenium Corporation have developed applications based on deep-sea genetic resources). Moreover, Verenium Corporation is engaged in collecting small samples of soil, water, sediment, leaf litter, or other materials from the environment in order to reduce the impacts of its activities on sensitive ecosystems, on the basis of the assumption that '[s]mall sample sizes provided abundant genetic samples for our collection of microbial gene libraries, which is estimated to contain over 2 million microbial genomes' (Verenium, 2014).

Such codes of conduct can contribute to responsible scientific behaviours, in addition to advancing our knowledge of the ecosystems studied. At the same time, they can contribute to developing capacity in marine scientific research through north–south cooperation (InterRidge), as well as furthering the potential of applications based on marine genetic resources within the private sector (Diversa/Verenium Corporation).

Recent developments in good practice related to scientific research on marine genetic resources include mechanisms for sharing of information through dedicated open-access internet-based databases. Examples of such databases are the United Nations University–Institute of Advanced Studies Bioprospector database (ATCM, 2008); the Gordon and Betty Moore Foundation Marine Microbial Genome Sequencing Project (J. Craig Venter Institute, 2014); and the Ocean Biogeographic Information System (OBIS), which is referred to in more detail in Chapter 12 of this book.

The identification and adoption of practical measures to further capacity-building, including through south–south cooperation, as well as the facilitation of access to samples of marine genetic resources and the transfer of technology, are desired developments related to marine genetic resources. Discussions on capacity-building measures which have taken place in the context of the UN General Assembly Ad Hoc Open-ended Informal Working Group to study issues relating to the conservation and sustainable use of marine biological diversity beyond areas of national jurisdiction (the 'Ad Hoc Open-ended Informal Working Group') and the UN Open-ended Informal Consultative Process on Oceans and the Law of the Sea (the 'ICP') seem to have had a limited concrete impact on the development of potential measures in this regard thus far.

9.3 Legal and policy aspects

A number of international instruments govern activities related to marine genetic resources, including the 1982 United Nations Convention on the Law of the Sea (UNCLOS), the Convention on Biological Diversity (CBD) and intellectual property rights instruments. This patchwork of instruments has rendered policy discussions in various forums relatively complex and subjected consideration of the conservation and sustainable use of marine biodiversity beyond areas of national jurisdiction to what may be called 'forum shopping'.

9.3.1 *The centrality of UNCLOS and the role of the CBD*

The 1982 United Nations Convention on the Law of the Sea (UNCLOS) provides the legal framework for all activities in the oceans and seas, including sustainable management of resources, marine scientific research, and the protection and preservation of the marine environment. While UNCLOS does not contain a specific mention of genetic resources or biodiversity, it addresses 'marine life' (Articles 1(1)(4) and 194(5)), 'natural resources' (Articles 56(1)(a), 77, 79(2), 145, 193, 194(3), 246(5), and 249(2)), 'living resources' (Preamble and Articles 1(1)(4), 21(1), 56(1), 61, 62, 69, 70, 71, 72, 73, Section II of Part VII, 123, 277, and 297(3)), and 'living organisms' (Article 77) in a number of ways from conservation, protection and preservation, to utilization and research. These terms can reasonably be understood as encompassing all biological resources in the oceans, including genetic resources.

A number of provisions of the CBD are also relevant, including Article 15 on access and benefit-sharing, as far as genetic resources within national jurisdiction are concerned, and Article 14 on impact assessments. The 2010 Nagoya Protocol, which aims at implementing the provisions of the CBD on access to genetic resources and the sharing of benefits arising out of their utilization, also applies to genetic resources within national jurisdiction.

Both UNCLOS and the CBD provide the legal framework for activities related to marine genetic resources from organisms found within national sovereignty (i.e. in internal waters, which are the waters situated on the landward side of the baselines; archipelagic waters, which are the waters enclosed by the archipelagic baseline; and the territorial sea, which are the waters extending up to 12 nautical miles from the baselines) or jurisdiction (i.e. the exclusive economic zone, which lies beyond and adjacent to the territorial sea, up to 200 nautical miles, and the continental shelf, which comprises the seabed and subsoil that extend beyond the territorial sea throughout the natural prolongation of its land territory to the outer edge of the continental margin, or to a distance of 200 nautical miles from the baselines where the outer edge of the continental margin does not extend up to that distance). However, divergent views continue to be held regarding the applicability of the CBD and its Nagoya Protocol to marine genetic resources of areas beyond national jurisdiction. As regards the components of biological diversity, the CBD's scope is set out in its Article 4(a), which limits its application to areas within the limits of national jurisdiction. However, some states argue that the CBD is also applicable to activities related to biological resources of areas beyond national jurisdiction, by virtue of Article 4(b). The latter provides that the CBD applies, in the case of processes and activities, regardless of where their effects occur, carried out under a Party's jurisdiction or control, within the area of its national jurisdiction or beyond the limits of national jurisdiction (United Nations, 2006, 2008, 2010; Wolfrum and Matz, 2000; Scovazzi, 2006; Proelss, 2008). In addition, while the Nagoya Protocol certainly applies to genetic resources from areas under national jurisdiction, the question of whether it may also apply to resources of areas beyond national jurisdiction or of Antarctica, remained among the issues under discussion until the last hours of the Protocol's negotiations (Buck and Hamilton, 2011). Ultimately, the scope of the Nagoya Protocol, as adopted, is limited to genetic resources within national jurisdiction. However, Article 10 of the Protocol leaves open the possibility for the future negotiation of a multilateral benefit-sharing mechanism, which could, if states so choose, provide the basis for a future benefit-sharing arrangement in regard to marine genetic resources of areas beyond national jurisdiction.

As regards UNCLOS, while states recognize that it provides the framework for all activities in the oceans and seas, including in relation to marine genetic resources beyond areas of national jurisdiction, divergent views continue to be held on the respective application of its Parts VII (high seas, which are all parts of the sea that are not included in the exclusive economic zone, in the territorial sea or in the internal waters of a state, or in the archipelagic waters of an archipelagic state) and XI (the Area, which is the seabed and ocean floor and subsoil thereof, beyond the limits of national jurisdiction), to marine genetic resources and the activities related thereto. A majority of developing countries hold the view that the common heritage of mankind, as set out under Part XI of UNCLOS, applies not only to mineral resources, but also to the biological resources of the area. Developed countries, generally, based on a literal interpretation of UNCLOS, are of the opinion that Part XI only encompasses mineral resources, while marine genetic resources fall under the regime of the high seas – the freedom of the high seas would therefore govern their collection and exploitation (cf. United Nations, 2008 and 2010). By and large, these positions reflect the respective capacity of each group of states to access, explore, and exploit those resources (Germani and Salpin, 2011).

Marine scientific research (MSR), which is governed by the provisions of Part XIII of UNCLOS and constitutes the usual first point of access to marine genetic resources *in situ*, is understood as encompassing both pure and applied research (Salpin and Germani, 2007), including research of direct significance for the exploitation of natural resources (UNCLOS, Article 246). All states, irrespective of their geographical location, and competent international organizations have the right to conduct MSR in the area and in the high seas (UNCLOS, Articles 256 and 257).

Assuming that a valuable marine compound had been identified and isolated, but could not be cultured in the laboratory in sufficient quantities for further research and subsequent application, one could wonder, in the absence of any explicit provision to that effect, whether the procurement of additional material *in situ* in the marine environment would still qualify as MSR, in particular if the quantities of material required were significant, or if other provisions applied. For instance, under the MSR regulations of some states, the collection of large amounts of fish would entail the deduction of the catch from the researching state's annual fishing quota, and most of the bioprospecting laws in place usually refer to small quantities of algae, animals, and microorganisms (Salpin, 2013).

Another issue for further consideration relates to the manner in which UNCLOS addresses commercial exploitation of genetic resources following identification of a lead compound through MSR. In that regard, it must be borne in mind that MSR, under Article 241 of UNCLOS, shall not constitute the legal basis for any claim to any part of the marine environment or its resources (Salpin and Germani, 2007). Salpin and Germani (2007) provide a discussion of the implications of this provision.

Particularly relevant in a context of limited capacity also, are the provisions of UNCLOS related to cooperation in MSR, which require states and competent international organizations, in accordance with the principle of respect for sovereignty and jurisdiction and on the basis of mutual benefit, to promote international cooperation in MSR for peaceful purposes (UNCLOS, Article 242); to create favourable conditions for the conduct of MSR in the marine environment (UNCLOS, Article 243); and to actively promote the flow of scientific data and information and the transfer of knowledge resulting from MSR, especially to developing states, as well as the strengthening of the autonomous MSR capabilities of developing states through, inter alia, programmes to provide adequate education and training of their technical and scientific personnel (UNCLOS, Article 244).

Part XIV of UNCLOS is entirely devoted to the transfer of marine technology. Under Part XIV, states are required to cooperate to promote the development and transfer of marine science and marine technology on fair and reasonable terms and conditions (UNCLOS, Article 266). They must also endeavour to foster favourable economic and legal conditions for the transfer of marine technology for the benefit of all parties concerned on an equitable basis. In so doing, they must have due regard for all legitimate interests, including the rights and duties of holders, suppliers, and recipients of marine technology (UNCLOS, Article 267). Of note, Articles 16 and 19 of the CBD, which address access to and transfer of technology and handling of biotechnology and distribution of its benefits, respectively, also provide a relevant context to promote transfer of technology related to marine genetic resources.

The nature of the marine environment, as constantly moving, and the complexity of the jurisdictional framework in the oceans are such that challenges to the smooth implementation of applicable legal instruments may exist in cases where the biological material is found in both areas within and beyond national jurisdiction (Salpin, 2013). The first of such challenges relates to biological material which is found both within the EEZ of one or more coastal states and in the high seas, such as the free-floating microbes of the pelagic zone or the genetic resources hosted by migratory species of fish, cetaceans, or mammals. Another challenge originates in the complexity of organisms or other biotic components that may depend on, or be associated with, sedentary species, such as sponge or corals, found on the portion of the continental shelf extending beyond 200 nautical miles, in accordance with Article 76 of UNCLOS, but which are not themselves necessarily sedentary in nature (Young, 2011; Anonymous, 2011; for recent discoveries of such symbiosis, see Young, 2011, and Anonymous, 2011).

Within 200 nautical miles, the coastal state has sovereign rights over resources, both on the continental shelf and in the superjacent territorial sea and the EEZ for the purposes of the exploration and exploitation of the resources (where an EEZ has been declared). However, beyond 200 nautical miles, while the sovereign rights of the coastal state apply to the continental shelf beyond 200 nautical miles, those rights do not apply to the superjacent water column – in other words, the high seas. Under UNCLOS, the prior consent of the coastal state is required for research on natural resources on its continental shelf. The natural resources of the continental shelf include the living organisms belonging to sedentary species, that is to say, organisms which, at the harvestable stage, are either immobile on or under the seabed, or are unable to move except in constant physical contact with the seabed or the subsoil (UNCLOS, Article 77). However, access to the non-sedentary organisms or compounds secreted by the sedentary species or closely associated with them, but found in the high seas, would not be subject to any prior consent requirement. The dichotomy between the legal boundaries established under UNCLOS and the biological and ecosystem characteristics of the marine environment is perplexing for the purposes of regulating the utilization of marine genetic resources (Salpin, 2013).

The policy and legal uncertainty concerning the regime applicable to marine genetic resources of areas beyond national jurisdiction may act as a deterrent to researchers and investors alike. In particular, the need for legal clarity for investors was highlighted at the eighth meeting of the United Nations General Assembly Informal Consultative Process on Oceans and the Law of the Sea in 2007, which focused its discussions on marine genetic resources in (United Nations, 2007a).

9.3.2 Intellectual property rights: incentives or obstacles?

From the standpoint of promoting innovation, in cases of privately funded research as well as university research and public–private partnerships, intellectual property protection and, in particular patents, may provide important incentives to promote investments in research and development and – through licensing – potential pathways towards technology transfer, product development, and commercialization. While several other types of intellectual property rights, including copyright, trade secrets, and *sui generis* database protection, play an important role in relation to marine genetic resources and related innovations, this chapter only considers in detail relevant aspects of patent law and policy.

Genetic resources, including marine genetic resources, can be modified by human intervention and take on characteristics that do not exist in nature. When these modifications result in a new biotechnological invention that involves an inventive step and is capable of industrial application, the invention may qualify for patent protection (Chiarolla, 2011; Eisenberg, 2003; Trevor, 2006). For instance, 'genetic patents' may comprise nucleic acids, nucleotide sequences and their expression products; transformed cell lines; vectors; as well as methods, technologies, and materials for making, using, or analysing such nucleic acids, nucleotide sequences, cell lines, or vectors (OECD, 2006).

Similarly to what has happened in other biotechnology areas, the application of molecular genetics and bioinformatics has transformed marine genetic resources into a promising source of appropriable information. The hereditary information which can be found in DNA, coupled with the identification of genes' functions (e.g. the way in which the latter code for proteins – a process also known as gene expression), may constitute the basis of patent applications that – if granted – can provide monopoly rights on the claimed inventions. Such inventions may include, inter alia: genes and gene sequences; cell lines; amino acids; proteins; viruses; and express sequence tags (ETSs), which are smaller segments of complementary DNA that are used as intermediate research tools. Since the inception of biotechnology patenting in the 1980s, the impact of these patents on research and innovation has been extremely controversial.

Besides the question of whether such inventions meet the three traditional statutory patentability requirements, certain inventions may not fall within the scope of patentable subject matter as defined by national law. This varies from one country to another. Many countries exclude from patentability such subject matter as scientific theories, mathematical methods, plant or animal varieties, discoveries of natural substances, methods for medical treatment (as opposed to medical products), and any invention where prevention of its commercial exploitation is necessary to protect public order and good moral or public health (WIPO,

2012). Nevertheless, patent offices in countries with advanced biotechnology capacity have routinely granted patents for gene-based inventions (as a matter of economic policy), while leaving courts to decide whether they should stand in the case of disputes.

The World Trade Organization (WTO) Agreement on Trade Related Aspects of Intellectual Property Rights (the TRIPs Agreement) requires all WTO member states to provide minimum standards of protection for a wide range of intellectual property rights. In so doing, the TRIPs Agreement incorporates provisions from several international intellectual property agreements administered by the World Intellectual Property Organization (WIPO). It also introduces a number of new obligations, including in relation to patents and trade secrets (Chiarolla, 2011). In terms of the rights conferred by patents and their duration, the TRIPs Agreement provides that a patent shall confer on its owner the right to prevent others from making, using, offering for sale, selling, or importing the patented product for a period of 20 years or more (TRIPs Agreement, Article 28). In the case of a process patent, the same rights extend at least to the product obtained directly by the patented process.

Article 27.1 of the TRIPs Agreement states: 'Patents shall be available for any inventions, whether products or processes, in all fields of technology, provided that they are new, involve an inventive step and are capable of industrial application.' This provision calls on WTO member states to provide patent protection for both products and processes, and forbids discrimination among different fields of technology. Thus, states are not allowed to take measures that would discriminate against the granting of patents, for instance, because the disclosed invention belongs to a particular technological domain (e.g. marine biotechnology).

Such a principle further requires that '... patents be available and patent rights enjoyable without discrimination as to the place of invention, the field of technology and whether products are imported or locally produced'. Thus, its general purpose is to protect right-holders against arbitrary policies that may undermine their rights. The geographical dimension of this principle focuses on prohibiting discrimination based on the place of the invention or the geographical location of the production facilities. Against this backdrop, the question of whether a WTO member country may be allowed to take measures that would exclude the granting of patents (or the enjoyment of patent rights) for inventions, which are based on marine genetic resources that are taken from areas beyond national jurisdiction, has as yet not arisen. However, such measures may be far reaching, since TRIPS Article 27.1 does not refer to the origin of the resources used for the invention, but rather the origin of the invention itself. In that regard, it must be borne in mind that once a marine genetic resource has been sampled from areas beyond national jurisdiction, the discovery and invention (in the patent sense) will likely not take place in areas beyond national jurisdiction, but in a laboratory on land, within national jurisdiction.

A potential norm that aims at establishing or preserving the public domain status of marine genetic resources from areas beyond national jurisdiction, if such were the policy objective, may involve an obligation on the concerned states to regulate (and subject to limits) the entitlements that researchers may have to file patents on marine genetic resources. The instruments that can be envisaged for enforcing these obligations could be of a contractual nature. In particular, they can take the form of contract clauses associated with the granting of research funding and/or authorizations or – when the material is received from *ex situ* sources – they can be inserted into Material Transfer Agreements (Krattiger *et al.*, 2006).

At the international level, the most relevant example of the above form of regulation, which entails the management of a global public good through a private law contract, is provided by the FAO International Treaty on Plant Genetic Resources for Food and Agriculture (ITPGRFA) and its Standard Material Transfer Agreement (SMTA). In particular, the ITPGRFA and its SMTA establish that recipients of PGRFA shall not claim 'any intellectual or other property rights that limit the facilitated access to the Material [i.e. PGRFA] ... or its genetic parts or components, in the form received from the Multilateral System' (ITPGRFA, Article 12.3(d), and SMTA, Article 6.2). However, the interpretation of this provision and of the limits it imposes on the private appropriation of plant genetic resources (and of their informational contents), which are shared within the multilateral system as public domain materials, remains controversial (Chiarolla and Jungcurt, 2011). Box 9.3 provides a short explanation of the multilateral system.

As anticipated, with regard to subject matter exclusions from patentability, WTO members are allowed to not grant patents for plants, animals, and essentially biological processes for their production. However, microorganisms and micro-biological or non-biological processes for the production of plants and animals must be protected (TRIPs Agreement, Article 27.3(b)). Therefore, while bearing the above exceptions in mind, under the TRIPs Agreement, countries are free to define subject matter exclusions that are broad enough to entirely avoid the granting of patents on marine organisms and on their genetic parts and components. On the other hand, countries are also free to protect biotechnological inventions that relate to such organisms, if they meet the patentability requirements of novelty, utility, and industrial applicability. For instance, in the case of gene sequences and the isolation of proteins, and in some instances, even micro-organisms in their original state, the mere isolation and characterization may be enough to satisfy the criteria for patentability, in particular if significant inventive ingenuity has been required to isolate and characterize them (Salpin and Germani, 2007).

Box 9.3
The multilateral system

The International Treaty on PGRFA provides an internationally agreed, legally binding framework for the conservation and sustainable use of crop diversity and the fair and equitable sharing of benefits, in harmony with the CBD (Chiarolla *et al.*, 2013). Within biodiversity, the Treaty defines a subset of genetic resources of particular importance for agriculture and food security – i.e. PGRFA – and it limits the scope of application of its norms to them. The International Treaty also establishes a Multilateral System of ABS (MLS) which consists of pooling selected crop and forage genetic resources from various countries (International Treaty Article 10.2). In particular, Annex I of the International Treaty lists the 64 crops and forages that are part of the MLS to ensure worldwide food security. These pooled resources are available under the facilitated access mechanism of the MLS only if access is requested for the purpose of utilization and conservation for research, breeding, and training for food and agriculture.

The material pooled in the MLS is governed by a set of common rules of access and benefit-sharing that states agreed upon and which were formalized in a standard contract called the Standard Material Transfer Agreement (SMTA) (Chiarolla, 2008). While the CBD promotes the development of a regime of contractual rules for the exchange of genetic resources that is based on bilateral contracts, access to PGRFA included in the MLS is done on the basis of the SMTA and hence does not require ad hoc negotiations between providers and recipients of PGRFA. This reduces transaction costs as 'access shall be accorded expeditiously, without the need to track individual accessions and free of charge, or, when a fee is charged, it shall not exceed the minimal cost involved' (International Treaty Article 12.3.b).

The International Treaty does not require a burdensome mechanism to track individual accessions, as providers of PGRFA do not have the obligation to keep track of all subsequent transfers of the material. However, reporting obligations for both providers and recipient are included in the SMTA in order to ensure that:

- some benefits flow back to the MLS when a product based on MLS materials is commercialized on the market; and
- to enable the functioning of dispute-settlement procedures.

Through these reporting obligations, in conjunction with the obligation to use the SMTA for any subsequent third-party transfer of PGRFA, the SMTA enables following the chain of transfers between individual providers and recipients of PGRFA at reduced costs.

States still enjoy relative flexibility in defining the extent to which genetic inventions, including inventions based on marine genetic resources, can constitute the subject matter of patent protection. For instance, Article 53(b) of the European Patent Convention (EPC) provides that plant or animal varieties or essentially biological processes for the production of plants and animals do not constitute patentable inventions (Chiarolla, 2011). However, while plant or animal varieties per se constitute unpatentable subject matter, Article 3.2 of the 1998 Biotechnology Directive provides that 'biological material which is isolated from its natural environment or produced by means of a technical process may be the subject of an invention even if it previously occurred in nature' (European Commission, 1998). Besides, the protection granted to biotechnological inventions may also encompass within its scope plant and animal varieties in accordance with Article 4.2, which states that plants and animals shall be patentable if the technical feasibility of the invention is not confined to a particular variety. In the USA, at least in theory, three categories of inventions are not patentable, namely: laws of nature, natural phenomena, and abstract ideas. In particular, the product of nature doctrine, which sets out the boundaries of patentable subject matter for genetic inventions under 35 USC Section 101, stands in the eye of the storm that surrounds the debate on patentability of DNA and its alleged stifling effects on biological innovation.[1]

While a detailed account of the economic theories that have been elaborated to justify the social function and coverage of the patent system is provided elsewhere in the literature (Chiarolla, 2011), it is useful to recall that some authors have argued that the world economy may '. . . not benefit from a general broadening and strengthening of patent protection' because, in many technology areas, overbroad patents 'entail major economic costs while generating insufficient additional social benefits' (Mazzoleni and Nelson, 1998). Among these technology areas, cumulative system technologies and science-based technologies figure prominently. One of the underlying concerns is that scientists and researchers are reluctant '. . . to conduct research and development where patents on genes exist because of fear of litigation' (KEI, 2012). This is particularly true for the increasing patenting of upstream results of basic research, including MSR. Because patents on 'upstream technologies' and, in particular, on isolated DNA that is identical to its natural counterparts, hold potential for limiting access to already available materials and research tools, patents may act as a barrier to further innovation. Therefore, unrestricted or facilitated access to basic information and discoveries as well as upstream research results, including those relating to marine genetic resources, may be critical for the progress of science, in general, and of marine scientific research, in particular (Kesselhein and Avorn, 2005).

Because of this, modern patent laws in most countries provide for some form of research exemption. A research exemption is an exception to the exclusive rights granted by a patent which allows researchers to undertake experiments on the patented invention with a view to discovering unknown effects or making improvements on the invention, without the prior consent of the patent holder

(WIPO, 2009). Examples can be found in European national patent laws, which generally provide some form of research exemption, especially with respect to activities whose results make improvements to a patented invention. However, research exemptions are not granted by the European Patent Convention and there is currently little harmonization and legal certainty regarding the extent to which marine genetic resources, which are covered by patent claims, may be used in the context of MSR without the patent holder's authorization. Another example is the Hatch Waxman Act of 1984, in the United States, which is the only statutory exemption in patent law, and is limited to experiments carried out on drugs or medical devices for the purpose of obtaining approval by the Food and Drug Administration (Drug Price Competition and Patent Term Restoration Act, 1984).

Article 30 of the TRIPs Agreement allows WTO member states to provide for research exemptions, which must meet three cumulative conditions, namely that they: (1) do not unreasonably conflict with the normal exploitation of the patent and (2) do not unreasonably prejudice the legitimate interests of the patent owner, (3) taking account of the legitimate interests of third parties. In particular, in the case *Canada – Patent Protection of Pharmaceutical Products*, the WTO Panel specifically considered the interpretation of the limiting conditions of Article 30 of the TRIPs Agreement (WTO, 2000).

The above provision is of a permissive nature and there are no minimum standards concerning the manner in which national patent laws should positively recognize a research exemption. Therefore, the formulations that have been adopted in various countries can greatly diverge. As a result, in most developed countries, the research exemption is construed in very narrow terms (Chiarolla, 2011).

In summary, because of the limited use that some countries may have made of both subject matter exclusions and the exceptions and limitations to patent rights, which are allowed under the TRIPs Agreement, newly discovered marine genetic resources may eventually be 'locked up' by patent monopolies. Instead, in the interests of humanity, the use of such resources and related knowledge would be usefully widely promoted, including through legal flexibilities, while respecting patent rights in inventions that embody genuinely new human-made constructs that are clearly different from their naturally occurring counterparts. To that end, a first necessary step requires landscaping existing patents with regard to marine genetic resources of areas beyond national jurisdiction, with a view to assessing their potential impact on scientific research and understating their economic value. This will help to enhance knowledge and understanding of relevant activities, including the extent and types of research on, and uses and applications derived from, marine genetic resources in areas beyond national jurisdiction.

9.3.3 Working hand in hand?

The relationship between the various applicable instruments, in particular UNCLOS and intellectual property instruments, is important. Pursuant to Article 311 of UNCLOS, which governs the relationship between UNCLOS and other conventions and international agreements, UNCLOS 'shall not alter the rights and obligations of States Parties which arise from other agreements compatible with this Convention and which do not affect the enjoyment by other States Parties of their rights or the performance of their obligations under this Convention'. In the present context, what this provision means, in substance, is that UNCLOS and the TRIPs Agreement are compatible, insofar that compliance with the TRIPs Agreement does not alter the exercise by a state of its rights and performance of its obligations under UNCLOS. The TRIPs Agreement is silent on the question of its relationship with international instruments, other than those relating to intellectual property. Whether compliance with the TRIPs Agreement affects the enjoyment by other state parties of their rights or the performance of their obligations under UNCLOS depends largely on: (1) whether patenting is considered as a claim to part of the marine environment or its resources (and hence runs counter to Article 241 of UNCLOS); (2) whether patenting is likely to interfere with the right to carry out MSR or any other activity in relation to marine genetic resources from the high seas and the Area (and hence runs counter to Article 240 of UNCLOS); and (3) whether the degree of confidentiality required prior to the filing for patents in order to safeguard the novel character of an invention is compatible with the requirement for dissemination and publication of data and research results (or runs counter to Articles 244 and 143.3(c) of UNCLOS) (Salpin and Germani, 2007).

The relationship between UNCLOS and the 2010 Nagoya Protocol should also be considered, since the latter foresees the possible establishment of a global multilateral benefit-sharing mechanism regarding genetic resources, for which it is not possible to grant or obtain prior informed consent in accordance with its Article 10 (Tvedt, 2011; Salpin 2013). It will thus be interesting to see which practical options for benefit-sharing are discussed with regard to these resources and whether they may provide a source of inspiration for the discussions at the General Assembly of the United Nations in relation to marine genetic resources of areas beyond national jurisdiction (Salpin, 2013).

9.3.4 Discussions at the General Assembly of the United Nations

Historically, the debate on scientific and policy aspects of bioprospecting in the deep seabed has been initiated in the context of the CBD (cf. Convention on Biological Diversity, 1995). Following consideration by the Convention's

Subsidiary Body on Scientific, Technical and Technological Advice (SBSTTA), including on the basis of a study on the relationship between the CBD and UNCLOS (Convention on Biological Diversity, 2003), prepared in cooperation with the Division for Ocean Affairs and the Law of the Sea of the Office of Legal Affairs of the United Nations, the issue was taken up, at policy level, at the United Nations General Assembly.

Marine genetic resources were the focus of discussions at the United Nations Open-ended Informal Consultative Process on Oceans and the Law of the Sea, at its seventh meeting in 2006. In testimony to the difficulty of the policy debate on marine genetic resources, the meeting could not agree on consensual elements to be forwarded to the General Assembly, as a result of divergent views. Previously, in 2004, in Resolution 59/24, the General Assembly had established an Ad Hoc Open-ended Informal Working Group, to study issues relating to the conservation and sustainable use of marine biological diversity beyond areas of national jurisdiction (hereinafter referred to as the Working Group) to: (a) survey the past and present activities of the United Nations and other relevant international organizations with regard to the conservation and sustainable use of marine biological diversity beyond areas of national jurisdiction; (b) examine the scientific, technical, economic, legal, environmental, socio-economic, and other aspects of these issues; (c) identify key issues and questions where more detailed background studies would facilitate consideration by states of these issues; and (d) indicate, where appropriate, possible options and approaches to promote international cooperation and coordination for the conservation and sustainable use of marine biological diversity beyond areas of national jurisdiction. Since the first meeting of the Working Group, in 2006, issues related to marine genetic resources beyond areas of national jurisdiction, in particular the applicable legal regime, including questions on the sharing of benefits, have been central to the discussions. At the 2008 and 2010 meetings of the Working Group, a number of delegations, while open to considering the legal regime, proposed to focus on practical measures to address existing implementation gaps and to enhance the conservation and sustainable use of marine genetic resources. Such measures included: the promotion of MSR and development of codes of conduct; mechanisms for cooperation, sharing of information, and knowledge resulting from research on marine genetic resources, including by increasing the participation of researchers from developing countries in relevant research projects; discussion of practical options for benefit-sharing, including options for facilitating access to samples; and consideration of relevant intellectual property aspects. While open to considering practical measures, other states continue to stress the importance of discussions on the legal regime (cf. United Nations, 2008 and 2010).

All issues under consideration by the Working Group, in particular marine genetic resources, including questions on the sharing of benefits; area-based management tools, including marine protected areas (MPAs); environmental impact assessments (EIAs); and capacity-building and technology transfer are now bundled into a package, as shown by the outcome of the 2011 meeting of the Ad Hoc Open-ended Informal Working Group, as endorsed by the General Assembly at its 66th session (Resolution 66/231). The General Assembly decided to initiate a process, within the Working Group, with a view to ensuring that the legal framework for the conservation and sustainable use of marine biodiversity in areas beyond national jurisdiction effectively addresses those issues by identifying gaps and ways forward, including through the implementation of existing instruments and the possible development of a multilateral agreement under UNCLOS. The process is mandated to consider, together and as a whole, marine genetic resources, including questions on the sharing of benefits, measures such as area-based management tools, including marine protected areas, and environmental impact assessments, capacity-building, and the transfer of marine technology. While reaffirming its central role relating to the conservation and sustainable use of marine biological diversity beyond areas of national jurisdiction, the General Assembly has noted the work of states and relevant intergovernmental organizations and bodies on these issues, and invited them to contribute, within the areas of their respective competence, to the consideration of these issues within the process it initiated in Resolution 66/231 (Resolution 68/70). In 2013, in the context of that process, as requested by the General Assembly, some of the aspects discussed by the Ad Hoc Open-ended Informal Working Group were further addressed in intersessional workshops held with a view to improving understanding of the issues and clarifying key questions as an input to the work of the Working Group (United Nations, 2012a, 2012b, and 2013a).

Political momentum for advancing discussions within the Working Group was provided at the highest political level in 2012 at the United Nations Conference on Sustainable Development, held in Rio. At the conference, states committed to address, on an urgent basis, building on the work of the Ad Hoc Open-ended Informal Working Group and before the end of the 69th session of the Assembly – that is before the end of September 2015, the issue of the conservation and sustainable use of marine biological diversity of areas beyond national jurisdiction, including by taking a decision on the development of an international instrument under UNCLOS (United Nations, 2012c). In order to prepare for the decision to be taken at the 69th session, the General Assembly, following the recommendations of the Ad Hoc Open-ended Informal Working Group at its sixth meeting (United Nations, 2013b), requested the Working Group, within its mandate established by Resolution 66/231 and in the light of Resolution 67/78, to make recommendations to the Assembly on the scope, parameters, and feasibility of an international instrument under UNCLOS. It also decided, to this end, that the Ad Hoc Open-ended Informal Working Group would meet for three meetings to take place from 1 to 4 April and 16 to 19 June 2014 and from 20 to 23 January 2015, with the possibility of the Assembly deciding that additional meetings will be held, if needed (Resolution 68/70).

9.4 Moving forward

9.4.1 Strengthening international cooperation

A number of proposals have been made, emanating from states and civil society, concerning the possible content of an international instrument under UNCLOS. Some of the core elements of such proposals include ways to strengthen international cooperation, the application of ecosystem approaches and a precautionary approach, the use of best available scientific information and environmental impact assessments, the establishment of marine protected areas, and mechanisms for sharing the benefits arising out of the exploitation of marine genetic resources of areas beyond national jurisdiction, as well as facilitating capacity-building and technology transfer (European Union, 2012; Hart, 2008; Greenpeace, 2008; Pew, 2012).

Strengthening the implementation of existing instruments also remains a high priority, as evidenced by the repeated calls of the General Assembly in its resolutions on oceans and the law of the sea and on sustainable fisheries, to implement UNCLOS and various instruments related, among others, to shipping, the protection and preservation of the marine environment, marine science, and sustainable fisheries. However, it is generally thought that the current cross-sectoral approach to the management of human activities in areas beyond national jurisdiction does not take into account sufficiently the cumulative impacts of those activities on the ecosystems. A key purpose of the development of a new instrument under UNCLOS would therefore be to promote and strengthen cross-sectoral cooperation for the conservation and sustainable use of marine biodiversity beyond areas of national jurisdiction.

9.4.2 The role of environmental impact assessments and strategic environmental assessments

Among the tools available to foster greater cooperation – and which are at the centre of discussions – are EIAs and SEAs. The actual or potential value of marine organisms used as sources of marine genetic resources should also be seen from the standpoint of ecological processes and ccosystem services provided by deep sea biodiversity, as indicated in Annex 2 of the CBD Voluntary Guidelines on EIA. In that regard, EIAs and SEAs, as a tool to implement an ecosystem approach, should be conceived as part of coherent policy-planning frameworks and include systematic follow-up procedures for the management of various threats to marine biodiversity in areas beyond national jurisdiction.

Impact assessments are currently required under a number of global instruments, including UNCLOS (Articles 205–206) and the CBD (Article 14), as well as regional instruments, including the Convention for the Protection of the Marine Environment in the North-East Atlantic (OSPAR Convention) and the Protocol on Environmental Protection to the Antarctic Treaty. At its 11th meeting, in October 2012, the CBD Conference of the Parties adopted voluntary guidelines for the consideration of biodiversity in EIAs and SEAs in marine and coastal areas (Convention on Biological Diversity, 2012). The voluntary codes of conduct for marine scientific research described in Section 9.1 also promote EIAs. EIAs are also being addressed in the context of the General Assembly Ad Hoc Open-ended Informal Working Group, in the light of challenges that exist for the application of EIAs in areas beyond national jurisdiction, including in relation to the responsible entity for carrying out the assessments, stakeholder consultations, and the authority overseeing such assessments and making the decision on whether an activity may proceed or not, based on the results of an assessment.

9.4.3 Sustainable management through area-based management tools

With regard to the conservation of marine genetic diversity, area-based management tools such as marine protected areas, fisheries closures, and special areas, can play an important role in the conservation and sustainable use of the resources, which are host to valuable genes (van Dover *et al.*, 2012). The development of networks of MPAs was called for at the World Summit on Sustainable Development in 2002. At the Rio+20 Conference, states reaffirmed the importance of area-based conservation measures, including marine protected areas, consistent with international law and based on best available scientific information, as a tool for conservation of biological diversity and sustainable use of its components (United Nations, 2012c). Area-based management tools, including marine protected areas, are sometimes perceived as an impediment to the freedom of the high seas, including navigation, fishing, and scientific research. Yet such tools, which do not necessarily imply that all activities are banned from taking place in a given area, also present opportunities as frameworks for strengthened cross-sectoral cooperation and the management of potential user conflicts.

Recent efforts towards the establishment of cross-sectoral MPAs include the establishment of six MPAs in areas beyond national jurisdiction in the OSPAR maritime area (Salpin and Germani, 2010), with attempts to involve other competent international organizations, including the International Seabed Authority, the International Maritime Organization, and relevant regional fisheries management organizations. There are also proposals for reserves for chemosynthetic ecosystems in the context of the ISA (ISA, 2011). The Sargasso Sea Alliance, a public–private partnership led by the government of Bermuda, aims, inter alia, at using existing regional, sectoral, and international organizations to secure a range of protective measures, including marine protected areas, for all or parts of the Sargasso Sea. The Sargasso Sea is host to floating mats of Sargassum seaweed which provide a habitat to many endemic species and are also being researched for their potential in various industrial, medical, and nutritional uses, including applications focused on inhibiting HIV infection, antibiotics, antifungals, and antifouling substances (Laffoley *et al.*, 2011).

While the role of area-based management tools, including MPAs, in the conservation and sustainable use of marine genetic resources is justified, very often, little or no information is available on the conservation status of species and organisms used as sources of marine genetic resources. Where information is available, the future prospects are gloomy. For example, there is evidence that psycrophiles of the Arctic that show high metabolic activity at very low temperatures might be under threat of extinction due to global warming (Arrieta *et al.*, 2010).

The identification of ecologically or biologically sensitive areas and vulnerable marine ecosystems, through undertaking the relevant processes under the CBD and the FAO, remains a scientific exercise which does not automatically entail the establishment of a protected area or fisheries closure. However, the identification of such areas is useful and provides a complementary tool to inform the selection and designation of appropriate tools for the conservation and sustainable use of marine biodiversity, including genetic resources.

9.4.4 Addressing benefit-sharing

From the standpoint of feasible approaches to dealing with benefit-sharing in relation to marine genetic resources, this chapter has emphasized that it is often difficult to establish whether a given marine genetic resource was collected within or beyond national jurisdiction. Therefore, the introduction of a mechanism that would allow the disclosure of, inter alia, the geographic coordinates of sample collection locations, would provide greater legal certainty for all those concerned with research and development. Such information should follow all samples or specimens of the collected material throughout the product development chain, including for specimens held by *ex situ* culture collections. In the latter case, this could be done, for instance, by quoting the specimens' unique identifier numbers, which should be linked to relevant documentation. Besides, such information should be readily available at any stage of research, development, pre-commercialization, and commercialization.

Interestingly, the recent renewal, in September 2013, of a new negotiating mandate for the World Intellectual Property Organization's Intergovernmental Committee on Intellectual Property, Genetic Resources, Traditional Knowledge and Folklore, in conjunction with the WIPO General Assembly's decision to postpone until 2014 whether to convene a diplomatic conference (for the adoption of new treaties for the protection of the above subject matter), signals that policy makers are still divided on the most appropriate legal and policy responses to stop 'biopiracy', including on the issue of disclosure of origins or legal provenance of genetic resources in patent applications (Chiarolla and Lapeyre, 2013). In particular, Article 4 of the Consolidated Document Relating to Intellectual Property and Genetic Resources (February 7, 2014) sets out a number of exceptions and limitations concerning the proposed patent disclosure requirements. There appears to be a convergence of views on excluding the application of this new instrument to genetic resources from areas beyond national jurisdiction.

If this were not the case, then the disclosure requirements would broadly promote synergies with the implementation of the 2010 Nagoya Protocol, in particular with the obligation to monitor and enhance transparency about the utilization of genetic resources. Therefore, relevant information, including information on the 'source' of marine organisms, which may have been collected within or beyond national jurisdiction, could be disclosed at, or notified to, the checkpoints established in accordance with Nagoya Protocol, Article 17, or to other relevant international authorities. If the disclosed sample collection locations were areas beyond national jurisdiction, the application of domestic access and benefit-sharing requirements (including prior informed consent and mutually agreed terms requirements) would obviously be excluded.

This question deserves particular attention in the light of international access and benefit-sharing standards established by other instruments, which include not only the CBD and its Nagoya Protocol, but also the FAO International Treaty on Plant Genetic Resources for Food and Agriculture and the WHO Pandemic Influenza Preparedness Framework for the Sharing of Influenza Viruses and Access to Vaccines and Other Benefits (Chiarolla, 2014, forthcoming). In the light of norms already established or under consideration in various forums with extensive experience in addressing issues related to marine genetic resources, one can wonder whether the further continuation of discussions at the General Assembly and its Ad Hoc open-Ended Informal Working Group may not lead, in the near future, to a strategic forum shifting towards the CBD in the context of the negotiations on a Global Multilateral Benefit-Sharing Mechanism under the Nagoya Protocol (Article 10), since both discussions are proceeding simultaneously.

9.4.5 Promoting capacity-building

From the point of view of capacity-building and development, a 1993 assessment by the World Bank, the latest of its kind, estimated that research institutes in the fields of marine biology and marine biotechnology related areas were scattered throughout the developing world, but that their capabilities varied widely. While many had the capability to perform rudimentary experiments in marine biology, only a few could take on complex projects. While marine biotechnology is a sector undergoing rapid expansion and several developing countries have now developed advanced technological capabilities, some conclusions of this assessment may still be valid, since the report notes that the capability required for research and development involving advanced biotechnologies is of an order of one or two magnitudes more demanding in terms of expertise and equipment than are classical investigations in biology and bioscience. It also appears that most developing countries lack the infrastructure and funding to perform the necessary activities to ensure continued viability of their gene bank collections (United Nations, 2007b).

The continued and strengthened participation of experts from developing countries in relevant activities (exploration, investigations, and observations) is a precondition to developing adequate capacity in marine sciences to study and research marine biodiversity, including in areas beyond national jurisdiction. In that regard, besides the capacity-building programmes offered by various international organizations, south–south cooperation, which increasingly complements north–south cooperation, has a critical role to play. The General Assembly has emphasized the need to focus on strengthening south–south cooperation as an additional way to build capacity and as a cooperative mechanism to further enable countries to set their own priorities and needs (United Nations, 2012b). Examples of south–south cooperation include the Zone of Peace and Cooperation of the South Atlantic, the Micronesia Challenge, the Caribbean Challenge, and the Coral Triangle Initiative.

There is also a need for international initiatives to address gaps in scientific information and data sharing related to marine biodiversity in areas beyond national jurisdiction. A major capacity challenge will be to set up adequate integrated monitoring systems for these areas. The scientific and technology frontier, because of the unique nature of these extreme environments, represents a considerable challenge to the operationalization of monitoring systems as well as to comprehensive research activities related to marine genetic resources. There is therefore a critical need to step up capacity-building efforts and international collaborative programmes and activities in marine scientific research, including the transfer of marine technology.

The assessment of past and current practices in research activities and applications shows that we have disposed of a number of pragmatic tools to resolve some of the challenges related to marine genetic resources of areas beyond national jurisdiction. While the policy discussions continue in the context of the General Assembly, including its Ad Hoc Open-ended Informal Working Group, to identify suitable benefit-sharing options, the conservation and sustainable use of marine genetic resources can benefit, as of now, from the application of a number of tools, including those related to international cooperation in marine scientific research, sustainable planning, and the implementation of an ecosystem approach to the marine environment illustrated in Chapter 6.

References

Allen, M. J. and Jaspars, M. (2009). Realizing the potential of marine biotechnology – Challenges and opportunities. *Industrial Biotechnology* **5**(2): 77–83.

Anonymous (2011). No sow's ear – species of crustacean makes silk underwater. *The Economist*, 19 November 2011. Available at www.economist.com/node/21538659.

Antarctic Treaty Consultative Meeting (ATCM) (2008). *An update on biological prospecting in Antarctica, including the development of the Antarctic Biological Prospecting Database*. ATCM Working Paper 11 (Kiev, Ukraine, 2008). Available at www.unep. org/dewa/Portals/67/pdf/ATCM31_wp011_e.pdf. (Accessed: 1 April 2014.)

Appeltans, W., Ahyong, S. T., Anderson, G., *et al.* (2012). The magnitude of global marine species diversity. *Current Biology* **22**(23): 2189–2202.

Aricò, S. and Salpin, C. (2005). *Bioprospecting of Genetic Resources in the Deep Seabed: Scientific, Legal and Policy Aspects*. Yokohama: United Nations University.

Arnaud-Haond, S., Arrieta, J. M., and Duarte, C. M. (2011). Marine biodiversity and gene patents. *Science* **331**: 1521, doi: 10.1126/science.1200783.

Arrieta, J. M., Arnaud-Haond, S., and Duarte, C. M. (2010). What lies underneath: Conserving the oceans' genetic resources. *PNAS* **107**(43): 18318–18324.

Brinckerhoff, C. (2012). *My myriad nightmare*. Available at www.pharmapatentsblog. com/2012/03/22/my-myriad-nightmare/. (Accessed: 2 May 2014.)

Buck, M. and Hamilton, C. (2011). The Nagoya Protocol on Access to Genetic Resources and the Fair and Equitable Sharing of Benefits Arising from their Utilization to the Convention on Biological Diversity. *Review of European Community and International Environmental Law* **20**(1): 47–61.

Census of Marine Life (2010). *About the Census of Marine Life*. Available at www.coml. org/about/. (Accessed: 1 April 2014.)

Chiarolla, C. (2008). Plant patenting, benefit-sharing and the law applicable to the FAO Standard Material Transfer Agreement. *The Journal of World Intellectual Property* **11**(1): 1–28.

Chiarolla, C. (2011). *Intellectual Property, Agriculture and Global Food Security: The Privatisation of Crop Diversity*. Cheltenham/Northampton, MA: Edward Elgar.

Chiarolla, C. (2014). Relations et enjeux des négociations relatives à l'accès aux ressources génétiques et connaissances traditionnelles associées et au partage des avantages issus de leur utilisation dans le contexte de la future entrée en vigueur du Protocole de Nagoya (IDDRI Studies, forthcoming).

Chiarolla, C. and Jungcurt, S. (2011). Outstanding Issues on Access and Benefit Sharing under the Multilateral System of the International Treaty on Plant Genetic Resources for Food and Agriculture. A background study paper by the Berne Declaration and the Development Fund.

Chiarolla, C. and Lapeyre, R. (2013). Biodiversity and Traditional Knowledge: How can they be protected? IDDRI Policy Brief N°13/2013, Paris, France, 4 pp.

Chiarolla, C., Louafi, S., and Schloen, M. (2013). Genetic Resources for Food and Agriculture and Farmers' Rights: An analysis of the relationship between the Nagoya Protocol and related instruments. In: Morgera E., Buck M., and Tsioumani, E. (eds.), *The Nagoya Protocol in Perspective: Implications for International Law and Implementation Challenges*. Leiden and Boston: Brill/Martinus Nijhoff.

Christoffersen, L. P. and Mathur, E. L. (2005). Bioprospecting ethics and benefits: A model for effective benefits-sharing. *Industrial Biotechnology* 1(4): 255–259.

Conley, J. M. and Makowski, R. (2003).Back to the future: Rethinking the product of nature doctrine as a barrier to biotechnology patents. *Journal of the Patent and Trademark Office Society* **85**: 301.

Convention on Biological Diversity (Text and Annexes) (1992), 1760 UNTS 79.

Convention on Biological Diversity (1995), Decision II/10 of the Conference of the Parties to the Convention on Biological Diversity – Conservation and Sustainable Use of Marine and Coastal Biological Diversity. UNEP/CBD/COP/2/19.

Convention on Biological Diversity (2003). Study of the Relationship between the Convention on Biological Diversity and the United Nations Convention on the Law of the Sea with regard to the Conservation and Sustainable Use of Genetic Resources on the Deep Seabed. Decision II/10 of the Conference of the Parties to the Convention on Biological Diversity. UNEP/CBD/SBSTTA/8/INF/3/REV.

Convention on Biological Diversity (2012). Decision XI/18, B of the Conference of the Parties to the Convention on Biological Diversity – Voluntary Guidelines for the Consideration of Biodiversity in Environmental Impact Assessments and Strategic Environmental Assessments in Marine and Coastal Areas. UNEP/CBD/COP/11/35.

Drug Price Competition and Patent Term Restoration Act (1984). Public Law 98-417, Section 505(j) 21 U.S.C. 355(j).

Eisenberg, R.S. (2003). Patent swords and shields. *Science* **299**(5609): 1018–1019.

European Commission (1998), Directive 98/44/EC on the Legal Protection of Biotechnological Inventions (the Biotechnology Directive), incorporated into the EPC by r. 23 b (1) EPC. OJL213 of 30.7.1998, at pp. 13–21.

European Union (2012). Possible Structure of an Implementing Agreement. Informal paper circulated at the fifth meeting of the Ad Hoc Open-ended Informal Working Group to study issues relating to the conservation and sustainable use of marine biodiversity beyond areas of national jurisdiction (on file with the authors).

Germani, V. and Salpin, C. (2011). The status of high seas biodiversity in international policy and law. In: *A Planet for Life – Oceans: The New Frontier*. Jacquet, P., Pachauri, R. K., and Tubiana, L. (eds.). Delhi: TERI Press, pp. 194–196.

Glowka, L. (2010). Evolving perspectives on the international seabed area's genetic resources: Fifteen years after the 'Deepest of Ironies'. In: *Law, Technology and Science for Oceans in Globalisation*. Vidas, D. (ed.). Leiden/Boston: Martinus Nijhoff Publishers/Brill, pp. 397–419.

Goffredo, S. and Lasker, H. R. (2008). An adaptive management approach to an octocoral fishery based on the Beverton-Holt model. *Coral Reefs* **27**(4): 751–761.

Greenpeace (2008). *Suggested Draft High Seas Implementing Agreement for the Conservation and Management of the Marine Environment in Areas beyond National Jurisdiction.* Available at www.greenpeace.org/international/en/publications/reports/suggested-draft-high-seas-impl/. (Accessed: 1 April 2014.)

Hart, S. (2008). *Elements of a Possible Implementation Agreement to UNCLOS for the Conservation and Sustainable Use of Marine Biodiversity in Areas beyond National Jurisdiction.* Gland: IUCN.

International Seabed Authority (2011). *Environmental Management of Deep-Sea Chemosynthetic Ecosystems: Justification of and Considerations for a Spatially-Based Approach.* ISA Technical Study No. 9. Kingston: ISA, 90 pp.

InterRidge (2014). *InterRidge Statement of Commitment to Responsible Research Practices at Deep-Sea Hydrothermal Vents.* Available at www.interridge.org/IRStatement. (Accessed: 2 May 2014.)

J. Craig Venter Institute (2014). *The Gordon and Betty Moore Foundation Marine Microbial Genome Sequencing Project.* Available at www.jcvi.org/cms/research/past-projects/microgenome/overview/. (Accessed: 1 April 2014.)

KEI (2012). *Brief of* Amicus Curiae *Knowledge Ecology International in Support of Petitioners/.* Available at http://keionline.org/node/1347. (Accessed: 1 April 2014.)

Kesselhein, A. S. and Avorn, J. (2005). University-based science and biotechnology products: Defining the boundaries of intellectual property. *Journal of the American Medical Association* **293**(7): 850–854.

Kitchingman, A. and Lai, S. (2004). Inferences on potential seamount locations from mid-resolution bathymetric data. *Fisheries Centre Research Reports* **12**(5): 7–12.

Knowlton, N. (2010). Citizens of the sea: Wondrous creatures from the Census of Marine Life. *National Geographic.* Washington DC: National Geographic, 176 pp.

Krattiger, A., *et al.* (eds.) (2006). Glossary. In *Intellectual Property Management in Health and Agricultural Innovation – A Handbook of Best Practices*, Oxford, UK: PIPRA & MIHR.

Laffoley, D. d'A., Roe, H. S. J., Angel, M. V., *et al.* (2011). *The Protection and Management of the Sargasso Sea: The Golden Floating Rainforest of the Atlantic Ocean. Summary Science and Supporting Evidence Case.* Sargasso Sea Alliance, 44 pp.

Leary, D. (2007). *International Law and the Genetic Resources of the Deep Sea.* Leiden/Boston: Martinus Nijhoff Publishers.

Leary, D. (2011). Marine genetic resources: The patentability of living organisms and biodiversity conservation. In: *A Planet for Life – Oceans: The New Frontier.* Jacquet, P., Pachauri, R. K., and Tubiana, L. (eds.). Delhi: TERI Press, pp. 183–193.

Leary, D., Vierros, M., Hamon, G., *et al.* (2009). Marine genetic resources: A review of scientific and commercial interest. *Marine Policy* **33**(2): 183–194.

Mazzoleni, R. and Nelson, R. (1998). The benefits and costs of strong patent protection: A contribution to the current debate. *Research Policy* **27**(3): 273–284.

Molinski, T. F., Dalisay, D. S., Lievens, S. L. *et al.* (2009). Drug development from marine natural products. *Drug Discovery* **8**: 69–85, doi: 10.1038/nrd2487.

Nagoya Protocol on Access to Genetic Resources and the Fair and Equitable Sharing of Benefits Arising from their Utilization to the Convention on Biological Diversity (Text and Annex) (2011). Montreal: Secretariat of the Convention on Biological Diversity: 25 pp.

OECD (2006). *Guidelines for the Licensing of Genetic Inventions.* Available at www.oecd.org/science/biotechnologypolicies/guidelinesforthelicensingofgeneticinventions.htm.

Oldham, P., Hall, S., Barnes, C. *et al.* (2014). Valuing the Deep: Marine Genetic Resources in Areas beyond National Jurisdiction. Department for Environment Food and Rural Affairs (Defra). Available at www.google.com/url%3Furl=http://randd.defra.gov.uk/Document.aspx%3FDocument%3D12289_ValuingTheDeepDEFRAFinalVersionOne28092014.pdf&rct=j&frm=1&q=&esrc=s&sa=U&ei=Z7lkVJ3sJ8KtadTTgO-gI&ved=0CCMQFjAC&usg=AFQjCNEIEM4k6QInueArdJS2iHsVlYS9ew. (Accessed 13 November 2014.)

Pew (2012). *Potential Elements of an UNCLOS Implementing Agreement.* Available at www.pewenvironment.org/news-room/other-resources/potential-elements-of-an-unclos-implementing-agreement-85899382831.

PharmaMar (2014). *Yondelis.* Available at www.pharmamar.com/yondelis.aspx. (Accessed: 1 April 2014.)

PR Newswire (2010). Azur Pharma announces the signing of a definitive agreement to acquire PRIALT(R) from Elan. Available at www.prnewswire.com/news-releases/Azur-Pharma-Announces-the-Signing-of-a-Definitive-Agreement-to-Acquire-PRIALTR-From-Elan-87341777.html. (Accessed: 1 April 2014.)

Proelss, A. (2008). Marine genetic resources under UNCLOS and the CBD. *German Yearbook of International Law* **51**: 417–446.

Salpin, C. (2013). The Law of the Sea: A before and an after Nagoya? In: *The Nagoya Protocol in Perspective: Implications for International Law and Implementation Challenges.* Morgera, E., Buck, M., and Tsioumani, E. (eds.). Leiden/Boston: Martinus Nijhoff Publishers/Brill, pp. 149–183.

Salpin, C. and Germani, V. (2007). Patenting of research results related to genetic resources from areas beyond national jurisdiction: The crossroads of the Law of the Sea and Intellectual Property Law. *Review of European Community and International Environmental Law* **16**(1): 12–23.

Salpin, C. and Germani, V. (2010). Marine protected areas beyond areas of national jurisdiction: What's mine is mine and what you think is yours is also mine. *Review of European Community and International Environmental Law* **19**: 174–184.

Scovazzi, T. (2006). Bioprospecting on the deep seabed: A legal gap requiring to be filled. In: *Biotechnology and International Law.* Francioni, F., and Scovazzi, T. (eds.). Oxford and Portland: Hart Publishing, pp. 81–97.

Tara Expeditions (2014). *Tara Expeditions.* Available at http://oceans.taraexpeditions.org/en/a-2-5-years-marine-and-scientific-expedition.php?id_page=1. (Accessed: 1 April, 2014.)

Thambisetty, S. (2002). Database access crucial for developing countries. *Nature Biotechnology* **20**: 775.

The Pharma Letter (2010). Azur Pharma buys global rights (excl-Europe) to Elan pain drug Prialt. Available at www.thepharmaletter.com/article/azur-pharma-buys-global-rights-excl-europe-to-elan-pain-drug-prialt. (Accessed: 1 April 2014.)

Trevor C. (2006). Responding to concerns about the scope of the defence from patent infringement for acts done for experimental purposes relating to the subject matter of the invention. *Intellectual Property Quarterly* **3**: 193–222.

Tvedt, M. W. (2011). *A Report from the First Reflection Meeting on the Global Multilateral Benefit-Sharing Mechanism.* FNI Report 10/2011. Lysaker: FNI, 18 pp.

United Nations Convention on the Law of the Sea (Text and Annexes) (1982). 1833 UNTS 3.

United Nations (2006). Report of the Ad Hoc Open-ended Informal Working Group to Study Issues relating to the Conservation and Sustainable Use of Marine Biological Diversity beyond Areas of National Jurisdiction. UN Doc. A/61/65.

United Nations (2007a). Report on the Work of the United Nations Open-ended Informal Consultative Process on Oceans and the Law of the Sea at its Eighth Meeting. UN Doc. A/62/169.

United Nations (2007b). Report of the Secretary-General on Oceans and the Law of the Sea. UN Doc. A/62/66/Add.2.

United Nations 2008. Letter dated 15 May 2008 from the Co-Chairpersons of the Ad Hoc Open-ended Informal Working Group to Study Issues relating to the Conservation and Sustainable Use of Marine Biological Diversity beyond Areas of National Jurisdiction addressed to the President of the General Assembly. UN Doc. A/63/79.

United Nations (2009). Report of the Secretary-General on Oceans and the Law of the Sea. UN Doc. A/64/66/Add.2.

United Nations (2010). Letter dated 16 March 2010 from the Co-Chairpersons of the Ad Hoc Open-ended Informal Working Group to Study Issues relating to the Conservation and Sustainable Use of Marine Biological Diversity beyond Areas of National Jurisdiction to the President of the General Assembly. UN Doc. A/65/68.

United Nations (2011). Letter dated 30 June 2011 from the Co-Chairs of the Ad Hoc Open-ended Informal Working Group to the President of the General Assembly. UN Doc. A/66/119.

United Nations (2012a). Letter dated 8 June 2012 from the Co-Chairs of the Ad Hoc Open-ended Informal Working Group to Study Issues relating to the Conservation and Sustainable Use of Marine Biological Diversity beyond Areas of National Jurisdiction to the President of the General Assembly. UN Doc. A/67/95

United Nations (2012b). Oceans and the Law of the Sea. UN Doc. A/RES/67/78.

United Nations (2012c). The Future We Want. UN Doc. A/RES/66/288.

United Nations (2013a). Intersessional workshops aimed at improving understanding of the issues and clarifying key questions as an input to the work of the Working Group in accordance with the terms of reference annexed to General Assembly Resolution 67/78: Summary of proceedings prepared by the Co-Chairs of the Working Group. UN Doc. A/AC. 276/6.

United Nations (2013b). Letter dated 23 September 2013 from the Co-Chairs of the Ad Hoc Open-ended Informal Working Group to the President of the General Assembly. UN Doc. A/68/399.

United States Government (2010). *Brief for the United States as* Amicus Curiae *in support of Neither Party*. Available at www.genomicslawreport.com/wp-content/uploads/2010/11/Myriad-Amicus-Brief-US-DOJ.pdf.

Van Dover, C. L., Smith, C. R., Ardron, J., *et al.* (2012). Designating networks of chemosynthetic ecosystem reserves in the deep sea. *Marine Policy* 36: 378–381.

Verenium (2014). *Advanced Enzyme Technology: Ethical Bioprospecting and World Citizenship*. Available at www.verenium.com/ourwork3.html. (Accessed: 2 May 2014.)

WIPO (2009). *Exclusions from Patentable Subject Matter and Exceptions and Limitations to the Rights*. SCP/13/3. Available at: www.wipo.int/edocs/mdocs/scp/en/scp_13/scp_13_3.pdf. (Accessed: 2 May 2014.)

WIPO. *Understanding Industrial Property*. Available at www.wipo.int/export/sites/www/freepublications/en/intproperty/895/wipo_pub_895.pdf. (Accessed: 29 March 2012).

Wolfrum, R. and Matz, N. (2000). The interplay of the United Nations Convention on the Law of the Sea and the Convention on Biological Diversity. *Max Planck Yearbook of United Nations Law* 4: 445–480.

WTO (2000). *Canada – Patent Protection of Pharmaceutical Products*. WT/DS114/R.

Yooseph, S., Sutton, G., Rusch, D. B., *et al.* (2007). The Sorcerer II Global Ocean Sampling Expedition: Expanding the universe of protein families. *PLoS Biology* **5**(3): 432–466.

Young, E. (2011). Yeti crab grows its own food. *Nature*, doi: 10.1038/nature.2011.9537.

Notes

* The views expressed in this chapter are those of the authors and do not necessarily represent those of the United Nations.

1 35 USC Section 101 states: 'Whoever invents or discovers any new and useful process, machinery, manufacture, or composition of matter, or any new and useful improvement thereof, may obtain a patent therefore, subject to the conditions and requirements of this title.' For a pivotal study on the application of the products of nature doctrine see John M. Conley and Roberte Makowski (2003). Back to the future: Rethinking the product of nature doctrine as a barrier to biotechnology patents, *Journal of the Patent and Trademark Office Society.* **85**: 301. Such boundaries were recently scrutinized in the test case *Association for Molecular Pathology* v. *Myriad Genetics, Inc.* In particular, on 13 June 2013, the US Supreme Court eventually held that 'a naturally occurring DNA segment is a product of nature and not patent eligible merely because it has been isolated, but synthetic complementary DNA ('cDNA') is patent eligible because it is not naturally occurring' (Brinckerhoff, 2012; US Government, 2010).

10

The assumption that the United Nations Convention on the Law of the Sea is the legal framework for all activities taking place in the sea

TULLIO SCOVAZZI

10.1 A commonly repeated statement

The resolutions that the United Nations General Assembly has adopted in past years on the subject 'Oceans and the Law of the Sea' emphasize 'the universal and unified character' of the United Nations Convention on the Law of the Sea (Montego Bay, 1982) (UNCLOS) and the need to maintain its integrity. The resolutions also reaffirm 'that the Convention sets out the legal framework within which all activities in the oceans and seas must be carried out' (Res. 68/70, 2013). Such a statement is often repeated by states and experts in international law of the sea. But does it fully correspond with the truth?

There is no doubt that the UNCLOS is a cornerstone in the process for the codification of international law. It has been rightly described as a 'Constitution for Oceans', 'a monumental achievement in the international community', 'the first comprehensive treaty dealing with practically every aspect of the uses and resources of the seas and the oceans', as well as an instrument which 'has successfully accommodated the competing interests of all nations' (Koh, 1983). The UNCLOS has many merits, which nobody could deny. However, the assumption that everything that occurs in the seas must necessarily fall under the scope of the UNCLOS, if this is what the words of the above mentioned resolutions are intended to mean, is far from being satisfactory, as this chapter will try to show.[1]

In fact, the UNCLOS, as any legal instrument, is linked to the time when it was negotiated and adopted (from 1973 to 1982 in this case) and to what was possible to achieve during the negotiations for the drafting of its text.

A first remark regarding this is that there are some evident gaps in the UNCLOS itself, because the states involved in the negotiations were not willing or able to address and solve a few thorny questions that were deliberately left vague. Here, the gaps can be filled by resorting to provisions of customary international law (regulation through customary international law). It may also happen that some

UNCLOS provisions are worded in too general terms, which lack sufficient precision. Where different interpretations of the relevant UNCLOS provisions are in principle admissible, state practice may be important in making one interpretation prevail (regulation through interpretation).

A second, and even more obvious, remark is that, while it provides a solid basis for the regulation of many matters, it would be illusory to think that the UNCLOS is the end of legal regulation in the field of law of the sea. Being itself a product of time, the UNCLOS cannot stop the passing of time. International law of the sea is subject to a process of natural evolution and progressive development which is linked to states' practice and can also affect the UNCLOS. In two cases, changes with respect to the original UNCLOS regime have been integrated into the UNCLOS itself through adoption of so-called implementation agreements (regulation through integration). In another case, because the UNCLOS regime was clearly unsatisfactory – this happens very seldom, but it cannot be excluded altogether – a new instrument of universal scope has been drafted to avoid the risk of undesirable consequences (regulation in another context).

10.2 Regulation through customary international law

The most evident gap in the UNCLOS can be found in Art. 74, relating to the delimitation of the exclusive economic zone, and Art. 83, relating to the delimitation of the continental shelf:

1. The delimitation of the exclusive economic zone between States with opposite or adjacent coasts shall be effected by agreement on the basis of international law, as referred to in Article 38 of the Statute of the International Court of Justice, in order to achieve an equitable solution.
2. If no agreement can be reached within a reasonable period of time, the States concerned shall resort to the procedures provided for in Part XV [Settlement of Disputes].
3. Pending agreement as provided for in paragraph 1, the States concerned, in a spirit of understanding and co-operation, shall make every effort to enter into provisional arrangements of a practical nature and, during this transitional period, not to jeopardize or hamper the reaching of the final agreement. Such arrangements shall be without prejudice to the final delimitation.
4. Where there is an agreement in force between the States concerned, questions relating to the delimitation of the exclusive zone shall be defined in accordance with the provisions of that agreement (Art. 74).

Art. 83 repeats Art. 74, the only change being 'continental shelf' instead of 'exclusive economic zone'.

The reading of Arts. 74 and 83 leaves the impression that, despite their lengthy content, they do not provide any clear substantive regime. By resorting to procedural means, they avoid tackling the main issue. Apart from specifying the procedures to which the states concerned are bound to resort to find a solution, Arts. 74 and 83 do not say very much on the question of delimitation. In particular, paragraph 1 does not indicate what the substantive rules are that become applicable if the states concerned do not reach an agreement. The reference to Art. 38 of the Statute of the International Court of Justice (ICJ), which specifies the category of rules that the ICJ must apply, does not provide any clear guidance on how to address the merits of the question, that is how to draw a boundary line on a map.

The indication of the objective of achieving an equitable solution seems also redundant, as any agreement which has been freely negotiated by the parties embodies, by definition, an equitable solution. Paragraph 4, which provides that, where there is an agreement in force for the states concerned, the agreement applies, somehow recalls the story of Monsieur de La Palice. Some substantive content can only be found in paragraph 3 which, however, has a provisional character. It can be understood as a confirmation of the general obligation of the states concerned to behave in good faith in order to reach a final agreement on delimitation. But it is impossible to infer from UNCLOS Arts. 74 and 83 any further guidance on how to give a final solution to the problem.

The vague content of Arts. 74 and 83 of the UNCLOS was due to practical reasons. During the negotiations, states involved in complex issues of maritime delimitation strongly opposed specific solutions which would have played in favour of their opposite or adjacent neighbouring states. Also states facing manifold issues of delimitation, depending on the characteristics of the coastlines and the different neighbouring states concerned, preferred a vague provision which would grant them enough flexibility to be able to play different games in different fields. It was unwise to force the situation by trying to set forth clear-cut solutions in the text of the UNCLOS. This explains the UNCLOS drafter's choice to leave the very controversial issue of delimitation unresolved. It was necessary to avoid the opening of a Pandora's box which could have precluded adoption of the Convention itself or its universal acceptance.

The stalemate of the UNCLOS is so evident that in the award rendered on 17 December 1999 on the *Eritrea–Yemen Arbitration (Second Stage: Maritime Delimitation)* case the following remark is made: 'In any event there has to be room for differences of opinion about the interpretation of articles [Arts. 74 and 83 of the UNCLOS] which, in a last minute endeavour at the Third United Nations Conference on the Law of the Sea to get agreement on a very controversial matter, were consciously designed to decide as little as possible. It is clear, however, that both articles envisage an equitable result.'

Today the normative gap left in the UNCLOS is being filled by international decisions, which since 1969[2] have been rendered by the ICJ or arbitral tribunals.

Such a notable body of international decisions shows the use of a number of 'methods' (such as equidistance, proportionality, reduced effect of islands, the shifting of the equidistance line, the drawing of a corridor) which, in the light of the circumstances relevant to each specific case, were found by courts to be appropriate for delimiting maritime jurisdictional zones in order to achieve an equitable solution. The methods in question have today consolidated into rules of customary international law.

Another gap in the UNCLOS is left by paragraph 6 of Art. 10. It states that the UNCLOS provisions on bays do not apply to so-called historic bays. This exception creates a legal vacuum, since the only conclusion that can reasonably be drawn from the UNCLOS is that historic bays exist. But nowhere in the UNCLOS is it specified what historic bays are and what other rules apply to them. Some elements – namely the exercise of state authority, the long-lasting duration of this exercise, acquiescence by other states and, although less frequently, the existence of vital interests by the coastal state – are referred to in international practice and doctrinal works as the constitutive elements of a historical title over marine waters.

10.3 Regulation through interpretation

The instance of the criteria for drawing straight baselines (Art. 7 of the UNCLOS) shows that sometimes state practice corresponds with rather elastic interpretations of UNCLOS provisions which could have a different meaning if understood in a literal manner. The word baseline designates the line from which the territorial sea and the other coastal zones are to be measured. A system of straight baselines is an exception to the rule according to which the normal baseline of the territorial sea is the low water mark along the coast. Instead, straight baselines are drawn in the sea and connect appropriate points on land. The text of the relevant paragraphs of Art. 7 reads as follows:

1. In localities where the coastline is deeply indented and cut into, or if there is a fringe of islands along the coast in its immediate vicinity, the method of straight baselines joining appropriate points may be employed in drawing the baseline from which the breadth of the territorial sea is measured. (...).
3. The drawing of straight baselines must not depart to any appreciable extent from the general direction of the coast, and the sea areas lying within the lines must be sufficiently closely linked to the land domain to be subject to the régime of internal waters.

Some redundancy may be perceived, as a deeply indented coast is necessarily cut into, and vice versa. But the idea behind the provision seems clear. Nature cannot be remade by changing in a radical way the shape of a state. What is allowed is to rectify by a geometrical device a manifestly irregular coastline. To simplify without altering: this is the basis of the straight baselines method.

However, the reading of Art. 7 leaves a feeling of uncertainty. While based on geography, the wording of the provision does not contain any geometrical and mathematical precision. A number of questions may be asked in this respect. When can a coastline be considered to be deeply indented and cut into? What should the ratio be between the length of the closing line of the indentation and the distance between this line and the most internal point of the indentation? At what distance from the coast should a fringe of islands be located to be considered in its immediate vicinity? Should this distance be measured between the coast and the closest island or the most external one? What should be the distance between the islands themselves in order to constitute a fringe? Could fringes located in a direction perpendicular to the coast, and not parallel to it, qualify for straight baselines? How is it possible to determine whether the drawing of straight base-lines departs to any appreciable extent from the general direction of the coast? How is the general direction of the coast itself to be determined (it is evident that any conclusions in this respect are greatly influenced by the scale of the map utilized)? In what cases can marine areas be considered as sufficiently closely linked to the land domain? No clear responses are given.

All the problems would have been solved if the UNCLOS had established a limit of maximum length for the segments of straight baselines. But this was not done and Art. 7 has been left vague despite some attempts to give more precision to its content. To infer maximum length, limits could be considered a distortion in the interpretation of a provision which has to be understood in conformity with its flexible nature.

Irrespective of the codified provisions, a practice has developed at the level of state legislation. It shows that several coastal states rely on rather elastic criteria in the determination of the drawing of their straight baselines systems. It is true that questionable instances may be found, but the criteria followed in average domestic legislations go far beyond the strict wording of the relevant UNCLOS provisions. In deciding on 16 March 2001 the case between Qatar and Bahrain on *Maritime Delimitation and Territorial Questions*, the International Court of Justice pointed out that 'the method of straight baselines, which is an exception to the normal rules for the determination of baselines, may only be applied if a number of conditions are met. This method must be applied restrictively' (paragraph 212 of the judg-ment). In spite of this warning, the practice of several states seems today definitely oriented towards a quite liberal interpretation and application of such provisions. Protests are in most cases limited to those made by a few countries or by neighbouring states which could be affected by straight baselines systems at the time of maritime boundary delimitations.

10.4 Regulation through integration

An evident instance of regulation through integration in the UNCLOS itself are the two UNCLOS implementation agreements which have been adopted so far. In this case, changes in the UNCLOS regime are introduced in a 'physiological' manner which does not entail a rupture in the UNCLOS system.

In 1994, several provisions of the most innovative part of UNCLOS, that is the part relating to the seabed beyond the limits of national jurisdiction (so-called 'Area') subject to the regime of common heritage of mankind, were changed to meet the hope for universal participation in the Convention. This was done by the Agreement Relating to the Implementation of Part XI of the UNCLOS, which was annexed to Resolution 48/263, adopted by the General Assembly on 17 August 1994. In fact the politically prudent label of an 'implementation agreement' is a euphemism for the word 'amendment', which would have been more correct from the legal point of view. The provisions of the 1994 Implementation Agreement and those of Part XI of the UNCLOS 'shall be interpreted and applied together as a single instrument' (Art. 2 of the Agreement). However, in the event of any inconsistency between the 1994 Implementation Agreement and Part XI of the UNCLOS, the provisions of the former shall prevail (incidentally, the fact that 21 states (as at May 2014) which are parties to the UNCLOS are not yet parties to the 1994 Implementation Agreement is a persistent matter of concern and raises almost inextricable questions of the law of treaties).

The trend to 'integrate' the UNCLOS regime with some relevant additional provisions has also occurred in the field of fisheries. The Agreement for the Implementation of the Provisions of the United Nations Convention of the Law of the Sea, of 10 December 1982, Relating to the Conservation and Management of Straddling Fish Stocks and Highly Migratory Fish Stocks, was opened for signature in New York on 4 December 1995. This treaty has one evident defect (which is the unbearable length of its title) and many merits. For instance, it includes detailed provisions on the precautionary approach as applied to fisheries (Art. 6 and Annex II), it establishes that a party may authorize a vessel to use its flag for fishing on the high seas 'only where it is able to exercise effectively its responsibilities in respect of such vessel' (Art. 18, para. 2) and, most notably, it brings an evident, but welcome, encroachment on the traditional principle of freedom of the high seas, as states which are not willing to comply with conservation and management measures can be excluded from fishing on the high seas: on the one hand, all states having a real interest in the fisheries concerned have the right to become members of a subregional or regional fisheries management organization or participants in such an arrangement (Art. 8, para. 3); on the other, only those states that are members of such an organization or participants in such an arrangement, or that agree to apply the conservation and management measures established by such an organization or arrangement, have access to the fishery resources to which those measures apply (Art. 8, para. 4). This important conse-quence was not spelled out in the UNCLOS.

10.5 Regulation in another context

In the case of underwater cultural heritage, where the UNCLOS regime is lacking or is clearly unsatisfactory, a new instrument of universal scope has been adopted to avoid the risk of unexpected and undesirable consequences.

The regime provided by the UNCLOS for underwater cultural heritage is fragmentary, being composed of only two provisions, included in different parts of the Convention, namely Art. 149[3] (in Part XI, 'The Area') and Art. 303[4] (in Part XVI, 'General provisions'). Moreover, the two provisions are in a conceptual contradiction one with the other. While Art. 149 is based on assump-tions that the heritage must be preserved and used for the benefit of mankind and that preferential rights are granted to certain states having a link with it, paragraph 3 of Art. 303 can be interpreted, at least in its English text, as a covert invitation to the looting of the underwater cultural heritage found on the contin-ental shelf. It gives priority to 'the law of salvage and other rules of admiralty', that is, to a body of rules that are intended in some common law countries to provide for the application of a first-come-first-served or freedom-of-fishing approach for the underwater cultural heritage. This can only serve the interest of private commercial gain, to the detriment of the public interest in the study and exhibition of the underwater cultural heritage for public interest. If this is the case, a state that has a cultural link with certain objects can be deprived of any means of preventing the pillage of its historical and cultural heritage.

For example, according to the United States Court of Appeals for the Fourth Circuit, in the decision rendered on 24 March 1999 in the case *R.M.S. Titanic, Inc.* v. *Haver*, the law of finds means that 'a person who discovers a shipwreck in navigable waters that has been long lost and abandoned and who reduces the property to actual or constructive possession becomes the property's owner'. In its turn, the law of salvage gives the salvor a lien (or right *in rem*) over the object.

The Convention on the Protection of the Underwater Cultural Heritage (CPUCH), adopted on 2 November 2001 within the framework of the UNESCO (United Nations Educational, Scientific and Cultural Organization), builds on the assumptions contained in Art. 149 UNCLOS and basically aims at preventing the risk of a freedom of fishing regime arising from Art. 303, para. 3, UNCLOS. It provides in general that states parties are bound to 'preserve underwater cultural heritage for the benefit of humanity' (Art. 2, para. 3) and that 'under-water cultural heritage shall not be commercially exploited' (Art. 2, para. 7). Although it does not totally ban the law of admiralty, including law of salvage and law of finds, the CPUCH regime has the practical effect of preventing all of the undesirable effects of the application of these kinds of rules.[5] As regards the heritage found in the continental shelf, the CPUCH sets forth a procedural mechanism which involves the participation of the states having a verifiable link to the heritage. It is based on a three-step procedure (reporting, consultations, urgent measures).

The CPUCH entered into force on 2 January 2009. While stating that the UNCLOS is not prejudiced,[6] in fact the CPUCH tries to bring a remedy to a very questionable aspect of the former. Sadly though, it appears that the sensible message coming from the CPUCH has not yet been fully appreciated, as it is today (May 2014) binding on only 46 states.

10.6 The question of marine genetic resources

The ongoing discussion on the regime of marine genetic resources found beyond the limits of national jurisdiction is an instance of a gap that exists in the UNCLOS.

10.6.1 The prospects for the exploitation of genetic resources in the deep seabed

While the prospects for commercial mining in the seabed beyond national jurisdiction are uncertain, the exploitation of commercially valuable genetic resources may in the near future become a promising activity to take place in the area.

The deep seabed is not a desert, despite extreme conditions of cold, complete darkness, and high pressure. It is the habitat of diverse forms of life associated with typical features, such as hydrothermal vents, cold-water seeps, seamounts, or deep water coral reefs. In particular, it supports biological communities that present unique genetic characteristics. For instance, some animal communities live in the complete absence of sunlight where warm water springs from tectonically active areas (so-called hydrothermal vents). Several species of microorganisms, fish, crustaceans, polychaetes, echinoderms, coelenterates, and molluscs have been found in hydrothermal vent areas. Many of these were new to science. These communities, which do not depend on plant photosynthesis for their survival, rely on specially adapted microorganisms able to synthesize organic compounds from the hydrothermal fluid of the vents (chemosynthesis). The ability of some deep seabed organisms to survive extreme temperatures (thermophiles and hyperthermofiles), high pressure (barophiles), and other extreme conditions (extre-mophiles) makes their genes of great interest to science and industry.

But what is the international regime applying to genetic resources in areas beyond national jurisdiction? In fact, neither the UNCLOS nor the 1992 Convention on Biological Diversity provide any specific legal framework in this regard. The factual implications of the question are pointed out in a document issued in 2005 by the Subsidiary Body on Scientific, Technical and Technological Advice (SBSTTA), established under the Convention on Biological Diversity (CBD, 2005).

First, only a few states and private entities have access to the financial means and sophisticated technologies needed to reach the deep seabed:

Reaching deep seabed extreme environments and maintaining alive the sampled organisms, as well as culturing them, requires sophisticated and expensive technologies. ... Typically, the technology associated with research on deep seabed genetic resources involves: oceanographic vessels equipped with sonar technology, manned or unmanned submersible vehicles; in situ sampling tools; technology related to culture methods; molecular biology technology and techniques; and technology associated with the different steps of the commercialization process of derivates of deep seabed genetic resources. With the exception of basic molecular biology techniques, most of the technology necessary for accessing the deep seabed and studying and isolating its organisms is owned by research institutions, both public and private. To date, only very few countries have access to these technologies. (CBD, 2005, paras. 12 and 13)[7]

Second, the prospects for commercial applications of bioprospecting activities seem promising:

Deep seabed resources hold enormous potential for many types of commercial applications, including in the health sector, for industrial processes or bioremediation. A brief search of Patent Office Databases revealed that compounds from deep seabed organisms have been used as basis for potent cancer fighting drugs, commercial skin protection products providing higher resistance to ultraviolet and heat exposure, and for preventing skin inflammation, detoxification agents for snake venom, anti-viral compounds, anti-allergy agents and anti-coagulant agents, as well as industrial applications for reducing viscosity. (CBD, 2005, para. 21)

The commercial importance of marine genetic resources is demonstrated by the fact that all major pharmaceutical firms have marine biology departments. The high cost of marine scientific research, and the slim odds of success (only one to two percent of pre-clinical candidates become commercially produced) is offset by the potential profits. Estimates put worldwide sales of all marine biotechnology-related products at US$100 billion for the year 2000. (CBD, 2005, para. 22)

Last, but not least, another important element to take into consideration is that the patent legislation of several states does not compel the applicant to disclose the origin of the genetic materials used:

Assessing the types and levels of current uses of genetic resources from the deep seabed proves relatively difficult for several reasons. First, patents do not necessarily provide detailed information about practical applications, though they do indicate potential uses. Moreover, information regarding the origin of the samples used is not always included in patent descriptions. (CBD, 2005, para. 22)

10.6.2 Common heritage of mankind versus freedom of the high seas

In 2006, the subject of the international regime for genetic resources in the deep seabed was discussed within the Ad Hoc Open-ended Informal Working Group to Study Issues Relating to the Conservation and Sustainable Use of Marine Biological Diversity beyond Areas of National Jurisdiction (hereinafter referred to as the Working Group), established by United Nations General Assembly Resolution 60/30 of 29 November 2005. Two opposite views were put forward by the states concerned.

Some states took the position that the UNCLOS principle of common heritage of mankind should be extended to also cover genetic resources. The mandate of the International Seabed Authority (ISBA), the body in charge of the organization and control of activities relating to mineral resources in the area, should also be extended accordingly:

Several delegations reiterated their understanding that the marine genetic resources beyond areas of national jurisdiction constituted the common heritage of mankind and recalled article 140 of the Convention, which provides that the activities in the Area shall be carried out for the benefit of mankind and that particular consideration should be given to the interest and needs of developing States, including the need for these resources to be used for the benefit of present generations and to be preserved for future generations. ... A number of delegations mentioned that the International Seabed Authority constituted an existing mechanism in this area and that consideration should accordingly be given to the possibility of broadening its mandate. (United Nations, 2006, para. 71)

Other states relied on the UNCLOS principle of freedom of the high seas, which would imply a right of freedom of access to, and unrestricted exploitation of, deep seabed genetic resources:

Other delegations reiterated that any measures that may be taken in relation to genetic resources in areas beyond national jurisdiction must be consistent with international law, including freedom of navigation. In their view, these resources were covered by the regime of the high seas, which provided the legal framework for all activities relating to them, in particular marine scientific research. These delegations did not agree that there was a need for a new regime to address the exploitation of marine genetic resources in areas beyond national jurisdiction or to expand the mandate of the International Seabed Authority. (United Nations, 2006, para. 72)

The Working Group held a second meeting in 2008. Again, very different views were expressed regarding the regime to be applied to marine genetic resources, repeating what had already taken place in 2006:

In that regard, divergent views were expressed on the relevant legal regime on marine genetic resources beyond areas of national jurisdiction, in particular whether those marine genetic resources were part of the common heritage of mankind and therefore fell under the regime of the Area, or were part of the regime for the high seas. (United Nations, 2008, para. 32)

The same, different positions were manifested during the 2010 meeting of the Working Group (United Nations, 2010, paras. 70–72).

Also, the United Nations Open-Ended Informal Consultative Process on Oceans and the Law of the Sea addressed the subject of marine genetic resources at its 2007 meeting. However, the meeting was unable to reach overall agreement on the regime that should apply to such resources.

This basic disagreement on the international regime of genetic resources leaves a sentiment of dissatisfaction. In fact, both divergent positions move from the same starting point:

The United Nations Convention on the Law of the Sea was recognized as the legal framework for all activities in the oceans and seas, including in respect of genetic resources beyond areas of national jurisdiction. (United Nations, 2008, para. 36)

Why do two groups of states, moving from the same starting point, namely that the UNCLOS is the legal framework for all activities taking place in the sea, reach two completely opposite conclusions with regard to the same question? A possible answer to the question is that the starting point is not as solid as it seems at first sight.

It is a matter of fact that the UNCLOS cannot regulate those activities that its drafters did not intend to regulate, for the simple reason that they were not foreseeable in the period when the Treaty was being negotiated. At that time, very little was known about the genetic qualities of marine organisms. For evident chronological reasons, the potential economic value of the marine genetic resources was not considered by the UNCLOS negotiators. When dealing with the special regime of the Area and its resources, the UNCLOS drafters had only mineral resources in mind.

This is fully evident from the plain text of the UNCLOS. The term 'activities' in the Area is defined as 'all activities of exploration for, and exploitation of, the resources of the Area' (Art. 1, para. 1). Art. 133 (a) defines the 'resources' of the Area as 'all solid, liquid or gaseous mineral resources *in-situ* in the Area at or beneath the sea-bed, including polymetallic nodules'. The UNCLOS regime of the common heritage of mankind does not include the non-mineral resources of the Area. However, for the same chronological reasons, the regime of freedom of the high seas does not apply to genetic resources either. While including rules on living (fish) and non-living (minerals) resources in areas beyond national jurisdiction, the UNCLOS does not provide any specific regime for the exploitation of marine genetic resources. The words 'genetic resources' or 'bioprospecting' do not appear anywhere in the UNCLOS. A legal gap exists in this regard. Sooner or later it should be filled (better sooner than later) through a regime which, to be consistent, should encompass under the same legal framework the genetic resources of both the Area and the superjacent waters.

However, not all of the UNCLOS should be left aside when envisaging a future regime for marine genetic resources beyond national jurisdiction. The scope of the regime of the Area is already broader than may be believed at first sight. Under the UNCLOS, the legal condition of the Area also has an influence on the regulation of activities that, although different from minerals and mining, are also located in that space. The regime of the Area already encompasses subjects which are more or less directly related to mining activities, such as marine scientific research (Art. 143 of UNCLOS), the preservation of the marine environment (Art. 145 of UNCLOS), and the protection of underwater cultural heritage (Art. 149 of UNCLOS). As far as the first two subjects are concerned, it is difficult to draw a clear-cut distinction between what takes place on the seabed and what in the superjacent waters.

While a specific regime for exploitation of genetic resources is lacking, the aim of sharing the benefits among all states can still be seen as a basic objective embodied in a treaty designed to 'contribute to the realization of a just and equitable international economic order which takes into account the interests and needs of mankind as a whole and, in particular the special interests and needs of developing countries, whether coastal or land-locked' (UNCLOS preamble). Also in the field of genetic resources, the application of the principle of freedom of the sea (that is the 'first-come-first-served' approach) leads to inequitable and hardly acceptable consequences. New cooperative schemes, based on provisions on access and sharing of benefits, should be envisaged in a future agreement on genetic resources beyond the limits of national jurisdiction. This is also in full conformity with the principle of fair and equitable sharing of the benefits arising out of the utilization of genetic resources set forth by Article 1 of the Convention on Biological Diversity and, more recently, by Article 10 of the Protocol on Access to Genetic Resources and the Fair and Equitable Sharing of Benefits Arising from their Utilization (CBD, 2010).[8]

Moreover, bioprospecting, which is currently understood as the search for commercially valuable genetic resources, can already be considered as falling under the UNCLOS regime of marine scientific research. The UNCLOS does not provide any definition of 'marine scientific research'. However, Article 246, which applies to the exclusive economic zone and the continental shelf, makes a distinction between two kinds of marine scientific research projects, namely those carried out 'to increase scientific knowledge of the marine environment for the benefit of all mankind' (para. 3) and those 'of direct significance for the exploration and exploitation of natural resources, whether living or non-living' (para. 5(a). This distinction supports the conclusion that, under the UNCLOS logic, research activities of direct significance for the purpose of exploration and exploitation of genetic resources also fall under the general label of 'marine scientific research'. There is an inextricable factual link between marine scientific research (either pure or applied) and bioprospecting. A research endeavour organized with the intent of increasing human knowledge may well result in the discovery of commercially valuable information on genetic resources. Also, bioprospecting is consequently covered by Article 143, para. 1, of the UNCLOS, which sets forth the principle that 'marine scientific research in the Area shall be carried out exclusively for peaceful purposes and for the benefit of mankind as a whole'.[9] This provision refers to any kind of marine scientific research and is not limited to research on mineral resources. Yet the reading of Article 143 in combination with Article 246 contradicts the assumption that there is an absolute freedom to carry out bioprospecting in the Area. States which are active in bioprospecting in this space are already bound to contribute to the benefit of mankind as a whole.

Article 143, para. 3, grants to states the right to carry out scientific research in the Area, but binds them to cooperate with other states and the ISBA in various fields, including dissemination of results. Also this provision refers to any kind of marine scientific research in the Area. Yet, the mandate of ISBA deserves close scrutiny, especially if it is to be understood, not only as an entity involved in marine mining activities in competition with others, but as the international organization which bears the main responsibility to realize a just and equitable economic order for the oceans and seas. Nothing prevents states from expanding the mining focus of the ISBA and granting to it some broader management competences in the field of genetic resources.

10.6.3 Possible future developments

New prospects emerged at the 2011 meeting of the Working Group. A number of states, both developed and developing, proposed the commencement of a negotiation process towards a new and third UNCLOS implementation agreement that could fill the gaps in the present regime of conservation and sustainable use of marine biological diversity in areas beyond national jurisdiction. A further process of regulation through integration in the UNCLOS was thus envisaged.

A new implementation agreement was already envisaged by certain states, in particular the member states of the European Union, during the 2008 meeting of the Working Group: 'Several delegations considered that an implementation agreement under the United Nations Convention on the Law of the Sea was the most effective way to establish an integrated regime and address the multi-plicity of challenges facing the protection and sustainable use of marine biodiver-sity in areas beyond national jurisdiction. These delegations suggested that such an instrument was necessary to fill the governance and regulatory gaps that prevented the international community from adequately protecting marine bio-diversity in the areas beyond national jurisdiction. It was proposed that such an instrument would address currently unregulated activities, ensure consistent application of modern ocean governance principles in sectoral management regimes and provide for enhanced international cooperation' (United Nations, 2008, para. 47).

While a general consensus on this proposal has not yet been achieved, a number of common elements are being developed among several states that were previ-ously putting forward divergent positions. The states participating in the 2011 meeting of the Working Group recommended that

(a) A process be initiated by the General Assembly, with a view to ensuring that the legal framework for the conservation and sustainable use of marine biodiversity in areas beyond national jurisdiction effectively addresses those issues by identifying gaps and ways forward, including through the implementation of existing instruments and the possible development of a multilateral agreement under the United Nations Convention on the Law of the Sea.

(b) This process would address the conservation and sustainable use of marine biodiversity in areas beyond national jurisdiction, in particular, together and as a whole, marine genetic resources, including questions on the sharing of benefits, measures such as area-based management tools, including marine protected areas, and environmental impact assessments, capacity-building and transfer of marine technology.

(c) This process would take place: (i) in the existing Working Group; and (ii) in the format of intersessional workshops aimed at improving understanding of the issues and clarifying key questions as an input to the work of the Working Group. (United Nations, 2011, Annex, para. 1)

At its 2012 meeting, the Working Group requested the United Nations Secretary-General to convene in 2013 two intersessional workshops on the topics of 'marine genetic resources' and 'conservation and management tools, including area-based management and environmental impact assessment'. The workshops were intended to improve understanding of the issues and clarify key questions in order to enable the United Nations General Assembly to make progress on ways to fulfil its mandate (United Nations, 2012, para. 1 and appendix). By Resolution 67/78, adopted on 11 December 2012, the UN General Assembly decided (para. 182) to convene the two workshops in May 2013 and recalled (para. 181):

that in 'The Future We Want' States committed to address, on an urgent basis, building on the work of the Ad Hoc Open-ended Informal Working Group and before the end of the sixty-ninth session of the General Assembly, the issue of the conservation and sustainable use of marine biological diversity of areas beyond national jurisdiction, including by taking a decision on the development of an international instrument under the United Nations Convention on the Law of the Sea.[10]

The workshops, which were held in May 2013 (United Nations, 2013), ensured a wide participation of scientists, industry, and non-governmental organizations and resulted in a well-informed presentation and discussion of the two topics. The workshop on marine genetic resources addressed the following subjects: 'Meaning and scope; extent and types of research, uses and applications; technological, environmental, social and economic aspects; access-related issues; types of benefits and benefit-sharing; intellectual property rights issues; global and regional regimes on genetic resources, experiences and best practices; impacts on and challenges to marine biodiversity beyond areas of national jurisdiction; exchange of information on research programmes regarding marine biodiversity in areas beyond national jurisdiction'.

By Resolution 68/70, adopted on 9 December 2013, the UN General Assembly requested, inter alia, 'the Working Group to make recommendations to the Assembly on the scope, parameters and feasibility of an international instrument under the Convention' (para. 198), in order to prepare the decision to be taken at the General Assembly 69th session (starting in September 2014).

In all these discussions, the possibility of a third UNCLOS implementation agreement is envisaged as a possible way to move forward, insofar as the existing instruments cannot fill the present governance and regulatory gaps and cannot provide the required specific regime. Rather than elaborations on theoretical questions and legal principles, such as freedom of the high seas or common heritage of mankind, what is needed in the near future is the consolidation of a general understanding on a number of 'commonalities' which could become the key elements in the 'package' for a future global regime for the conservation and sustainable use of marine biodiversity in areas beyond national jurisdiction. This package could include rules on a network of marine protected areas, environmental impact assessment, marine genetic resources, including access to and sharing of benefits from them, as well as capacity-building and technology transfer.

References

Convention on Biological Diversity (CBD) (2005). Status and Trends of, and Threats to, Deep Seabed Genetic Resources beyond National Jurisdiction, and Identification of Technical Options for their Conservation and Sustainable Use. Doc. UNEP/CBD/SBSTTA/11/11 of 22 July 2005.

CBD (2010). Protocol on Access to Genetic Resources and the Fair and Equitable Sharing of Benefits Arising from their Utilization (Nagoya, 2010). Montreal: Convention on Biological Diversity.

Koh, T. R. B. (1983). 'A Constitution for the Oceans'. In: *The Law of the Sea – Official Text of the United Nations Convention on the Law of the Sea with Annexes and Index*. New York: United Nations, pp. xxxiii–xxxvii.

Scovazzi, T. (2001). The evolution of international law of the sea: New issues, new challenges. *Recueil des Cours*. **286**: 39–244.

United Nations (2006). Report of the Ad Hoc Open-ended Working Group to Study Issues Relating to their Conservation and Sustainable Use of Marine Biological Diversity beyond Areas of National Jurisdiction. Doc. A/61/65 of 20 March 2006.

United Nations (2008). Joint Statement of the Co-Chairpersons of the Ad Hoc Open-ended Informal Working Group. Doc. A/63/79 of 16 May 2008.

United Nations (2010). Letter Dated 16 March 2010 from the Co-Chairpersons of the Ad Hoc Open-ended Informal Working Group to the President of the General Assembly. Doc. A/65/68 of 17 March 2010.

United Nations (2011). Letter Dated 30 June 2011 from the Co-Chairpersons of the Ad Hoc Open-ended Informal Working Group to the President of the General Assembly. Doc. A/66/119 of 30 June 2011.

United Nations (2012). Letter Dated 8 June 2012 from the Co-Chairpersons of the Ad Hoc Open-ended Informal Working Group to the President of the General Assembly. Doc. A/67/95 of 13 June 2012.

United Nations (2013). Intersessional workshops aimed at improving understanding of the issues and clarifying key questions as an input to the work of the Working Group in accordance with the terms of reference annexed to General Assembly Resolution 67/78: Summary of proceedings prepared by the Co-Chairs of the Working Group. Doc. A.AC/276/6 of 10 June 2013.

United Nations General Assembly (UNGA) (2012). Resolution 66/288 of 11 September 2012.

Notes

1 A number of considerations made in this paper are based, with the necessary updating, on Scovazzi (2001).

2 The first relevant judgment was rendered by the International Court of Justice on 20 February 1969 in the *North Sea Continental Shelf* cases (*Federal Republic of Germany* v. *Denmark*; *Federal Republic of Germany* v. *Netherlands*). The last one is the judgment by the same court of 27 January 2014 on the *Maritime Dispute* between Peru and Chile.

3 'All objects of an archaeological and historical nature found in the Area shall be preserved or disposed of for the benefit of mankind as a whole, particular regard being paid to the preferential rights of the State or country of origin, or the State of cultural origin, or the State of historical and archaeological origin.'

4 '1. States have the duty to protect objects of an archaeological and historical nature found at sea and shall co-operate for this purpose. 2. In order to control traffic in such objects, the coastal State may, in applying article 33 [= the contiguous zone], presume that their removal from the sea-bed in

the zone referred to in that article without its approval would result in an infringement within its territory or territorial sea of the laws and regulations referred to in that article [= customs, fiscal, immigration or sanitary laws and regulations]. 3. Nothing in this article affects the rights of identifiable owners, the law of salvage and other rules of admiralty, or laws and practices with respect to cultural exchanges. 4. This article is without prejudice to other international agreements and rules of international law regarding the protection of objects of an archaeological and historical nature.'

5 'Any activity relating to underwater cultural heritage to which this Convention applies shall not be subject to the law of salvage or law of finds, unless it: (a) is authorized by the competent authorities, and (b) is in full conformity with this Convention, and (c) ensures that any recovery of the underwater cultural heritage achieves its maximum protection' (Art. 4).

6 'Nothing in this Convention shall prejudice the rights, jurisdiction and duties of States under international law, including the United Nations Convention on the Law of the Sea. This Convention shall be interpreted and applied in the context of and in a manner consistent with international law, including the United Nations Convention on the Law of the Sea' (Art. 3 CPUCH).

7 'A limited number of institutions worldwide own or operate vehicles that are able to reach areas deeper than 1,000 meters below the oceans' surface, and can therefore be actively involved in deep seabed research' (CBD, 2005, para. 16).

8 'Parties shall consider the need for and the modalities of a global multilateral benefit-sharing mechanism to address the fair and equitable sharing of benefits derived from the utilization of genetic resources and traditional knowledge associated with genetic resources that occur in transboundary situations or for which it is not possible to grant or obtain prior informed consent. The benefits shared by users of genetic resources and traditional knowledge associated with genetic resources through this mechanism shall be used to support the conservation of biological diversity and the sustainable use of its components globally.' While the Nagoya Protocol does not apply to areas beyond national jurisdiction, it could become a source of inspiration. Another relevant instrument could be the International Treaty on Plant Genetic Resources for Food and Agriculture, concluded in 2001 under the auspices of the Food and Agriculture Organization (FAO).

9 Art. 241 UNCLOS is also relevant in a discussion on the legal condition of the genetic resources of the deep seabed. It provides that 'marine scientific research activities shall not constitute the legal basis for any claim to any part of the marine environment or its resources'.

10 'The Future We Want' is the outcome document adopted by the United Nations Conference on Sustainable Development, held in Rio de Janeiro in 2012 (so-called 'Rio+20 Conference') (UNGA, 2012).

11

The legal regime of outer space in light of the Law of the Sea

VIVIANA IAVICOLI

11.1 Space law at its inception and analogy with other regimes

Outer space is an area encompassed among spaces beyond the limits of national jurisdiction and for this reason its legal regime is often associated with those applicable to other common spaces, such as the high seas and the deep seabed, and to Antarctica. The legal regime concerning outer space, the Moon, and the other celestial bodies does not draw inspiration from a specific part of international law; rather it deals with a mixture of multiple analogies (Peterson, 1997; DeSaussure, 1992; Kopal, 1992).

During the 1980s, space law was frequently defined as a special branch of international law. However, this definition cannot be considered more than a *ratione materiae* criterion, since space law was born in the context of international law, and its applicability to space activities has been confirmed by all of the relevant United Nations general treaties and declarations. Currently, it is commonly also presumed to cover domestic laws and the laws of the European Union concerning space activities. Therefore, space law can be defined as a 'specific and coherent set of rules applicable to activities of states and other subjects, including private subjects, in outer space or related to outer space activities' (Marchisio, 2006).[1]

The inception of space activities marks the beginning of space law. On 13 December 1958, the General Assembly of the United Nations set up an Ad Hoc Committee as its subsidiary body, with the aim to deal with space matters. Between 1959 and 1962, the Committee for Peaceful Uses of Outer Space (Ad Hoc Committee for Peaceful Uses, then COPUOS) released three important resolutions: Resolution 1472 (XIV) (12 December 1959) and Resolution 1721 (XVI) (20 December 1961), both concerning international cooperation in the peaceful uses of outer space; and Resolution 1962 (XVIII) (13 December, 1963) concerning the legal principles governing the activities of states in the exploration and uses of outer space. Although not binding, these resolutions reflected the opinion of a relevant number of states on basic principles of space law *in statu nascendi*. States complied with these from the beginning without any substantial objection in their implementation (*opinio juris*). Since these principles were soon affirmed within the international community, a few years later, it was possible to codify them as binding norms in the Treaty on Principles Governing the Activities of States in the Exploration and Use of Outer Space, including the Moon and Other Celestial Bodies (27 January 1967, 610 UNTS 250, Outer Space Treaty – OST).

11.2 Rejection of the legal approach of air law

Long before the inception of the space era, which has conventionally been defined as commencing at the launch of the Soviet satellite Sputnik I, the first space object launched into orbit, in October 1957, Vladimir Mandl had already remarked, in 1928, that the special features of outer space required basic principles devoted to ruling the activities carried out in this area (Kopal, 1992). During the 1950s, the choice of possible analogies within the already existing international law, the law of the sea and Antarctica, whose Treaty of Washington was concluded on 1 December 1959 and entered into force on 23 June 1961 (402 UNTS 71), were taken into consideration along with related implications on the future space law legal regime. Looking at experienced antecedents can be an easier solution for lawmakers, but in any case it could not have entailed a simple process of adapta-tion to a new discipline, since a number of key space-related issues required innovative solutions (Lachs, 2010).

Analogies with air law might have been a natural option, since the location of the space area is just above the air layer surrounding the Earth. Air and space areas are connected, but there is not a physical border between them (*délimitation horizontale*), and the law did not establish any definition accordingly.

Since air law recognizes the sovereignty of the subjacent state, air space, which falls under national jurisdiction, a possible analogical approach to outer space would have led to the fragmentation of the cosmic space into areas subject to different sovereignties, thus conditioning the use of outer space to the express permission of the national state.

Furthermore, the political context of the Cold War played a fundamental role in the birth of *corpus iuris*. Between the 1950s and 1960s, the conquest of space manifested itself as a struggle for power among the principal antagonists of the Cold War. This situation explains why, from the beginning, by tacit agreement of both superpowers, space affairs were dealt with by the United Nations in the framework of the debate on peaceful uses of outer space. Military activities carried out in space, such as the deployment of rockets carrying nuclear weapons and satellites for reconnaissance, were also crucial in the rejection of the legal approach of air law, which would have imposed restrictions to the military uses of outer space. On the other hand, aviation was unable to maintain the control exercised by individual states on space areas, which made a comparison with air law useless.

It is noteworthy that since the launch of the first Sputnik, no state had objected to the flying over of satellites above its own territory. The effect of this acquiescence was that states did not require authorization for satellites to orbit above their territory. This presumed the establishment of a free common area, not subject to national jurisdiction, therefore the freedom to use outer space.

It has rightly been noticed that '[t]he absence of any objection from other States meant that the orbiting of satellites around the Earth was not a privilege but a right given to all nations' (DeSaussure, 1992). A claim of sovereignty made in 1976 by some equatorial countries on the segments of their respective overarching geostationary orbits was issued in the Bogotá Declaration. However, it provoked strong opposition by other states.[2]

Essentially, the debate concerning the analogical approach with air law ended in 1961, when the government of the Soviet Union accepted some preliminary key issues embodied by United Nations Resolution 1721 A (XVI) (International Cooperation in the Peaceful Uses of Outer Space, 12 December 1959), followed by the Declaration adopted by Resolution 1962 (XVIII). Resolution 1721 stated for the first time that the activities carried out in outer space and celestial bodies should be in compliance with international law, including the Charter of the United Nations, and proclaimed the freedoms of exploration and use for peaceful purposes in outer space and celestial bodies, which are not suitable for national appropriation (typical space activities are the launch of satellites, experiments in outer space, invention in outer space, production of satellite data in outer space or on the Earth, broadcasting activities, Earth observation, and the use of intercontinental ballistic missiles and of anti-satellite weapons. Cf. Hobe, 2009). Such basic principles were definitively provided for by the 1967 OST, in particular by Article I.2, and have fundamental implications for the legal regime of outer space.

The enjoyment of freedoms of exploration, use for peaceful purposes, and the non-appropriation principle are interconnected, since the latter constitutes, at the same time, a precondition and a limit to those freedoms themselves. Claims of sovereignty could not arise from any activity carried out in space since Article II stipulates: 'Outer space, including the Moon and other celestial bodies, is not subject to national appropriation by claim of sovereignty, by means of use or occupation, or by any other means', and the non-appropriation principle should grant the less technologically capable countries the opportunity to access and use outer space, on the basis of non-discrimination.

Additionally, the sovereignty of the state cannot be extended to the air spaces located above either the high seas, or the continental shelf, and the exclusive economic zone (EEZ) for which the coastal state maintains limited sovereign rights. Similarly to subjacent high seas, the international air spaces are free zones beyond national jurisdiction.

For security reasons, Article 12 of the Chicago Convention on Air Law (Convention on International Civil Aviation of 7 December 1944, entered into force on 4 April 1947 (15 UNTS 2959)) stipulates that the enforceable regulations on these areas are within the auspices of the International Telecommunication Union (ITU).[3] For air traffic control and military surveillance, some parts of them are the objects of different agreements (Air Defense Identification Zones – ADIZ) (Lyall and Larsen, 2009).

Thus the legal regime of outer space has some analogy with that of international air space, especially in regard to the principles of non-appropriation, free use, and international cooperation. It should be noted that the definition of outer space and the delimitation between air space and outer space remains unsettled and that this is an item still present on the agenda of COPUOS. On the other hand, the lack of delimitation thus far has not involved serious consequences in terms of peaceful use of space activities. That being said, commercialization of the latter is likely to spur new developments in the coming years.

11.3 The analogy with the law of the sea

11.3.1 *The* res communis omnium: *the high seas and the OST*

The freedom to exercise space activities, intended as exploration and use, granted in space law, presumes the establishment of a common area. This approach has received support among governments, including both neutral and non-aligned, in debates within the United Nations since the first session of the COPUOS in 1958 (Peterson, 1997). The antecedent of the law of the high seas is more appropriate than air law because of the customary freedoms of navigation and exploration characterizing these areas beyond the limits of national jurisdiction. Both the high seas and space indeed concern areas not subject to national jurisdiction, but the respective legal regimes have evolved in different contexts.

The law of the high seas was shaped over a long period dating back to the 17th century, when Grotius, in his treatise *Mare Liberum*, theorized that the high seas were a free area to all states for traditional activities, for passage, and use without preventive notification or authorization. These freedoms played a fundamental role in the development of international trade and were essential in the governance of these maritime zones, according to the *res communis omnium* principle.

The Geneva Convention codified the law of the sea only relatively recently, in 1958 (Convention on the High Seas of 29 April 1958, entered into force on

23 June 1961(450 UNTS 11)). The enjoyment of the customary freedoms provided for by the Geneva Convention on the High Seas establishes the availability of this part of the sea to all states for navigation, fishing, laying submarine cables and pipelines, and flying over (Geneva Convention, Art. 87). The United Nations Conventions on the Law of the Sea (UNCLOS), concluded on 10 December 1982 and entered into force on 16 November 1994 (1833 UNTS 396), expressly added the freedoms to construct artificial islands and other installations (UNCLOS, Arts. 58 and 60).

Conversely, space law was from the beginning the object of conventional regulation. Its fundamentals were negotiated, principally by the superpowers, over a short period, given the urgency of the situation; at the time, outer space was, from the legal standpoint, a *tabula rasa*. The general consensus of the states within the United Nations Ad Hoc Committee (Marchisio, 2005) allowed the adoption of fundamental resolutions covering general principles, some of which were followed by consistent practice (UN General Assembly Resolution 1721 of 20 December, 1961). These principles were incorporated into the OST, which was adopted by the General Assembly on 19 December 1966 and entered into force on 10 October 1967.

The subsequent space agreements, negotiated under the auspices of the United Nations, further developed and integrated the general principles of the OST.[4] The period covering the late 1950s to the late 1970s, when the fundamentals of space law were fleshed out and embedded in binding treaties, identifies the first phase of space law, defined also as the 'law making phase' (Marchisio, 2005).

Although the legal regime of space is encapsulated by all the space-related treaties, the cornerstone is the OST, to which more than one hundred states are part (see www.oosa.unvienna.org/oosa/en/SpaceLaw/treatystatus/index.html). The regime enshrined in the OST establishes the architecture of space law and provides the legal framework of the *res communis omnium*, stipulating fundamental principles already developed and consolidated in the law of the high seas, such as the freedom of exploration and use, the applicability of international law to space activities, the non-appropriation principle, and the respect of the environment (Back Impallomeni, 1983).

No article stipulates *expressis verbis* the regime of *res communis omnium*, but this can be deduced from the whole and coherent framework of the Treaty, which proclaims the freedom for all states to explore and use outer space and that these activities should be carried out for the benefit and in the interests of the whole of mankind (Art. I); that outer space, including the Moon and celestial bodies, is not subject to national appropriation by any means (Art. II); that international law is applicable to outer space activities (Art. III), which shall be peaceful (Art. IV); and that states shall avoid harmful contamination in carrying out space activities (Art. IX).

The freedoms of exploration and use of outer space and celestial bodies enunciated by the Treaty are granted for the benefit and in the interests of all countries (Art. I.1) irrespective of their degree of economic and scientific development, and as a consequence these activities are 'the province of all mankind'. The freedoms of exploration and use are intended to serve as a limitation of the absolute freedom to carry out space activities, which should be performed on an ethical basis for the advantage of the whole of mankind ('province of all mankind'). The presence in the OST of Article I, paragraph 1, is because of the need for space law to be legitimized. Unlike the law of the sea, which is essentially based on old customs, space law, at the time, concerned only principles *de lege ferenda* (Kerrest, 2004b).

Article I.1 raises some questions in regard to how to implement this ethical clause and whether it also entails access to space as a precondition for the freedoms of exploration and use (Hobe, 2009; Del Ville, 2009). Strictly speaking, it is debatable whether access to space deals with a faculty or a right granted to all. Opponents of such an opinion remark that the Treaty simply 'consacre une liberté d'accès à l'espace extratmosphérique et non un droit d'accès' (allows for freedom of access to outer space rather than a right to access), but should rather be an unconditional faculty to access to outer space. 'Toutefois en aucune cas le principe de liberté des activités spatiales ne signifie que tous les Etats, quelque soit leur stade de développement doivent être aptes, d'un point de vue technologique, à accéder à l'espace extratmosphérique. Le principe de liberté d'accès à l'espace extratmosphérique et d'exploration ne constitue pas un droit effectif d'exploration pour tout Etat demandeur et non doté de capacité spatiales' (In any event, the principle of freedom of use does not signify that all states, regardless of their degree of development, will be technologically able to access outer space. The principle of freedom of access to outer space does not automatically imply a right to exploration for any states who wish to, but which may not possess the necessary technological capabilities) (Del Ville, 2009).

The gap between states with and without space capabilities should therefore be solved through the principle of cooperation, which is at the basis of space law and should be intended as a remedy to inequality, especially in terms of technological discrepancies between states. Indeed space missions in particular require international cooperation since they frequently rely on the technological and financial capacities of several states. The principle of cooperation has been implemented through multilateral and bilateral agreements and especially with the participation of countries lacking space capabilities with international organizations such as Intelsat, Inmarsat, ESA, etc.

However, the most prominent examples of cooperation are efforts in projecting, building, and managing the International Space Station (von der Dunk and Brus,

2006). In this context, while different forms of cooperation are stipulated among partners (USA, Russia, ESA, Japan, Canada), based primarily on the Intergovernmental Agreement (IGA – Agreement among the Government of Canada, Governments of Member States of the European Space Agency, the Government of Japan, the Government of the Russian Federation, and the Government of the United States of America Concerning Cooperation on the Civil International Space Station, 29 January 1998, entered into force on 28 March 2001), they do not envisage any form of cooperation with developing countries.

The Declaration on International Cooperation in the Exploration and Use of Outer Space for the Benefit and in the Interest of All States, Taking into Particular Account the Needs of Developing Countries, adopted by the General Assembly in 1996 (Resolution 51/122 of 13 December, 1996) draws some guidelines for fostering cooperation programmes between states with space capabilities and those without. This Resolution, issued at the end of long discussions, is based, fundamentally, on two considerations according to which states are free to determine their cooperation and to choose the most efficient and appropriate means of collaboration in the scientific, technological, and economic fields.

The International Code of Conduct of Outer Space Activities (CoC), adopted by the EU Council in 2008 (European Council Doc. 17175/08, Council Conclusions and Draft Code of Conduct for Outer Space Activities, 17 December 2008, Annex II) and currently underway at the international level, which – although it was born in a different context – can be considered as the development of the basic principles established in the OST, expressly mentions among the freedoms, the access to space: any other freedom in space without access is indeed unfeasible (Marchisio, 2012).

Unlike UNCLOS, which includes 'exploration' and 'exploitation' of the area under part XI, and establishes a separate regime for scientific research, which falls under Article 143, the OST does not provide any definition of the terms 'exploration' and 'use'. As a matter of fact, the concepts of exploration and use in the OST are extensively understood to cover all current civil uses of space. The term 'exploration' is usually intended to relate to an activity with scientific purpose, notwithstanding that the difference between the exploration and research activity may not necessarily be a coincidence. Exploration and scientific research are different activities which may overlap to a limited extent, but exploration is a more general term which may or may not include scientific research (Hobe, 2009).

As far as the uses of outer space are concerned, these include both non-economic and economic; the commercialization of space activities falls into the latter category, also in the light of subsequent practice and the 1996 Declaration on International Cooperation in the Exploration and Use of Outer Space for the

Benefit and in the Interest of all States, Taking into Particular Account the Needs of Developing Countries.

Furthermore, the draft version of the OST as a supplementary means of interpreting the Treaty, confirms this wide interpretation of the meaning of the term, and this opinion is widely supported (Hobe, 2007).

With regard to the activities of exploration and use, it has to be stressed that there is a different meaning of the term 'exploration' under Part IX of UNCLOS, which includes exploitation in the definition of the activities in the area and does not include the regime of marine scientific research, for which a separate regime is established under Article 143 of UNCLOS. In this context, scientific research enjoys an autonomous status in respect of exploration and exploitation of resources, which are regulated by the whole of Part XI, while exploration is connected with commercial interests.

In this regard, in the context of space law, the common benefit clause and its interpretation are subject to debate, which is centred on the issue of 'whether the benefit must be shared in a practical sense, perhaps including technological transfer, or whether the requirements are met simply by the activities being beneficial in a generalized way' (Lyall and Larsen, 2009). The former interpretation discourages space-faring countries and private investors from risking resources in space activities, while the latter, which should be read in the light of the significant impacts and spillover deriving from the application of technology developed and knowledge gathered in the context of space programmes, discriminates against those countries without space capabilities, because of their lack of knowledge and technology to access space.

The first article of the Treaty is closely connected with Article II, which provides for an absolute prohibition of national appropriation of outer space, the Moon, and other celestial bodies by claim of sovereignty, by use, occupation, or by any other means. The appropriation of such areas would not be compatible with the *res communis omnium* nature of these commons and excludes any analogy with the practice put in place in the framework of international law following the discoveries of new lands. Indeed, during the fifteenth and sixteenth centuries, the colonial states acquired extensive territories by occupation, having considered that they did not belong to anybody (*res nullius*), and thus susceptible to appropriation by whomever through the effective control coupled with manifested intention to establish sovereignty.

On the contrary, *res communes* are to be intended as a property for all, not suitable to be the object of private rights, since they identify '*une appartenance à tous en indivision*' ([property] which belongs to all and is indivisible) in order to better guarantee the collective rights (Kiss, 1982). According to Roman law, which extended the state sovereignty *usque ad sidera*, some useful or essential assets for the life of human beings, such as water, air, and the sea, were not suitable for occupation and accordingly everyone was entitled to enjoy them: they could not be appropriated by anyone and consequently nobody could prevent the right of others to utilize them. This entails that impairment of the freedoms of others constitutes the limit of these freedoms (Paliouras, 2014). Outer space, the Moon, and other celestial bodies are common spaces belonging to the international community, so that every state can utilize, but not appropriate them.[5]

The expression 'province of mankind' is not intended to be connected with the territory, but with the exploration and use, so it is referred to rather as the 'responsibility over a territory', a certain competence given to humanity as far as the activities in outer space and on celestial bodies are concerned (Kerrest, 2001). Therefore only humanity should be entitled to regulate exploration and use, and no entity is permitted to do this without the agreement of humanity as a whole, although in international law this does not enjoy a legal status (Kerrest, 2001). This interpretation makes reference to a different and enlarged concept of humanity, which in the law of the sea is vested in the International Seabed Authority (ISBA).

Respect for the environment in regard to space activities is addressed by the OST through Article III, which stipulates their conformity with international law and thus also with international environmental law, and through Article IX, which formulates only very general principles concerning cooperation, mutual assistance, non-harmful interference, non-contamination, and consultation.

It is important to point out that the integrity of the space environment is not the core concern of space law, since the concerns of the space community have always been Earth-oriented. Such anthropocentric orientation has been declining in recent years but persists in some matters (i.e. the case of the Safety Framework for Nuclear Power Applications in Outer Space, adopted in 2009).

It is noteworthy that although the OST was concluded in 1967, Article IX of the Treaty is fully in line with later developments in international environmental law. In particular, Principle 21 of the 1972 Stockholm Declaration on Human Environment, the cornerstone of international environmental law, and Principle 2 of the Rio Declaration on Environment and Development establish that states have a duty to avoid the contamination of areas beyond national jurisdiction, among which lie outer space and the Moon and other celestial bodies. The duty of preventing transboundary environmental damage has become a customary rule of international law. At the time of the Cold War, this principle was essentially connected with military activities in outer space. In particular, nuclear explosions and tests in the atmosphere, outer space, and underwater were already prohibited by the 1963 Non Proliferation Test Ban Treaty (Treaty Banning Nuclear Weapons Tests in the Atmosphere, in Outer Space and Under Water, concluded in Moscow on 5 August 1963 and entered into force on 10 October 1963 – 480 UNTS 43).

In carrying out space activities, by means of the applicability of international law, states should take into consideration the corresponding interests and rights of other states and avoid interference or measures which might threaten the activities of others. This conduct requires states to act with due diligence and high standards of care vis-à-vis the other states and in respect of the global environment, since space activities are per se ultra-hazardous and produce environmental pollution.

Article IX, which does not define the concept of harmful contamination, identifies two risks: forward contamination, the introduction to space and celestial bodies of terrestrial materials or living organisms; and back contamination, which, conversely, is the introduction to Earth or its atmosphere of extraterrestrial agents. Although back contamination has not produced risks so far, both risks are frequently faced through application of the 'Planetary Protection Policy', which materializes and implements the necessary due diligence. The Planetary Protection Policy is a set of technical rules elaborated by a scientific body not involved with political matters. The Committee on Space Research (COSPAR), set up in 1958, formulated the Policy at the beginning of the 1960s and updated it in subsequent years. It provides a high international standard of procedures and accepted guidelines, fully in line with the wording of Article IX of the OST.

11.3.1.1 State practice with regard to the Outer Space Treaty

The legal regime of space law, negotiated within the COPUOS and concluded under the auspices of the United Nations, is based on the five UN treaties on space. As previously stated, beyond the mentioned 1967 OST, subsequent space agreements further developed and integrated the general principles incorporated in earlier agreements.

The legal regime of space set forth by the OST also encompasses the applicability of international law to all space activities (Article III), the peaceful uses of outer space and celestial bodies (partial demilitarization of outer space and total demilitarization of the Moon and other celestial bodies – Article IV), the international responsibility of states for space activities carried out by their national entities, either governmental or non-governmental (Article VI), the international liability of the launching state for damage caused by space objects (Article VII), the registration of space objects (Article VIII), and the protection of the environment (Article IX).

It has to be stressed that principles I–IV stipulated in the OST are generally accepted and have therefore passed into customary international law so that they are binding for all states (Lyall and Larsen, 2009). The large number of ratifications (102 states are party to the Treaty – cf. United Nations Doc. A/AC.105/1045 of 23 April 2013) and the presence of its fundamental principles in all the agreements and documents concerning space law coupled with the practice of states make this Treaty truly successful. Furthermore it has guaranteed the peaceful development of space activities so far.

Nevertheless the prevailing opinion on the matter does not draw a balance between the fundamental principle of international cooperation and the ethical provision enshrined in the Treaty in Article I.3 and the practice. Indeed the regime of freedom intrinsic to the *res communis omnium* is 'inherently favourable towards a laissez-faire approach', which leaves room for pragmatic and opportunistic reasons, since the perception is rather founded on individualism than 'on a community-oriented basis' (Paliouras, 2014).

The concept of use, which includes both commercial and non-commercial purposes of space activity, is frequently the object of ambiguous interpretation, since the *res communis* principle allows the economic exploitation of space to one's advantage to the extent that it does not impair the respective freedoms of others (Back Impallomeni, 1983; Paliouras, 2014).

The logic of the market, especially in connection with the development of the commercialization and privatization of space activities, prevails on the principle of common benefit arising from space activities, stipulated in the OST, according to which benefits should be granted to all countries on the basis of equal opportunity. The materialization of this principle should be found in the fact that the space-faring countries should help those states lacking space capabilities to participate in space activities for the sharing of derived benefits. This principle is usually deemed to have been implemented through the participation of developing countries in international organizations and in bilateral and multilateral agreements. Scientific research and observation of the Earth in areas of natural disasters are good examples of such cooperation.

However, 'affirmer que les Etats coopèrent dans le domaine des activités spatiales pour satisfaire le prescrit de l'article 1ᵉʳ, 3ᵉ alinéa, du Traité de 1967 serait faire prévue d'un optimisme juridique exagéré. Les Etats coopèrent tout simplement parce qu'ils trouvent leur compte' (saying that states cooperate in the area of space activities to meet the obligations of Article 1(3) of the 1967 Treaty would reflect a sense of excessive legal optimism. Quite simply, states cooperate when they find it convenient to do so) (Mayence, 2008).

With the aim of clarifying the rationale of the common benefit clause, the General Assembly adopted, in 1996, the Declaration on International Cooperation in the Exploration and Use of Outer Space for the Benefit and in the Interest of all States, Taking into Particular Account the Needs of Developing Countries. According to these recommendations, states are free to determine all aspects of their participation in international cooperation, exploration, and use of outer space on an equitable and mutually acceptable basis. Furthermore, international cooperation should be conducted in ways that are considered most effective and appropriate by the concerned states, taking into special consideration the needs of countries at all levels of development for technical assistance and rational and efficient allocation of financial and technical resources. However the Declaration cannot go any further towards imposing mandatory conditions.

A meaningful example of the current situation concerns data generated by the remote sensing activities of states, which follows the dynamics of a profit-oriented sector. The commercialization and privatization of some international organizations, such as Inmarsat and Intelsat, paved the way to the understanding that the concept of common benefits has to be interpreted in the light of the commercialization and privatization process, which serves the market economy. Thus outer space has become a resource available for exploitation by all who have the capacity to do so, and the advantage of the space-faring countries has been identified as benefiting all other countries. It follows that the benefits for those states lacking space capabilities can only arise from international cooperation, 'without any obligation upon the space-faring countries to enter *into* agreements' (Hobe, 2009).

The Sub-Commission on the Ethics of Outer Space, set up by the UNESCO World Commission on the Ethics of Scientific Knowledge and Technology (COM-EST), formulated some recommendations on these matters and, inter alia, highlighted the role of ethical principles in space law which 'must come before law and not the reverse', although 'contrary opinions must be taken into account' (COM-EST, 2003). As previously stressed, the ethical principles (freedom of use, common benefits, etc.) are legal principles and they were included within space treaties as a counterpart to space freedoms.

11.3.2 The common heritage of mankind: the deep seabed and the Moon

The *res communis omnium* is considered to be the foundation of the concept of the common heritage of mankind (CHM). Strictly speaking, leaving aside the freedom which characterizes the *res communis omnium*, the fundamental feature which distinguishes the common heritage of mankind from the *res communis* is the obligation to share the benefits (equitable sharing) derived from the exploitation of the natural resources found beyond the limits of national jurisdiction.

Paliouras submitted that at the basis of the common heritage of mankind there would be a *res communis humanitatis* principle (Paliouras, 2014). Under this theoretical framework, common property requires common management, so that the exploitation of resources and deriving benefits needs to be distributed 'equitably' among states. It follows that the common heritage of mankind can be accomplished only through the establishment of a conventional organization for the exploitation of resources and distribution of the arising benefits, which, in the case of the law of the sea, is structured in the model of the ISBA. Thus the Moon

Agreement 'converted' the Moon and celestial bodies from the '*res communis omnium* into *res communis humanitatis*' (Paliouras, 2014).

Besides the Moon landing of the Apollo 11 mission in 1969, which raised expectations of exploration and use of the Moon, other factors led to a re-examination of the traditional principles included in the *res communis omnium* and elaboration of a common heritage of mankind in space law.

Technological evolution, in particular, made exploitation of the natural resources on the Moon potentially feasible (Baslar, 1998). Additionally, since the *res communis omnium* de facto allows for application of the first-come-first-served principle and access to limited resources is not endless, this has induced a better implementation of the principles of justice and equity stipulated by Article I.1 of the OST. Under the influence of theories of the new international economic order (NEO), the debate has been focused on balancing opportunities among states and respecting their individual interests (Baslar, 1998).

In its first formulation, the concept of the common heritage of mankind was included in the Declaration on Principles Governing the Sea-bed and Ocean Floor, and the Subsoil Thereof, beyond the Limits of the National Jurisdiction (United Nations Resolution 2749 (XXV) of 17 December 1970 – A/RES/2749), and was later incorporated in Article 136 and Annex VII of the 1982 UNCLOS. Detailed provisions on the basic elements of the common heritage of mankind are included in Part XI of the same convention, which was later negotiated and adopted by the UN General Assembly on 17 August 1994. This part has to be interpreted by Parties to the OST in light of the implementing agreement, finalized in New York in 1994. The seabed and the ocean floor and the subsoil beyond the limits of national jurisdiction are declared the common heritage of mankind. Article 136 enunciates that the area and its resources are the common heritage of mankind and that all rights over resources are vested with the ISBA, the organization which acts on behalf of all mankind.

Contextually to the long negotiations of the 1982 UNCLOS, the notion of the common heritage of mankind was put forward by Ambassador Cocca (Argentina) in the context of space law. The Agreement Governing the Activities of States on the Moon and Other Celestial Bodies, usually identified as the Moon Agreement, was indeed concluded on 18 December 1979, just a few years before the UNCLOS, when the concept of the common heritage of mankind was still debated in the framework of the law of the sea. The Moon Agreement entered into force on 11 July 1984.[6]

The Moon Agreement fundamentally reiterates the principles affirmed in the OST, all but Article 11. Article 1.1 widens the scope of the Moon Agreement to the other celestial bodies within the solar system, other than the Earth (Art. 1.1), except insofar as specific legal norms enter into force with respect to any of these planets. Reference to the Moon includes orbits around or other trajectories around it. However the Agreement fails to draw a definition of the Moon and celestial bodies.

According to Article 2, the activities carried out on the Moon, including the exploration and use, are free and governed in compliance with international law and the Charter of the United Nations. The formulation of this article is clearly shaped on the basis of Article III of the OST, since its primary aim is to affirm that space activities on the Moon do not fall in a *vacuum juris*, but within the scope of international law and its developments and the Charter of the United Nations as a part of international law (Hobe *et al.*, 2013).

Similarly Article 4, which is formulated as an ethical principle, echoes Article I of the OST and establishes that the exploitation and use of the Moon shall be the province of mankind, and that these activities shall be conducted for the benefit of and in the interests of all countries and present and future generations, and to promote higher standards of living, in accordance with the United Nations Charter. These expressions are connected and must be interpreted in the light of the common heritage principle affirmed in Article 11. The principles of non-appropriation and non-occupation, enshrined in Article 11.2, are essential for implementation of the notion of the common heritage of mankind, along with the destination only for peaceful purposes and demilitarization of the Moon and other celestial bodies, provided for by Article 3, taking into consideration preservation of the environment.

Article 6 is devoted to the freedom of scientific investigation. Again this provision evokes Article 1 with respect to its ethical formulation ('[t]here shall be freedom of scientific investigation on the Moon by all States Parties without discrimination of any kind, on the basis of equality and in accordance with international law'), but it widens the scope of the Agreement beside the general freedoms of exploration and use provided for by the OST. Only in the context of scientific research is it possible to remove minerals from the lunar soil in the limited quantities needed for this aim (second paragraph). It follows that states which have removed samples have the duty to make them available to the scientific community for investigations.

The ban on the use of natural resources *in situ* ('in place') (Art. 11.3) is instead connected with the non-appropriation principle provided by Article 11.2, which prohibits the national appropriation of the Moon by any claim of sovereignty, by means of use or occupation or any other means. The resources which are *in situ*, that is, located in their natural status in the surface or subsurface of the Moon, are not susceptible to appropriation until they are incorporated in the soil or subsoil, so they fall within the scope of Article 11.3. Their removal should instead make them usable and subject to the regime dictated by Article 11.7 of the Moon Agreement (common heritage of mankind).

In this regard, the Moon Agreement pays special attention to the appropriation of natural resources and to the possible acquisition of property rights over these goods. The exploitation of mineral resources turns around this fundamental aspect. Commentators follow a different line of reasoning to demonstrate the commercial exploitation of natural resources of the Moon, as encompassed or excluded in the concept of use in compliance with space treaties.

In this vein, Paliouras (2014) distinguishes between the exhaustible (e.g. minerals) and the inexhaustible (e.g. cosmic rays) resources, so that the latter case 'could hardly be prohibited'. He then denies any interpretation which 'would render any appropriation of space natural resources impermissible' since it is based on a wrong assumption, which confounds 'the exercise of territorial sovereignty over a given area from the sovereign right to exploit the natural resources'.

As previously mentioned, if the general provision of the OST (Art. II – common benefit clause) and subsequent practice allow the use for the benefit of mankind, the subsequent practice recorded so far does not help to clarify the question at stake, since it only concerns the commercial exploitation of orbits and frequencies of outer space. The provision of Article II is reiterated in the Moon Agreement (Art. 4), which adds that the uses should be taking place with due regard to inter-generational interest (the interests of present and future generations), 'as well as to promote higher standard of living and condition of economic and social progress and development' in the light of Article 11.7, which imposes the equitable sharing of resources through the organizational mechanism to be defined (Art. 11.5).

According to the general rules for the interpretation of treaties, stipulated by the 1969 Vienna Convention on the Law of the Treaties (Art. 31.3, 1135 UNTS 331, 1980), each expression has to be connected with the other provisions of the same treaty.[7] The Moon Agreement enunciates that 'the Moon and its natural resources are the common heritage of mankind, which finds its expression in the provisions of this Agreement, in particular in paragraph 5 of this article'. The reference to paragraph 5 is intended to highlight that any interpretation of the common heritage of mankind must be in the light of the principles as embedded in the Moon Agreement.

Nevertheless, the concept of the common heritage of mankind as adopted in the context of the law of the sea constitutes a helpful model in international law, a relevant antecedent, and a basis for comparison for space law, since both concepts are related to the management of natural resources lying beyond national jurisdiction and were conceived under the influence and in the context of NEO. Unlike the UNCLOS, the Moon Agreement does not provide a legal framework to govern the management of natural resources of the Moon but mentions only the purpose of an international regime, to be established, but its set-up postponed 'as such exploit-ation is about to become feasible' (Art. 11.5). This prospective development remains unsettled given the scarcity of ratifications of the Agreement, especially by the countries who have the potential capability to exploit the Moon.[8] Adoption of the Agreement by the General Assembly through consensus reflects the opinion of states on the general features of the common heritage regime of exploitation of lunar resources at the time.

The purpose of the international regime, outlined in the Treaty, is for the orderly and safe development of the natural resources of the Moon (Art. 11.7a); their rational management (Art. 11.7b); the expansion of opportunities in the use of these resources (Art. 11.7c); and an equitable sharing by all state parties of the benefits deriving from these resources 'whereby the interests and needs of the developing countries, as well as the efforts of those countries which have contrib-uted either directly or indirectly to the exploration of the Moon, shall be given special consideration' (Art. 11.7d).

With reference to common and essential elements which can be deduced as a general standard for the exploitation of natural resources according to the common heritage approach, Kiss recalls that on different occasions the notion of trust has been evoked to indicate a legal mechanism devoted to the management and conservation of goods (Kiss, 1982). Therefore, humanity should be considered as the trustee, namely, the depositary of the natural resources of the whole planet, with an obligation to transfer them as completely as possible to future generations. The inter-generational preservation of the environment, indirectly dealt with by Article IX of the OST, is also stressed by the Moon Agreement in a specific provision (Art. 7.1).

In this context, the rational use of resources makes reference to the equitable sharing of the benefits, and that imposes special consideration of the needs of developing countries. According to the common heritage of mankind principle, the owner of the resources should be considered mankind as a whole, entitled by the international community to authorize the exploitation and appropriation of the common resources.

In the context of the law of the sea, the ISBA has been conceived as an international organization, acting on the behalf of all states, and accordingly, it is intended to represent humanity as a whole. The regime for the exploitation of the seabed beyond national jurisdictions is based on a 'parallel system', according to which it can be exploited both by the Enterprise and by commercial operators. The Enterprise acts as the operational body of the ISBA and has the legal capacity provided for by statute as envisaged by Annex IV to the 1994 Agreement to carry out direct activity in the area (Art. 170 of UNCLOS).

The common heritage of mankind principle contained in the Moon Agreement, which at the time of its inclusion was very much debated, is again at stake because of the manned return to the Moon planned mostly by space agencies (USA, Russia, Canada, China, India, Japan, ESA) to conduct scientific investigation and to test the potentiality of the natural resources, which actually are limited. It deals especially with oxygen and water which could potentially be utilized in a space colony and helium-3. Thus the feasibility of the exploitation of the Moon does not seem to be an imminent prospect.

11.3.2.1 The Moon Agreement and the debate on the exploitation of lunar resources

Leaving aside Article 11, most of the principles enunciated in the Moon Agreement are basically a re-elaborated version of the provisions of the OST, so that they are generally accepted (Arts. I–IV of the OST have passed into customary law). Article 11 constitutes at the same time, novelty and evolution with respect to the previous space treaties, but also the most controversial point.

A huge legal literature has developed over the years, especially in recent times, which analyses the controversial points from different perspectives, taking into consideration the conventional framework provided by the UNCLOS as the fundamental model that realizes the common heritage principle.

Space law is experiencing opposition to the Agreement by states, especially industrialized countries, which are supposed to have the potentiality to finance exploitation of the Moon, and whose efforts should also be recognized in the light of the text of the Agreement: 'the interests and needs of the developing countries, as well as the efforts of those countries which have contributed either directly or indirectly to the exploration of the Moon, shall be given special consideration' (Art. 11.7d).

Again this situation evokes the adoption of the UNCLOS, which was originally opposed by the most advanced countries because of the inclusion of the common heritage of mankind principle. To overcome this deadlock, the Implementing Agreement in Part XI relating to the area had to be negotiated with the aim of mitigating the obligations incumbent on developed states, stipulating a 'rapprochement' to the free market rules.

During a session of the New Delhi Conference in 2002, the International Law Association discussed the possibility of either improving the Moon Agreement or discarding it. Reference to the law of the sea was inevitable. But the process of amending multilateral treaties appears to be equally as difficult as the process of formulating a new treaty. The suggestion of the General Rapporteur, von der Dunk, to replace the concept of common heritage of mankind with 'common concern of mankind' was not retained. The Rapporteur suggested to 'save the

Agreement by introducing amendments', but other scholars thought it appropriate to renegotiate a new Agreement, with the aim of reaching a new consensus among states (von der Dunk, 2002). The debate was particularly heated and some opinions diverged greatly.

In 2008, a Joint Statement on the Benefits of Adherence to the Agreement Governing the Activities of States on the Moon and Other Celestial Bodies (United Nations Doc. A/AC.105/C.2/L.272) was issued by seven states parties to the Agreement. The Statement brings to light inter alia that: '[t]he Moon Agreement does not preclude any modality of exploitation, by public or private entities, or prohibit the commercialization of such resources, provided that such exploitation is compatible with the principle of a common heritage of mankind'.

The ownership of extraterrestrial property is indeed not relevant to the profitability of extraction of resources. Some authors shift the focus towards the uses of extraterrestrial resources, rather than on the claim of property ownership. Thus the focus is on the profitability of the resources even in the absence of property rights. Various examples concerning private entities extracting resources from property that does not belong to them are taken into consideration (e.g. grazing leases on public land, offshore oil platforms, logging rights, etc.) (Setsuko, 2004; Tennen, 2004).

It is worth noting that the law of the high seas allows the freedom of all states, including nationals subject to the state's jurisdiction, to exploit the marine resources located in the water column beyond national jurisdiction. It follows that this freedom corresponds with a freedom of exploitation.

In terms of the exploitation of natural resources, the Moon Agreement only bans the appropriation of resources *in situ*, which are incorporated in the soil or subsoil of the Moon in their natural status. For this reason, the acquisition of the valid legal title of property after their extraction from the soil is generally deemed admissible.

Conversely, there is no doubt that state responsibility for activities carried out by its nationals (private individuals and legal persons, governmental and non-governmental entities), as provided for by Article VI of the OST, guarantees compliance with the OST. Any activity must be conducted under the authorization and continuing supervision of the 'appropriate' state, which should enact national legislation. However, in the absence of territorial sovereignty, domestic law could not be applicable and indeed it is applicable only to space objects by means of registration. As a consequence, the state of registry maintains jurisdiction and control over the registered space object (OST, Art. VIII).

Furthermore, it is important to also stress that any possible solution must be in compliance with the legal principles and criteria of sustainable development and the environmental approach of international law, indicated by the 1972 Stockholm Declaration and 1992 Rio Declaration on Environment and Development (Ferrajolo, 2011; Tronchetti, 2010). An environmental approach is necessary to carry out any activity on the Moon in conformity with international environmental law and Article IX of the OST.

In the context of the law of the sea, the environmental approach has been reaffirmed by the International Tribunal for the Law of the Sea in case no. 17 'Responsibilities and obligations of States sponsoring persons and entities with respect to activities in the Area' (request for Advisory Opinion submitted to the Seabed Disputes Chamber, www.itlos.org/index.php?id=109&L=0). Although it deals with an advisory opinion with limited legal value, this opinion provides interesting indications on the functioning of the CHM principle in the area (Andreone, 2012). The opinion of the tribunal has a paradigmatic value and paves the way for application of the UNCLOS, in particular, its Part XI and the 1994 Implementing Agreement concerning the application of Part XI in the area (Andreone, 2012). The tribunal identifies specific states' obligations concerning the exploitation of the resources of the area.

Since the Moon Agreement is only in force among a few states with limited space capabilities, the OST, which does not regulate the prospective exploitation of lunar resources, binds the large majority of countries. In the absence of an ad hoc international regime for the exploitation of natural resources, the OST remains the Treaty in force among more than one hundred countries and the extraction of minerals would be regulated on the basis of the first-come-first-served rule. Without new developments, the prospective exploitation of the lunar soil and subsoil will likely not be regulated in accordance with the common heritage approach.

11.4 Antarctica

The legal regime of Antarctica, the groundwork for which was established by the Treaty of Washington (1 December, 1959), was negotiated during the years 1958 and 1959. It provides interesting analogies and solutions, especially in regard to questions raised pertaining to the conduct of human activities on the Moon and other celestial bodies. The law of Antarctica has also influenced the space regime, since Antarctica deals with terra firma, solid ground, as is the case for the Moon and planets.

The status of Antarctica, governed by the 1959 Treaty of Washington, freezes the claims of certain states over the territory, which is 'a part of the international domain without any territorial sovereign' (Peterson, 1997). This provides the rationale for avoiding territorial claims and provides an alternative to territorial sovereignty.

In 1983, the countries claiming parts of Antarctica began negotiations for finalizing a Convention on the Regulation of Antarctic Mineral Resources (CRAMRA) with the aim of establishing a mineral regime. This Convention, concluded in 1988, has never entered into force. It was replaced by a Protocol on Environmental Protection to the Antarctic Treaty (Madrid Protocol), which entered into force in January 1991 and which bans all mining activity in Antarctica until 2048. It has proclaimed the Antarctic continent as a nature reserve, devoted to peace and science.

The harsh environment, the unexplored territory and its resources, and the difficulties in defending the region from a military perspective are conditions shared by both Antarctica and outer space. Because of the coexistence of the non-appropriation principle and the control retained over humans and installations, along with the freedoms of exploration and scientific research not subject to interference by other states, the demilitarization and prohibition of nuclear explosions can be evoked as a useful model. This is confirmed by the opinion, expressed by President Eisenhower of the United States in 1960, that the Treaty of Washington should be used as a model for the new space regime (Lyall and Larsen, 2009).

Most of the legal solutions provided for Antarctica were indeed utilized as analogical antecedents to the Moon Agreement. Some key elements of the Antarctic Treaty, such as peaceful cooperation for scientific research and environmental protection, have inspired the legal regimes of the oceans and outer space (Tuerk, 2004). The analogy with Antarctica has allowed the superpowers to avoid transfer of the arms competition of the Cold War to the Moon by providing an example of the non-militarization of celestial bodies.

The use of the analogy with Antarctica in the case under consideration has been described as such: 'Though neither analogy was mentioned explicitly in the UN debates, the superpowers consensus on military activity was consistent with mixing the available analogies to generate a set of rules congruent with their current perceptions of interest. By creating the vacuum of space like the high seas and celestial bodies like Antarctica, the superpowers could secure both the freedom to pursue military uses of near-Earth space and the non-extension of arms competition to celestial bodies that they described' (Peterson, 1997).

Furthermore, the solution to retaining ownership of and jurisdiction over the stations and equipment on the Moon and over all persons sheds light on the parallels that emerge from the texts of both agreements, including the mutual inspection system. Besides the provisions that ruled an absolute ban on placing objects carrying nuclear weapons or weapons of mass destruction into the Moon's orbit (Art. 3) and the installation of military bases and fortifications, the Moon Agreement reaffirms the freedom of exploration and use (Art. 4 and Art. 5), and stresses the freedom of scientific investigation (Art. 6), while Article 9 allows for the establishment of manned and unmanned stations on the Moon, provided that they do not impede the free access of personnel, vehicles, and equipment of other states to all areas of the Moon and that these installations shall be open to other states parties to the Agreement, upon prior notification (Art. 12).

11.5 Towards a long-term sustainability of space activities

The Moon Agreement marked the ending of the 'law making' phase of the COPUOS and opened a second phase, which deals with non-binding legal rules. It deals with a set of declarations of principles on specific matters, such as the 1982 Principles Governing the Use by States of Artificial Earth Satellites for International Direct Television Broadcasting; the 1986 Principles Relating to Remote Sensing of the Earth from Outer Space; the 1992 Principles Relevant to the Use of Nuclear Power Sources in Outer Space; and the 1996 Declaration on International Cooperation in the Exploration and Use of Outer Space for the Benefit and in the Interest of All States, Taking into Particular Account the Needs of Developing Countries.

The current phase can be identified as the third, because the COPUOS contribution to space law is focused on removing obstacles that hamper the implementation of space treaties and the debate on many pressing issues. Most of these fall within the domain of the sustainability of space activities, one of the major and most urgent issues at stake (United Nations Doc. A/AC.105/C.1/2012/CPR.1). It implies an environmental approach to space activities, which are sources of consistent pollution either on the Earth and/or within specific orbits.

As far as the current development of technology is concerned, space law has not had to deal with the dangers connected with living resources – which to date, have not yet been discovered – but rather with issues concerning space debris and, more specifically, their mitigation and removal. Space debris can be very dangerous to space navigation in the event of collisions with other space objects, and a new approach to the issue requires not only new technologies and concepts but also solution of a number of legal issues concerning, among others, the definition of space debris, which is connected with the definition of space objects for the purposes of removal; liability, jurisdiction, and control over space debris; national space legislation for facilitating active debris removal; and, last but not least, the establishment of mechanisms and procedures to settle disputes. This issue is one of the three pillars for the sustainability of outer space activities, which includes the technical set of rules adopted for space debris by the Interagency Debris Committee in 2002, the International Code of Conduct (CoC) aimed at ensuring space sustainability, safety and security, and the measures recommended by the Group of Governmental Experts on Transparency and Confidence Building Measures in Outer Space Activities (GGE).

The chapter concerning the sustainability of space activities, debated within the COPUOS, which has set up an ad hoc working group, is not limited to space debris issues, but also covers space weather, space traffic management, and other issues. That being said, the sustainable approach of the COPUOS still remains limited in terms of applicability to the Earth-oriented approach

11.6 Conclusions

At its inception, space law was influenced by the concern of the international community about the eventual militarization and weaponization of outer space and by the urgency to avoid activities in this context being undertaken in a *vacuum juris*. Thus space law has looked at other areas of international law, in particular, the law of the sea as a reproducible model with respect to many solutions, and through the absence of state sovereignty in space, and shaping the freedoms of exploration and use (although marked by functional limits).

The end of the Cold War rekindled fears that the scale of investment in space activities would have dwindled in the absence of an incentive for private capital to replace the role of states. Although it is assumed that during the law making phase, space activities were largely due to the states' initiatives – and indeed the drafters of space treaties saw little immediate need to focus attention on regulating private activities – these were envisaged by Article VI of the OST, which charges states with international responsibility for the activities carried out in outer space, including the Moon and other celestial bodies, by private, both natural and juridical persons, governmental agencies or non-governmental entities. As a part of this responsibility, Article VI specifies that private activities require authorization and continuing supervision by the 'appropriate' state and therefore this entails states providing an adequate legal framework.

Article VI of the OST constitutes the only provision dealing with the private sector, but it is a starting point for the development of national space legislation, which currently engages most of states' efforts. Private space sectors will be essential to the future colonization and exploitation of the Moon and celestial bodies, but a clear legal framework governing the activities of exploitation and mining of lunar resources and the connected regime, concern for protecting and preserving the environment, and dispute resolution are fundamental *conditio sine qua non* to allow the future development of space activities on the Moon and Mars, which are the new frontiers of space missions.

Again space law looks at the analogical antecedents, such as the deep seabed, lying beyond the limits of national jurisdiction and the setting up of the ISBA, which realizes the common heritage of humankind. Similarly to the situation produced before introduction of the amendments provided by Part XI of the UNCLOS and by the Implementing Agreement, space law is in deadlock, since the Moon Agreement draws balances which satisfied the ambitions of the developing countries in the 1980s, but which fail to realize the economic return on the important investments needed for exploitation of the Moon's resources. However the scarcity of ratification, even by developing countries, shows that the Moon Agreement is unsatisfactory for all countries, both developed and developing.

If it is not clear what the future will be for this Agreement, given the number of questions it raises (Hobe *et al.*, 2013), including which states should be the negotiators, surely a prospective international regime governing the exploitation of natural resources is needed. The Joint Statement on the benefits of adherence to the Moon Agreement by States Parties to the Agreement (United Nations Doc. A/AC.105/C.2/L.272 of 3 April, 2008), previously mentioned, is particularly meaningful in this regard.

The law of the sea shows that a compromise between opposite interests is possible, although this does not necessarily entail adoption of the same models established under the UNCLOS.

In the meantime, the OST remains the legal framework of reference, the applicability of which has so far been limited to the use of outer space and has not served for the exploitation of natural resources on the celestial bodies. The commercial exploitation of limited natural resources on the Moon and other celestial bodies should be governed according to a prospective legal framework which should prevent the 'first-come-first-served' principle and be able to at least implement the benefits established by Article I of the OST.

If implementation of the legal framework envisaged by the Moon Agreement does not fulfil the expectations of the majority of the countries, it deals with a known dynamic, which opposes developed and developing countries accessing and sharing the benefits arising from limited resources. UNCLOS provides an invaluable example in this respect.

The solution must be sought within the context of space law. If it is clear that each field of law presents its own peculiarities, the various parts of international law influence each other, and this does not necessarily imply an automatic acquisition, in a given part of international law, of the antecedents elaborated in another.

At the end of its work on the fragmentation of international law, the International Law Commission reached the conclusion that any *lex specialis* or self-contained regime cannot operate in absolute autonomy with no connections with other special regimes and separated from the general international law.[9]

References

Andreone, G. (2012). *Chronique de la jurisprudence, Annuaire du Droit de la Mer*: pp. 689–705.

Back Impallomeni, E. (1983). *Spazio Cosmico e Corpi Celesti nell'Ordinamento Internazionale*. Padova: CEDAM.

Baslar, K. (1998). *The Concept of the Common Heritage of Mankind in International Law*. The Hague/Boston/London: Nijhoff.

COMEST (2003). Third session, 1–3 December Rio de Janeiro. Proceedings: 21. http://unesdoc.unesco.org/images/0013/001343/134391e.pdf

Del Ville, P. (2009). Réflexions sur le principe de non appropriation de l'espace extramosphérique et des corps célestes. *Revue française de Droit Aérien et Spatial* **250**(2): 137–160.

DeSaussure, H. (1992). The freedom of outer space and their maritime antecedents. In: *Jasentuljiana, Space Law: Development and Scope*. Westport, CT: Praeger.

von der Dunk, F. (2002). 70th Conference of the International Law Association, Space Law Committee New Delhi.

von der Dunk, F., and Brus, M. (eds.) (2006). *The International Space Station: Commercial Utilisation from a European Legal Perspective*. Leiden/Boston: Nijhoff.

Ferrajolo, O. (2011). Il Trattato 'incompiuto'. L'Accordo sulla Luna del 1979 e le altre norme internazionali rilevanti per l'uso delle risorse naturali nello spazio esterno. In *Studi in Onore di Claudio Zanghì, Vol. IV, Diritto dello Spazio e Miscellanea*. Torino: Giappichelli: 51–66.

Hobe, S. (2007). Adequacy of the current legal and regulatory framework relating to the extraction and appropriation of natural resources in outer space. *Journal of Space Law* **32**: 115–130.

Hobe, S. (2009). Art. I, OST. In: *Cologne Commentary on Space Law (CoCoSL)* Vol. I, Hobe, S., Schmidt-Tedd, B. and Schrogl, K.-U. (eds.). Cologne: Carl Heymanns Verlag: 25–43.

Hobe, S. (2013). Moon Agreement Future Perspectives. In: *Cologne Commentary on Space Law (CoCoSL)* Vol. II, Hobe, S., Schmidt-Tedd, B., and Schrogl, K.-U. (eds.). Cologne: Carl Heymanns Verlag: 423–426.

Kerrest, A. (2001). Outer Space: Res Communis, Common Heritage or Common Province of Mankind? Notes for a lecture in the Nice 2001 ECSL Summer Course, ECSL: 16–26.

Kerrest, A. (2004a). Exploitation of the resources of the high seas and Antarctica: Lesson for the Moon? Issues in IISL/ECSL Space Law Symposium 2004 on New Developments and the Legal Framework Covering the Exploitation of the Resources of the Moon (United Nations Doc. A/AC.105/C.2/2004/CPR.11, 1 April 2004), 33–38.

Kerrest, A. (2004b). The Evolution of the Basic Principles vis-à-vis the Evolution of Space Activities and Technologies: Frictions and Future. ECSL Summer Course, Terni.

Kiss, A. (1982). La notion de patrimoine commun de l'humanité. In Hague Academy of International Law, *Recueil des Cours de*. vol. 175, II: 99–256.

Kopal, V. (1992). Evolution of the doctrine of space law. In: N. Jasentuliyana (ed.). *Space Law: Development and Scope*. New York: Praeger: 17–32.

Lachs, M. (2010). *The Law of Outer Space – An Experience in Contemporary Law-Making by Manfred Lachs*, reissued on the occasion of the 50th anniversary of the International Institute of Space Law. Masson-Zwaan, T., and Hobe, S. (eds.). Leiden/Boston: Nijhoff Publishers.

Lyall, F., and Larsen, P. (2009). *Space Law: A Treatise*. Aldershot: Ashgate.

Marchisio, S. (2005). The evolutionary stages of the legal subcommittee of the United Nations Committee on the Peaceful Uses of Outer Space (COPUOS). *Journal of Space Law* **31**(1): 219–242.

Marchisio, S. (2006). The Law of Outer Space: How to Define It? 15th ECSL Summer Course on Space Law and Policy, Noordwijk, CDROM.

Marchisio, S. (2009). Art. IX. In: *Cologne Commentary on Space Law (CoCoSL)* Vol. I, Hobe, S., Schmidt-Tedd, B., and Schrogl, K.-U. (eds.). Cologne: Carl Heymanns Verlag: 169–182.

Marchisio, S. (2012). The legal dimension of the sustainability of outer space activities. In: *Outer Space Activities: International Legal Aspects. X Annual All-Russian Scientific Conference on Current Problems of Modern International Law, commemorating Professor Igor P. Blischenko*. Abashidze, A., Zhukov. G., and Solntsev, A. (eds.). 13–14 April 2012, Moscow: PFUR: 157–177.

Mayence, J.-F. (2008). Gouvernance et coopération dans le domaine des activités spatiales à travers quarante années de droit de l'Espace. In: *Droit de l'Espace : Télécommunications, Observation, Navigation, Défense, Exploration*. Brussels: Larcier: 47–67.

Paliouras, Z. A. (2014). The non-appropriation principle: The *grundnorm* of international space law. *Leiden Journal of International Law* **27**: 37–54.

Peterson, M. (1997). The use of analogies in developing outer space law. *International Organization* **51**(2): 254–274.

Setsuko, A. (2004). Commentary on emerging system of property rights in outer space. *UN Workshop on Space Law, Action at the National Level, Proceedings*: 310–316.

Tennen, I. (2004). Commentary on emerging system of property rights. *UN workshop on Space Law, Action at the National Level, Proceedings*: 342–347.

Tronchetti F. (2010). The commercial exploitation of the natural resources of the moon and other celestial bodies: What role for the Moon Agreement? In: *Proceedings of the International Institute of Space Law*. Prague: 614–624.

Tuerk, H. (2004). The negotiation of the Moon Agreement. In: *Space Law Symposium 2009, 30th Anniversary of the "Moon Agreement": Retrospect and Prospect* (available at www.unoosa.org/pdf/pres/lsc2009/symp00.pdf, last visited on 6 November 2014): 491–499.

Notes

1 Marchisio (2006) highlights the context in which space law was born, that is, international law and illustrates why space law cannot be 'submitted to special rules, different to those applicable to the other's States activities and behaviours'.

2 Declaration of the First Meeting of Equatorial Countries, Bogotá, 1976, made by Brazil, Colombia, Congo, Ecuador, Indonesia, Kenya, Uganda, and Zaire.

3 Article 12 of the 1944 Chicago Convention on International Civil aviation states: 'Over the high seas, the rules in force shall be those established under this Convention. Each contracting State undertakes to insure the prosecution of all persons violating the regulations applicable'.

4 Agreement on the Rescue of Astronauts, the Return of Astronauts and the Return of Objects Launched into Outer Space (Rescue Agreement of 22 April, 1968, entered into force on 3 December, 1968 (672 UNTS 119)); 1972 Convention on International Liability for Damage Caused by Space Objects (Liability Convention of 29 March, 1972, entered into force on 1 September, 1972 (961 UNTS 1879)); 1975 Convention on Registration of Objects Launched into Outer Space (Registration Convention of 14 January, 1975, entered into force on 15 September, 1976 (1023 UNTS 15)); and 1979 Agreement Governing the Activities of States on the Moon and Other Celestial Bodies (Moon Agreement of 18 December, 1979, entered into force on 11 July, 1984 (1363 UNTS 3)).

5 With respect to the modes of acquisition of sovereign titles over territory, it is worth noting a distinction between areas where the exercise of effective control (*corpus occupandi*) is physically and legally possible and areas where the sovereignty cannot be acquired by means of occupation (Paliouras, 2014). According to this opinion, the high seas fall in the former category, since these are areas unsusceptible to a minimum degree of effective control and therefore 'have *ipso facto* the status of *res communis omnium*'. Outer space, 'by direct analogy to the high seas', had been *res communis omnium* and not *terra nullius* even before the entry into force of the Outer Space Treaty. Thus, the status of *res communis omnium* of outer space would follow from its nature, not

susceptible to be occupied, and therefore 'to forming a part of any state's territory' (Paliouras, 2014).

6 Lyall and Larsen (2009) remark that in reality the common heritage of mankind 'goes a long way back' and the authors make reference to discussions begun early in the 19th century on the Spitsbergen and Svalbard Archipelago. Furthermore, they also indicate numerous documents taking into consideration the common heritage of mankind in relation to matters other than the high seas and space law.

7 Article 31.1 of the 1969 Vienna Convention on the Law of the Treaties: 'A treaty shall be interpreted in good faith in accordance with the ordinary meaning to be given to the terms of the treaty in their context and in the light of its object and purpose.'

8 In January 2012, only the following states had ratified the Moon Agreement: Australia, Belgium, Lebanon, Mexico, Pakistan, Peru, Kazakhstan, Albania, Algeria, and Andorra.

9 See the works of the International Law Commission (ILC) on the Fragmentation of International Law. See also the Conclusions on the Work of the Study Group on the Fragmentation of International Law: Difficulties Arising from the Diversification and Expansion of International Law, 2006: http://untreaty/un.org/ilc/texts/instruments.

12

Towards sustainable oceans in the 21st century

SALVATORE ARICÒ, GUNNAR KULLENBERG, AND BILIANA CICIN-SAIN

12.1 Introduction

Problems faced by the oceans are interconnected. The state parties to the 1982 United Nations Convention on the Law of the Sea (UNCLOS) declared that they were 'conscious that the problems of ocean space are closely interrelated and need to be considered as a whole' (UNCLOS, 1982, Preamble). Therefore, we should tackle the problems of the oceans in an integrated manner. Opportunities offered by the oceans should equally be interconnected. State parties also recognized 'the historic significance of this Convention as an important contribution to the maintenance of peace, justice and progress for all peoples of the world' (UNCLOS, 1982, Preamble) and that the achievement of its goals[1] would 'contribute to the realization of a just and equitable international economic order which takes into account the interests and needs of mankind as a whole and, in particular, the special interests and needs of developing countries' (UNCLOS, 1982, Preamble). Thus, we should bear in mind that efforts aimed at making a better use of the oceans in the broader framework of UNCLOS should be aimed ultimately at the maintenance of peace and sustainable development in an equitable manner for all humankind.

More than fifteen years after the entering into force of UNCLOS, it is legitimate to ask oneself whether the belief 'that the codification and progressive development of the law of the sea achieved in this Convention' (UNCLOS, 1982, Preamble) has indeed contributed 'to the strengthening of peace, security, cooperation and friendly relations among all nations in conformity with the principles of justice and equal rights and will promote the economic and social advancement of all peoples of the world, in accordance with the Purposes and Principles of the United Nations as set forth in the Charter' (UNCLOS, 1982, Preamble). Our preliminary assessment is: yes, but more could be done and better.

The findings, considerations, and recommendations contained in the ten substantive chapters – Chapters 2 to 11 – in this book deal with achievements and failures in our relation with the oceans, conflicts over space, the degradation of marine biodiversity and ensuing impact on its life-supporting role, diminished ocean resources and inequitable access to them, climate change and related impacts on ocean life, peoples and properties, inequity in the distribution of scientific and technological capacity to study, monitoring and legislating on ocean-related matters, and considerations on whether the framework provided by the current law of the sea regime is still adequate.

New scientific and societal challenges related to the oceans emerge on a regular basis. An example in this regard is provided by current discussions on geo-engineering the climate, an option that is now gaining scientific, policy, and public attention, while also raising important environmental, ethical, social, and political challenges. The deliberate large-scale manipulation of the Earth's climate aimed at mitigating the effects of climate change through, for example, ocean fertilization would have direct consequences on the functioning of the ocean. The assumption is that adding nutrients such as nitrogen or iron into the oceans would increase phytoplankton activity, thus enhancing the ocean's capacity to store CO_2 in the deep sea, a process referred to as the 'biological pump'. It is difficult to assess the impacts of large-scale testing of geo-engineering as these may generate unintended consequences for the Earth system, and this should be taken fully into account in the context of current discussions on policy aspects of geo-engineering. Moreover, associated ethical and legal questions include impacts on the most vulnerable populations, i.e. those already being affected by climate change; responsibility over deciding whether or not geo-engineering techniques should be applied; the potential additional CO_2 emissions induced by the application of geo-engineering on countries that are unwilling to reduce their carbon emissions; and issues related to how to deal with liability and compensation (UNESCO–SCOPE–UNEP, 2011). Furthermore, as illustrated in Chapter 2 in this book, the ocean takes a rather long time to assimilate new CO_2, and because the rate of increase in fossil fuel emissions is increasing, the degree of perturbation is increasing, and the ocean is having a harder time keeping up with the increase in carbon emissions in the atmosphere. The uncertainty in the science prevents us from embarking on hazardous potential solutions to the problems faced by the oceans.

The primacy of the law of the sea is recognized in many forums, as illustrated in Chapters 2 and 9, and partly also in Chapter 10 of this book. At the same time, activities such as routine scientific observations and data collection, not covered by relevant provisions in UNCLOS, are recognized to be of common interest to all countries and to have universal significance (cf. Chapter 8); many provisions under, for example, the Convention on Biological Diversity (CBD) and its Nagoya Protocol on Access to Genetic Resources and the Fair and Equitable Sharing of Benefits Arising from their Utilization to the Convention on Biological Diversity, are relevant and can provide solutions to dealing with unresolved issues, as exemplified by issues related to marine genetic resources in areas beyond national jurisdiction (cf. Chapters 9 and 10); and, more generally, there might be a need to develop new international rules to address new subjects in the context of the current legal order, as evolution by interpretation might not always provide adequate responses to dealing with such new subjects, and evolution by codification might then be required (cf. Chapter 10).

While the debate on policy and legal aspects of how to deal with current and new problems of the oceans, especially in areas beyond national jurisdiction, are pursued in the context of relevant institutions and processes (cf. Chapters 9 and 10, in particular), there are already efforts in place to pilot-test new approaches to dealing with such new problems. In this regard, the Global Environment Facility (GEF) has explicitly included the subject of marine areas beyond national jurisdiction in its portfolio. A GEF Programme on Sustainable Fisheries Management and Biodiversity Conservation in the Areas beyond National Jurisdiction (ABNJ) has been developed, which will benefit from a $50 million investment from the GEF. The Programme foresees a capacity development element aimed at advancing multistakeholder discussions of global and regional level perspectives and experiences on issues related to sustainable management of fisheries, other resource uses, and biodiversity conservation in marine areas beyond national jurisdiction through the assessment of current knowledge and the identification and dissemination of best practices. It is foreseen that the GEF Programme on marine areas beyond national jurisdiction will assist in the identification of recommendations on how to address gaps in knowledge and information, and capacity in areas beyond national jurisdiction (see www.fao.org/cofi/33199-02b19a0956b086b1a64430e7a73205051.pdf for an overview of the Programme).

Voluntary initiatives aimed at promoting collaboration among relevant stakeholders, including scientists, and access to information and data on the status and trends of the oceans as well as participation in decision-making by the public at large, include scientific cooperation programmes, voluntary codes of conduct for scientific activities, and for the commercial application of discoveries and participatory mechanisms, as described in particular in Chapters 6 and 9 of this book.

Another example of such voluntary endeavours is provided by the Ocean Biogeographic Information System (Box 12.1), which stems from the 10-year programme on a Census of Marine Life.

While the developments presented above are promising, including information and data on marine biodiversity free-of-access to all, a sense of urgency remains. There is a critical need to be fully aware and to act in relation to the demonstrated impacts and related effects of human-induced changes in the oceans, especially in relation to climate change and to the biodiversity crisis, as convincingly presented

Box 12.1

The Ocean Biogeographic Information System
(see www.iobis.org/)

The Ocean Biogeographic Information System (OBIS) is the world's largest online system for absorbing, integrating, and accessing data about life in the ocean. It provides open access to information on the diversity, distribution, and abundance of marine species, and aims at assisting decision-makers to sustainably manage the ocean's living resources.

OBIS emanates from the decade-long Census of Marine Life (www.coml.org), and is now integrated into the International Oceanographic Data and Information Exchange (IODE) Programme of UNESCO's Intergovernmental Oceanographic Commission (IOC).

As of June 2014, OBIS provides 38.8 million geo-referenced species observations of 115,000 marine species, from the Poles to the Equator, from the surface of the ocean to the deepest trenches and from bacteria to whales.

However, our knowledge of marine biodiversity still remains very incomplete. Not only one- to two-thirds of all marine species still remain undiscovered, the majority of the marine species that are known are only known from a single observation. The median number of observations of a species in OBIS is 4, and most data are from vertebrates and other large animals (mostly fish, mammals, and birds), while it is known that most of the species are smaller invertebrates (crustacea, worms, molluscs). The coastal waters in the Northern Hemisphere are the best-sampled, but ~95% of the vast midwaters (the open ocean, Earth's largest habitat by volume) is mostly unexplored. In addition, historical records (most data are from post 1960) and long-term time series (>10 years) are scarce.

OBIS contributes to the protection of marine ecosystems by assisting in identifying marine biodiversity hot spots and large-scale ecological patterns. The system is used extensively by the research community and plays a crucial role in providing scientific and technical guidance and data and information for the identification of Ecologically or Biologically Significant Marine Areas under the CBD. Among the areas mentioned, some are renowned for containing 'hidden treasures', such as the Sargasso Sea, the Tonga Archipelago, and key corals sites off the coast of Brazil. This work is conducted in the framework of the UN Strategic Plan for Biodiversity 2011–2020, and in particular Aichi Biodiversity Target 11, to conserve and sustainably manage at least 10% of coastal and marine areas by 2020, as agreed upon by the Conference of the Parties to the CBD in Nagoya in 2010.

OBIS will also contribute to regular assessments under the Intergovernmental Platform on Biodiversity and Ecosystem Services (IPBES), and to the World Ocean Assessment.

in Chapters 3, 4, and 5 of this book. There is also a need to fully realize the economic and societal values of the ecosystem services provided by the oceans (cf. Chapters 2 and 7).

In this regard, a salient, time-relevant, and impactful ocean policy agenda in the years to come appears to be critical to solve the problems of the oceans.

12.2 How well have we been doing? Progress since the 1992 Rio Summit on Sustainable Development

Cicin-Sain *et al.* (2011) have analysed in a comprehensive manner the progress made on ocean commitments from the 1992 Earth Summit and the 2002 World Summit on Sustainable Development.

The authors state that Ecosystem-Based Management/Integrated Coastal Management (EBM/ICM) is now widely accepted internationally and is being applied both at the national as well as regional levels, including through Large Marine Ecosystems (LMEs) projects. Future efforts should focus on improvements in institutional decision-making processes and on enhancing integration of this approach in the UN system, including in relation to areas beyond national jurisdiction. In addition to institutional considerations, effective implementation of EBM/ICM has suffered from lack of data on ecosystem structure and functioning, and from limited funding.

In the context of the 1995 Global Programme of Action for the Protection of the Marine Environment (GPA), efforts to reduce the impacts of land-based activities on the marine environment are being pursued. Over 70 countries have developed National Programmes of Action in support of implementation of the GPA, and regional programmes to mitigate marine pollution have been developed in the context of the Regional Seas Programmes. However, degradation of the marine environment as a consequence of sewage, excess of nutrients, marine litter, and alteration and destruction of physical habitats continues. Progress in the implementation of GPA provisions has been affected by a lack of public education and awareness, limited political will, inadequate financial and human resources, and the fragmentation of legal and institutional arrangements, as well as lack of or limitations in compliance, enforcement, and reporting mechanisms.

In terms of efforts aimed at advancing the protection of biodiversity, including through the designation of Marine Protected Areas (MPAs), the authors note that the global goal of 10% MPA coverage by 2012 is far from being met, while marine biodiversity continues to be lost. Designation and management of MPAs has to be conducted in the broader EMB/ICM framework, and the lack of standardized data and reporting on MPAs should be noted.

While the special circumstances and needs of Small Island Developing States (SIDS) are now widely acknowledged, implementation of the provisions of the 1995 Barbados Plan of Action and the 2005 Mauritius Strategy of Implementation for SIDS are lagging behind, mainly due to lack of institutional and financial support. Human capacity development for SIDS is also required, and concerns related to the impacts of climate change on SIDS needs to be reflected adequately in international commitments related to reducing fossil fuel emissions. Progress has been made in relation to MPA designation and increased efforts in EBM/ICM.

Efforts related to sustainable fisheries and aquaculture have resulted in the development by over 90% of FAO member states of fisheries management plans, while some 80% of them have taken steps to develop or implement plans to address illegal, unregulated, and unreported (IUU) fishing. However, overexploited, fully exploited, and depleted fish stocks have continued to increase. Reducing overcapacity through the International Plan of Action for the Management of Fishing Capacity provides an opportunity for progress in this regard. Implementation of UN General Assembly resolutions on mitigating the impacts of bottom trawling on vulnerable ecosystems remains a major impediment. Future needs include the need to lessen and ultimately eradicate destructive fishing practices to reduce perverse incentives which contribute to overfishing and IUU fishing.

The authors report that addressing critical uncertainties for the management of the marine environment, including from the perspective of climate change, requires further enhancement of systematic observations, including in the context of the Global Ocean Observing System (GOOS), and of systematic data exchange between nations (as provided for by OBIS). Developing countries continue to lack the capacity needed to participate in and benefit from systematic studies and observations of the oceans, and technology transfer, education, and training should be further promoted to overcome this barrier.

In relation to capacity development, major efforts by UN organizations, educational institutions, and multilateral and bilateral donors are ongoing, but the lack of adequate measurement and monitoring of the results attained makes an assessment of the impacts of such efforts difficult. There is a need to integrate capacity development efforts into long-term development strategies. In this regard, it is worth noting the current efforts of IOC in developing a Global Ocean Science Report, a proposal currently under consideration by the IOC member states. The Global Ocean Science Report would provide the type of tracking tool necessary to measure and monitor progress towards capacity development, in particular, the level of investment in ocean science, infrastructure, and human and institutional capacity.

12.3 The Rio+20 UN Conference on Sustainable Development and the ocean agenda in the years to come

The Rio+20 United Nations Conference on Sustainable Development Outcome Document contains a number of references and commitments related to the oceans (United Nations, 2012; Cicin-Sain *et al.*, 2012).

The Rio+20 Outcome Document reiterates the importance of an ecosystem approach to the management of activities impacting the marine environment. It calls for states to take action to reduce the incidence and impacts of marine pollution, including through the implementation of the relevant conventions of the International Maritime Organization (IMO) and through the GPA, and also calls for significant reduction in marine debris, to prevent harm to coastal and marine environments by 2025. It recognizes the importance of area-based measures, including MPAs, to conserve biodiversity and to mainstream the socio-economic benefits of biodiversity, and of the important role played by indigenous and local communities in relation to the conservation and sustainable use of biodiversity.

The Rio+20 Outcome Document contains several commitments related to the need to reduce overfishing and IUU fishing, including intensification of efforts to restore stocks to maximum sustainable yield levels, the need to address by-catch and destructive practices, protection of vulnerable ecosystems, application of environmental impact assessment, and elimination of subsidies in the context of the World Trade Organization (WTO). Capacity is also to be developed and provided for improved monitoring, control, surveillance, compliance, and enforcement systems. Regional Fisheries Management Organizations (RFMOs) were encouraged to undergo independent review to increase transparency and accountability. It also called for strengthening international research on ocean acidification.

In relation to SIDS, the Rio+20 Conference called for a third international conference on SIDS in 2014 and for strengthened support by the UN system to the implementation of the SIDS agenda. A new multi-year strategy and renewed political commitment for SIDS is expected to emerge late this year. The Rio+20 Conference also called for support to improve national capacity to manage and realize the benefits of sustainable resource use, including in relation to improved market access for small-scale, artisanal, and women fishworkers, and indigenous peoples and communities.

The Rio+20 Outcome Document also refers to states' support to the Intergovernmental Science-Policy Platform on Biodiversity and Ecosystem Services (IPBES) and the World Ocean Assessment (the latter, in the context of the Regular Process for Global Reporting and Assessment of the State of the Marine Environment, including Socio-economic Aspects).

The Outcome Document also stresses the need to consider the precautionary approach in ocean fertilization activities.

In relation to areas beyond national jurisdiction, the Rio+20 Conference called for a decision at the 69th session of the UN General Assembly in 2014 on the development of an international instrument under UNCLOS for the conservation and sustainable use of marine biodiversity in areas beyond national jurisdiction.

12.4 The need for education

It is stipulated in Agenda 21, Chapter 36.3 that '[e]ducation is critical for promoting sustainable development and improving capacity of the people to address environment and development issues' (United Nations, 1992). The continued validity of this stipulation is evident from the chapters in this book and is further elucidated in this section.

12.4.1 The coastal area and the EEZs

The coastal zone space is one of the richest zones on Earth. It has been examined as a possible 'prelude to conflict' (Goldberg, 1994). Following the endorsement of the United Nations Conference on Environment and Development (UNCED), in 1992, of integrated coastal zone management as the paradigm for addressing concerns related to coastal areas, the analysis of these concerns calls for an interdisciplinary and system-oriented approach. These concerns include the rising population in coastal areas, which is still ongoing, the tourism and recreation industries, still increasing and a major source of income in many developing and poor areas, as well as a source of conflict with other uses such as transportation, shipping, aquaculture, fishing, energy production, and mining, and the role of the oceans in waste management.

In many regions, pressure on coastal areas has led to large-scale land reclamation with displacement of communities, loss of natural habitat, and serious impacts on biodiversity. Estuaries are severely impacted, as are mangrove forests, deltas, coral reefs, and seagrass beds, which are disappearing in some regions of the world (cf. Chapter 4); this situation, in turn, leads to a severe decrease in available seafood, which may have serious implications for human health and security.

Legal instruments, regulations, and norms for the various uses of the coastal zone have been introduced in many areas, together with requirements for environmental impact assessments. This is coupled with the increasing importance of the coastal zone in national economies (cf. Kullenberg, 2010). However, the implementation and enforcement of such laws and norms are often unsatisfactory. This has been demonstrated by the impacts of natural and man-made disasters such as hurricanes, tsunamis, oil drilling and shipping accidents, and food contamination from uncontrolled waste disposal. Over past decades, conflicts have expanded in many regions of the world due to increasing potential for further offshore oil, gas, food, and mineral exploitation. These concerns, which are exacerbated by the expected impacts of climate change, can lead to nationalistic and political complications and conflicts.

In order to help to avoid devastating conflicts and national political unrest, education and proactive information from communities, populations, and decision-makers is required. Education should aim at enhancing knowledge about the whole of the coastal zone and shelf sea areas, including economic aspects, the inter-dependencies of uses and users, and the economic and social consequences of sound management and of mismanagement of disaster risks. Often communities are well aware of environmental and resource conditions and interdependencies. Their need is education regarding norms, legal instruments, laws, and how to achieve communication with authorities and other communities. This will help enforcement and application of the instruments so as to achieve functioning law and order with proper protection of community interests, including improved equity of distribution and use of resources. In the globalized economy, resources and uses are often controlled outside the local and subregional environment and their communities. This situation needs to be balanced through enhanced local and subregional knowledge about existing rights, laws, norms, and requirements for the sustainability of resources, environment, infrastructure, human life protection, and security. Education needs to include information about procedures that give communities an appropriate voice in decisions over the use of coastal resources, the distribution of incomes, and on industrial and infrastructure developments. The need for related education has been demonstrated repeatedly in connection with disaster risk prevention and reduction. Law and order includes practical matters related to daily life and survival, and these need to be put into the hands of ordinary people through adequate education and skill development.

The coastal areas and the EEZs are multipurpose zones, hence education should also be multipurpose. Users and managers should be made aware of the diversity of marine activities, their interdependencies, and increasingly important roles in global, regional, and national economies, and thus for peace and comprehensive human security. When practised over any extended period of time, it appears that any ocean use – fishing, aquaculture, oil and gas drilling, construction and mineral resources exploitation or mining, energy extraction, desalination, tourism development, transportation and ports, habitats and land reclamation, and many other land-based activities generating pollution and other impacts at the coast – is related to or influenced by all other uses of ocean and coastal spaces due to their interdependencies, environmentally and socially (Mann Borgese, 1998). These realizations are the basis for the Integrated Coastal Management paradigm and the Ecosystem-based Management and Large Marine Ecosystems approaches. Achievement and application of these all require a practically oriented education and skill development involving practitioners and their experiences. Education, including proactive provision of information about associated legal and policy instruments and their applications in practice, should lead to achievement of good governance based on accepted law and order (global conventions, regional agreements, and harmonized national laws).

In this regard, Marine Spatial Planning (MSP) has emerged as a very useful approach and tool to promote awareness and active participation among multiple stakeholders in planning processes related to access to and management of ocean areas and resources therein. About 35 countries are currently developing MSP approaches, and nine already have government-approved plans covering their territorial seas and/or exclusive economic zones (EEZs). Altogether, these plans now cover about 9% of the world's EEZ surface area; within the next 10 years, about a third (low estimate) or half (high estimate) of the EEZs will be covered by a marine spatial plan (IOC, personal comm.).

12.4.2 Areas beyond national jurisdiction

The technological developments during the Second World War laid the foundation for increased abilities to exploit the seabed and ocean resources. This triggered associated extension of national jurisdictions, which in turn led to gradual renewal of the law of the sea, with internationally agreed national jurisdictions to include the EEZs (cf. Kullenberg *et al.*, 2012). Further scientific and technological developments have continued to expand the exploitation of marine resources well beyond national jurisdictions. The discovery in the 1970s of deep-sea vents triggered an increasing interest in the use of deep-sea resources for pharmaceutical and medical research, food resources, fossil fuel, and mineral exploitation purposes (cf. Chapter 9 in this book). These developments go in parallel with enhanced technological capacities and economic interests, which are limited mostly to a few leading nations. This situation has stimulated renewed interest in adequate governance of marine areas beyond national jurisdiction (e.g. Vierros *et al.*, 2012).

There is a strong need for education regarding the impacts of human exploration and exploitation of this under-protected zone of the planet. It should address concerns for food resources, climate regulation, biodiversity, and biogeochemical cycles of fundamental importance for maintaining life on Earth. The biodiversity in these areas faces serious threats from resources exploitation, impacts of climate change, and ocean acidification (Vierros *et al.*, 2012; cf. also Chapters 3 and 4).

However, neither the climate nor the life-support system can be enclosed under national jurisdiction or as national property. The climate change problem, food security, unbalanced resource utilization, distribution of wealth, and lack of equity must also be addressed at global level through global instruments.

Incentives need to be established for governments, populations, and various industrial and economic interests to act in accordance with agreed guidelines, norms, and conventions of global coverage. Many such instruments exist, including UNCLOS and those developed through the 1992 UNCED process, sufficient to take action, although requirements for some supplements have been identified (Gjerde *et al.*, 2008; cf. also Chapter 2). The incentives for implementation and enforcement are evident in the socio-economic necessity to address the survival and livelihood issues identified by the challenges. These include biodiversity loss, threats to habitats and ecosystems, food security, water and energy sustainability, as well as distribution, equity, and space concerns. Practical procedures have been developed through lessons learnt in areas under national jurisdiction (cf. Chapters 2, 6, 8, and 9). However, in order to achieve sufficient consensus for governments to take action, education is needed. Several education, training, and skills development initiatives have been undertaken over the last 50 years through intergovernmental and non-governmental organizations at global and regional level (see e.g. Terashima, 2004; Kullenberg, 2004; Karlin *et al.*, 2010). Essential lessons learnt are the need for persistence in the efforts and the availability of a common framework for these activities in the Law of the Sea Convention (Kullenberg *et al.*, 2012).

12.4.3 The whole ocean

The United Nations Convention on the Law of the Sea (cf. Chapters 2, 9, and 10; in this book, UNCLOS is also seen from the perspective of possible analogies with the space law regime, as presented in Chapter 11) is the most comprehensive instrument available for achieving adequate governance of the open ocean, based on two principles: (i) that 'the problems of ocean space are closely interrelated and should be considered as a whole', and (ii) the common heritage of humankind principle (which only applies to the area). The definition of this notion given by Dr Pardo in 1967 (see e.g. Kullenberg *et al.*, 2012) includes: only peaceful use, freedom of access and use with regulation for the purpose of conserving the heritage and avoiding discrimination of the rights of others, responsibility for misuse, and equitable distribution of benefits from exploitation of the heritage. This notion is thus much different from the notion of common property, which implies right to use (and misuse). The two principles should be taken together, thus providing for integration and comprehensiveness. The Convention promotes an order for peace and comprehensive human security (Mann Borgese, 1998; Kullenberg, 2010). The premise of consensus underlying UNCLOS is that the ocean must be used for the benefit of all and not only in the interests of a few maritime powers (Anand, 2002; Chua *et al.*, 2008). The traditional freedom of the seas must be complemented by a proactive enforcement and possible expansion of the scope of application of these two principles.

Together with other instruments, UNCLOS provides the basis on which to build the required education and capacity of populations regarding the rules and regulations which can help them to register and communicate about their rights to marine resources, the adequate protection of these resources, and their sustainable use and equity of distribution, including the open ocean and the coastal zone. This notion is made operational, inter alia, through the IOC Criteria and Guidelines on the Transfer of Marine Technology (IOC, 2005), which consistently with Part XIV of UNCLOS should enable all parties concerned to benefit, on an equitable basis, from the area of marine sciences.

Education will lay the foundation for sufficient political will to enable implementation and enforcement. Education can provide for information relating to procedures, practical matters, and arguments to achieve a voice in decision-making on uses of the regime beyond national jurisdiction. This can include how to enhance the use and strength of the United Nations system. Education could be lifelong and start in schools. In this process the United Nations Educational, Scientific and Cultural Organization (UNESCO) and the United Nations University (UNU) would seem to have both a mandate to take the initiative and a large responsibility.

12.5 What can be learned from the findings and considerations in this book?

The findings of research and observations, as well as modelled projections, confirm that the world's ocean is under stress. Business as usual cannot continue. This applies, inter alia, to climate change and its adverse effects, the excessive overloads of nutrients due to increased pressure on coastal areas, land reclamation, habitat fragmentation, and overuse of resources. Ocean acidification also deserves a special mention, in the light of the interconnection with non-ocean governance processes, namely the UN Framework Convention on Climate Change (IGBP, SCOR and IOC, 2013). Our governance set-up does not seem adequate to deal with the multiple and often conflicting uses of ocean areas and the resources therein, nor do we apply systematically comprehensive valuation methods when assessing the true value of the oceans for humankind.

The changes being experienced by the oceans are unprecedented, from the perspectives of time and space scales, as exemplified by the fact that a new ocean is forming under our own eyes in the Arctic area of our planet. Much uncertainty does exist in relation to the resilience of the marine environment to human-induced impacts. At the same time, these impacts are already visible and quantifiable and now have direct consequences for human property, health, food security, and sustainability in general.

Understanding the complexity of ecosystems and their capacity to adapt and respond at multiple spatial and temporal scales is a difficult endeavour, and human actions as well as responses, in terms of governance and management, add another dimension of complexity, as humans at multiple scales (from individuals to nations) also respond to change.

The necessity to take into account the interactions among the ocean, the atmosphere and the land, the ecosystems, and human society is evident, as human activities have clear and measurable effects on the oceans, and as, without a healthy ocean, there will be no healthy life on Earth. The ocean system must be treated as a whole, as stipulated in UNCLOS. Application of the ecosystem approach calls for the need to adopt a 'mountain to ocean trench' approach to management of the coasts and seas (as stated in Chapter 6).

Changes induced by global warming and ocean acidification will have direct and indirect consequences on marine ecosystems and human life. Because of the high, not only social but also economic, importance of the services provided by marine habitats and ecosystems, including coastal, open ocean, and deep-sea habitats and ecosystems, the global economy is likely to suffer from the continuous degradation of these areas. In fact, a high level of unpredictability is associated with the effects of climate change on marine ecosystems and their capacity to deliver services important for human well-being. Moreover, the effects of human impacts on the ocean can be mutually reinforcing, and cumulative impacts suggest the need to act very promptly so as to reduce our pressure on the marine environment. To this end, we first need to be capable of assessing the real value brought by the oceans, in a methodologically sound manner and so also to reflect intangible values. Assessing values and benefits and implementing appropriate measures for the conservation and sustainable use of marine biodiversity appear therefore as imperatives to achieving the sustainability of the oceans in this very century.

Marine resources novel to science, namely, genetic resources, have a potential value to humans due to possible applications in the areas of health and industry. The related legal, regime, and policy frameworks may need further elucidation and to further evolve so as to accommodate developments related to the application of findings of research on organisms which are the source of compounds presenting novel properties. Emerging issues associated with the oceans, such as those related to marine genetic resources, indicate that, in addition to questions related to the conservation and sustainable use of marine biodiversity, new equity questions on access to and the sharing of benefits arising from the utilization of resources in marine areas beyond national jurisdiction also arise.

A 'culture of care' should guide our actions (Child, 2009), not just a culture of economic efficiency (cf. Chapter 7 in this book). We have a moral responsibility to ensure inter-generational equity too; this entails a longer-term responsible attitude vis-à-vis the time-scale of management choices, the need to ensure flexibility in exercising intellectual property over new discoveries not only vis-à-vis less scientifically and technologically advanced nations, but also vis-à-vis future generations, and the need to avoid commodification of nature.

Perceptions and approaches do play a role in the way we tackle our relationship with the oceans. A more holistic approach based on an ecosystem approach to ocean areas and resources is needed – one which also recognizes the importance of indigenous and local knowledge. The ecosystem approach has been codified, based on past successful experiences of a harmonious relationship of peoples with the ocean (although its application around the world has been very uneven), and should now be implemented extensively, including through the application of tools which benefit from an increasing recognition in relevant policy forums, such as the tool of marine spatial planning.

The history of the UNCLOS and its provisions and institutions have created a situation whereby '. . . today, the rules set out in the LOS Convention correspond to customary international law, unless (as regards specific provisions) the contrary is proven' (Treves, 2010). At the same time, evolution in the norms is sometimes necessary, hence the challenges to ensure that the legal instruments of which we dispose today can evolve so as to accommodate the new developments with which the oceans are currently faced.

The reality of routine observations, such as in relation to the collection of meteorological information, which are not explicitly reflected in relevant provisions of the UNCLOS, have not resulted in restrictions on conducting observations which tend to be considered to be in the common interest of all countries and to have universal significance, also because they guarantee developing countries' participation in the observational activities on an equitable basis. Quite on the contrary to the very positive collaborative climate faced by international scientific cooperation for the purpose of conducting observations, defining approaches and modalities to access and for sharing the benefits arising out of the exploitation of resources in areas beyond national jurisdiction, including from the point of view of the conduct of related marine scientific research activities, is proving extremely difficult.

This is why the current debate on the adequacy of UNCLOS in the light of the international access and benefit-sharing standards established by other instruments is so alive. Current in-built spatial limitations related mainly to territorial claims of coastal nations seem to hamper the realization of a truly equitable regime to deal with resources in areas beyond national jurisdiction, which requires bridging approaches related to areas within, with those outside, national jurisdiction.

12.6 Towards sustainable oceans in the 21st century

Some tools do exist that further our understanding of the oceans and promote their sustainable management and improve ocean governance.

Freestone (2009) has suggested a set of ten complementary principles, most of which have already been agreed upon at the international level, in order to serve as an integrated modern basis for further developing the system of high seas governance. They include: (1) conditional freedom of the high seas; (2) protection and preservation of the marine environment; (3) international cooperation; (4) science-based approach to management; (5) public availability of information; (6) transparent and open decision-making process; (7) precautionary approach; (8) ecosystem approach; (9) sustainable and equitable use; and (10) responsibility of the states as stewards of the global marine environment.

Ecosystem-based management and site-based action continue to be tested and promoted. Approaches such as the designation of biosphere reserves under UNESCO's Man and the Biosphere Programme have triggered bold initiatives such as the recent pledge by the government of the Maldives to designate the entire country and its exclusive economic zone (EEZ) as a biosphere reserve by 2017. This pledge entails challenging technical work in relation to the feasibility of applying the biosphere reserve status beyond the territory of a state, i.e. not only to the territorial sea; but in itself it clearly indicates that countries are increasingly cognizant and eager to utilize some of the practical tools at our disposal today.

Cicin-Sain *et al.* (2012) point out the following major initiatives to advance the oceans agenda in the near future: the reform and enhancement of UN-Oceans; the Global Partnership for Oceans; the Sustainable Ocean Initiative; the Yeosu Declaration on the Living Ocean and Coast; and the World Marine Assessment.

Capitalizing on the mandates, expertise, and experiences of individual UN specialized agencies, programmes, and funds active in the area of the oceans, and ensuring effective coordination of UN activities on oceans have all been promoted through UN–Oceans, the UN interagency coordination mechanism on ocean matters. Recently the UN General Assembly decided on an evaluation of UN–Oceans and requested UN–Oceans to submit to the General Assembly's attention terms of reference for its work, with a view to reviewing the mandate of UN–Oceans and enhancing transparency in its reporting to member states.

Efforts for effective and synergistic inter-agency coordination by UN organizations involved in ocean affairs in the context of UN–Oceans are complemented by initiatives aimed at enhancing the active participation of countries in the oceans agenda. One of such significant initiatives is the Global Partnership for Oceans, which brings together more than one hundred countries, non-governmental organizations, civil society groups, and the private sector, which are committed to the cause of ensuring healthy oceans for the benefit of peoples (see http://www.globalpartnershipforoceans.org). The Sustainable Ocean Initiative (see http://www.cbd.int/marine/doc/soi-brochure-2012-en.pdf) under the CBD, which promotes partnerships in support of the Aichi Biodiversity Targets under the United Nations Strategic Plan for Biodiversity 2011–2020 (see www.cbd.int/sp/targets/), is an example of another of such initiatives.

The Global Ocean Forum (see www.globaloceans.org/content/about-gof), established in 2001 and involving ocean leaders from over 110 countries, continues to provide a much needed forum for multistakeholder policy analyses and dialogues to advance the global ocean agenda.

The current process on a post-2015 development agenda underpinned by a set of Sustainable Development Goals (SDGs) may culminate in the adoption by governments of a stand-alone SDG on attaining the conservation and sustainable use of marine resources, oceans, and seas. This goal, albeit less ambitious than the one put forward at the beginning of the work of the UN General Assembly Open Working Group on the SDGs, in the second quarter of 2013, which used to refer to healthy, productive, and resilient oceans and seas (and prosperous and resilient peoples and communities), may act as a leverage for strengthening the political and policy enabling framework for action in support of oceans.

The momentum generated by the Rio+20 Conference and other related developments, the urgency of solving the problems of the oceans for the benefit of humankind as a whole in the broader context of UNCLOS, and the fact that we already have at our disposal some scientific and policy tools to solve the problems with which the oceans are faced, suggests that we can, in principle, achieve a sustainable management of the oceans in this very century. Humankind has always depended on the oceans for its subsistence. We cannot afford the continued degradation of the life-support system on which our very existence depends.

In this context, two key processes provide us with a historic opportunity to make our relation with the oceans last: the climate change negotiations, which must urgently reach a successful outcome; and the possible development of an international instrument under UNCLOS to deal with the conservation and sustainable use of marine biodiversity in areas beyond national jurisdiction. We cannot and should not miss this chance.

Acknowledgements

The authors wish to acknowledge the contributions of Maud Borie, Ward Appeltans, and Julian Barbière, who have provided information and inputs to specific sections of this chapter.

References

Anand, R. P. (2002). Non-European sources of Law of the Sea. In: *Proceedings of Pacem in Maribus 2000, XXVIII*, compiled by Wesnigk, J., and Ralston, S. (eds.). Halifax: International Ocean Institute, pp. 9–20.

Child, M. F. (2009). The Thoreau ideal as a unifying thread in the conservation movement. *Conservation Biology* **23**: 241–243.

Chua, T.-E., Kullenberg, G., and Bonga, D. (eds.) (2008). *Securing the Oceans: Essays on Ocean Governance – Global and Regional Perspectives.* Quezon City: GEF–UNDP–IMO Regional Programme on Building Partnerships in Environmental Management for the Seas of East Asia (PEMSEA) and the Nippon Foundation, 770 pp.

Cicin-Sain, B., Balgos, M., Appiott, J., *et al.* (2011). *Oceans at Rio+20: How Well Are We Doing in Meeting the Commitments from the 1992 Earth Summit and the 2002 World Summit on Sustainable Development? Summary for Decision Makers.* Global Ocean Forum, available at www.globaloceans.org/sites/udel.edu.globaloceans/files/Rio20-SummaryReport.pdf, last accessed on 18 January 2014.

Cicin-Sain, B., Balgos, M., Appiott, J., *et al.* (2012). *Global Ocean Forum Activities in Advancing the Rio+20 Outcomes. Summary of a Multi-stakeholder Brainstorming Session on Oceans Post Rio+20, November 12, 2012.* Washington: World Bank, available at www.globaloceans.org/sites/udel.edu.globaloceans/files/SummaryRio20BrainstormingSessionHeldAtTheWorldBank-November12-2012.pdf, last accessed on 18 January 2014.

Freestone, D. (2009). Modern principles of high seas governance: The legal underpinnings. *International Environmental Policy and Law* **39**(1): 44–49.

Gjerde, K. M., Dotinga, H., Hart, S., *et al.* (2008). *Regulatory and Governance Gaps in the International Regime for the Conservation and Sustainable Use of Marine Biodiversity in Areas beyond National Jurisdiction.* Marine Series No.1. Gland: IUCN.

Goldberg, E. D. (1994). *Coastal Zone Space: Prelude to Conflict?* IOC Ocean Forum 1. Paris: UNESCO Publishing, 138 pp.

IGBP, IOC and SCOR (2013). *Ocean Acidification Summary for Policymakers – Third Symposium on the Ocean in a High-CO_2 World.* Stockholm, Sweden: International Geosphere-Biosphere Programme, available at www.igbp.net/download/18.30566fc6142425d6c91140a/1385975160621/OA_spm2-FULL-lorez.pdf, last accessed on 18 June 2014.

Intergovernmental Oceanographic Commission (IOC) (2005). *IOC Criteria and Guidelines on the Transfer of Marine Technology.* Paris: UNESCO, available online at www.jodc.go.jp/info/ioc_doc/INF/139193m.pdf, last accessed on 18 June 2014.

Karlin, L. N., Oliounine, I., and Sychev, V. I. (eds.) (2010). *Proceedings of International Conference: 50 Years of Education and Awareness-Raising for Shaping the Future of the Oceans and Coasts.* St Petersburg: RSHU, 642 pp.

Kullenberg, G. (2004). Marine resources management: Ocean governance and education. In: *Ocean Yearbook No. 18.* Chircop, A., and McConnell, M. (eds.). University of Chicago Press, pp. 578–599.

Kullenberg, G. (2010). Human empowerment: Opportunities from ocean governance. *Ocean and Coastal Management* **53**: 405–420.

Kullenberg, G., Azimi, N., and Otsuka, M. B. (2012). Ocean governance education and training: Perspectives from contributions of the International Ocean Institute. In: *Ocean Yearbook No. 26.* Chircop, A., Coffen-Smout, S., and McConnell, M. (eds.). Leiden: Martinus Nijhoff Publishers, pp. 623–654.

Mann Borgese, E. (1998). *The Oceanic Circle. Governing the Seas as a Global Resource.* UNU Press, 240 pp.

Terashima, H. (2004). The importance of education and capacity-building programs for ocean governance. In: Chircop, A., and McConnell, M. (eds.). Ocean Yearbook, vol. 18, University of Chicago Press, pp. 600–611.

Treves, T. (2010). The development of the law of the sea since the adoption of the UN Convention on the Law of the Sea: Achievements and challenges for the future. In: *Law, Technology and Science for Oceans in Globalisation – IUU Fishing, Oil Pollution, Bioprospecting, Outer Continental Shelf.* Vidas, D. (ed.) Leiden/Boston: Martinus Nijhoff Publishers.

UNESCO–SCOPE–UNEP 2011. *Engineering the Climate. Research Questions and Policy Implications. UNESCO–SCOPE–UNEP Policy Briefs Series No. 14. Paris: UNESCO.*

United Nations Convention on the Law of the Sea – Text and Annexes (1982). 1833 UNTS 3.

United Nations (1992). *The United Nations Programme of Action from Rio.* New York: United Nations, 294 pp.

United Nations (2012). UN General Assembly Resolution 66/288 'The Future We Want', available at http://daccess-dds-ny.un.org/doc/UNDOC/GEN/N11/476/10/PDF/N1147610.pdf?OpenElement, last accessed on 18 June 2014.

Vierros, M., McDonald, A., and Aricò, S. (2012). Oceans and sustainability: The governance of marine areas beyond national jurisdiction. In: *Green Economy and Good Governance for Sustainable Development: Opportunities, Promises and Concerns.* Puppim de Oliveira, J.A. (ed.). UNU Press, pp. 221–244.

Note

1 These can be derived from the Preamble to UNCLOS: facilitate international communication, promote the peaceful uses of the seas and oceans, the equitable and efficient utilization of their resources, the conservation of their living resources, and the study, protection and preservation of the marine environment.

Index

Printed in the United States
By Bookmasters